Advances in Industrial Control

Series editors

Michael J. Grimble, Glasgow, UK
Michael A. Johnson, Kidlington, UK

More information about this series at http://www.springer.com/series/1412

Martha Belem Saldivar Márquez
Islam Boussaada · Hugues Mounier
Silviu-Iulian Niculescu

Analysis and Control of Oilwell Drilling Vibrations

A Time-Delay Systems Approach

Springer

Martha Belem Saldivar Márquez
Laboratoire des Signaux et Systèmes
(L2S, UMR CNRS 8506)
CNRS–Centrale Supélec-Université Paris-Sud
Gif-sur-Yvette
France

and

Facultad de Ingeniería (CA DSyC)
Universidad Autónoma del Estado de México
Toluca de Lerdo
Mexico

Islam Boussaada
Laboratoire des Signaux et Systèmes
(L2S, UMR CNRS 8506)
CNRS–Centrale Supélec-Université Paris-Sud
Gif-sur-Yvette
France

and

Institut Polytechnique des Sciences Avancées
(IPSA)
Ivry-sur-Seine
France

Hugues Mounier
Laboratoire des Signaux et Systèmes
(L2S, UMR CNRS 8506)
CNRS–Centrale Supélec-Université Paris-Sud
Gif-sur-Yvette
France

Silviu-Iulian Niculescu
Laboratoire des Signaux et Systèmes
(L2S, UMR CNRS 8506)
CNRS–Centrale Supélec-Université Paris-Sud
Gif-sur-Yvette
France

ISSN 1430-9491 ISSN 2193-1577 (electronic)
Advances in Industrial Control
ISBN 978-3-319-15746-7 ISBN 978-3-319-15747-4 (eBook)
DOI 10.1007/978-3-319-15747-4

Library of Congress Control Number: 2015932243

Mathematical Subject Classification Code: 34H15, 34H05, 93C20, 34K40, 74H45

Springer Cham Heidelberg New York Dordrecht London

Printed on acid-free paper

Springer International Publishing AG Switzerland is part of Springer Science+Business Media
(www.springer.com)

Series Editors' Foreword

The series *Advances in Industrial Control* aims to report and encourage technology transfer in control engineering. The rapid development of control technology has an impact on all areas of the control discipline. New theory, new controllers, actuators, sensors, new industrial processes, computer methods, new applications, new philosophies…, new challenges. Much of this development work resides in industrial reports, feasibility study papers and the reports of advanced collaborative projects. The series offers an opportunity for researchers to present an extended exposition of such new work in all aspects of industrial control for wider and rapid dissemination.

The operation of drilling is an important one for the production of oil and gas and is the key process in the very large production and exploration industry. Drilling takes place in a wide range of geographical and climatic environments operating both onshore and offshore. More recently the process of hydraulic fracturing (fracking) of oil- and gas-bearing strata has opened up even more swathes of the environment as potential oil and gas production areas. Exploration and production companies seek cost-effective operations; society, meanwhile, demands safe operation with minimal environmental and, ideally, no ecological impact. Making drill-strings safe contributes to these societal conditions and to their effective and efficient operation. A phenomenon that can cause poor operational performance and even develop into the catastrophic failure of the well is the presence of vibrations in the drill-string: torsional vibrations, axial vibrations and lateral vibrations. In very simple terms, operators at the surface are trying to control and direct the motion of the drill bit that is often thousands of metres below and at the end of a very long rotating jointed rod. This is a spatially distributed system that is modelled fundamentally by partial differential equations (PDEs); it represents a very challenging modelling and control problem.

Although spatially distributed systems modelled by PDEs have been studied by control theorists, and even form the basis of an undergraduate laboratory experiment involving a flexible beam, it is not often that monographs appear treating an important real-world industrial process that is a spatially distributed system. And it is this that makes this monograph *Analysis and Control of Oilwell Drilling*

Vibrations: A Time-Delay Systems Approach by Martha Belem Saldivar Márquez, Islam Boussaada, Hugues Mounier and Silviu-Iulian Niculescu such an interesting and valuable contribution to the control field.

The monograph is very sharply focused and reports on the fundamental modelling, analysis and control of well drill-strings. The authors' objectives are to understand and then control the various detrimental vibrations that can arise in drill-strings. Instead of approaching the modelling through lumped-parameter models and analogues, the authors take the fundamental approach and follow the PDE modelling route as a foundation on which to construct their research study. The monograph is noteworthy for a number of reasons:

i. The existing industrial strategies for vibration control for drill-strings are documented and discussed.
ii. The model devised by the authors is used to gain insight about these often empirical industrial control approaches.
iii. Using the models devised, a number of new control strategies are proposed, analysed, and simulated.
iv. A good comparative assessment of all the control strategies is given starting with a clear statement of the metrics against which the assessment is to be made. The conclusions of this comparison are drawn in the final chapter of the monograph.

This monograph will interest readers from the academic control community, both academics and researchers. It will also find a readership in the industrial control engineering community as it demonstrates how to model, analyse and control spatially distributed systems, so it has transferable technical content. It is a very welcome addition to the *Advances in Industrial Control* monograph series.

Readers seeking insight into oil and gas exploration and production processes might also be interested in another recent *Advances in Industrial Control* series monograph: *Dynamics and Control of Mechanical Systems in Offshore Engineering* by Wei He, Shuzhi Sam Ge, Bernard Voon Ee How and Yoo Sang Choo (ISBN 978-1-4471-5336-8, 2013). This monograph focuses on the units that make up the offshore oil and gas delivery system. Technically the monograph reports the modelling and control of distributed parameter systems, and uses robust adaptive control methods.

Industrial Control Centre, Glasgow, Scotland, UK M.J. Grimble
M.A. Johnson

Preface

Vibrations in mechanical systems are oscillations occurring without being intentionally provoked. They often have detrimental effects on the system performance and may cause premature wear of the system components, underperforming processes, and could even involve security problems, such is the case in aircraft wings; which in the worst case scenario, excessive vibration causes the aircraft to crash.

In oilwell drillstring systems, vibrations constitute an important source of economic losses; drill bit wear, pipes disconnection, borehole disruption and prolonged drilling time, are only some examples of consequences associated with drilling vibrations.

Extensive research effort on the modeling and control of drilling systems has been conducted in the last century. Before the sixties, investigations were focused on material strength of the drillstring components, but the trends have since changed to emphasize on its dynamic behavior [136]. In 1960, Bailey and Finnie of *Shell Development Company* conducted the first analytical and experimental study on torsional and axial drilling vibrations [20].

Since then, numerous approaches for modeling and control have been proposed. The most popular control techniques are listed below:

- *Steering torque feedback system* [127]. The underlying idea of the torque feedback method is to adjust the velocity provided by the rotary table according to torque variations; hence, propagating waves are dampened at the top extremity instead of being reflected back to the drillstring. The major shortcoming of this strategy is that it requires an accurate measurement of the torque, which, in practice, can be difficult to obtain.
- *Soft Torque Rotary System* (*STRS*) [138]. This method is an improved version of the feedback torque technique. It avoids the task of measuring the drill string torque by computing it through the motor current. A proper tuning of the controller allows reducing drillstring vibrations.

- *Proportional-Integral-Derivative* (*PID*)-*control* [225]. This is a simple strategy to avoid the stick-slip phenomenon. PID controller gains are obtained through an appropriate stability analysis. A drawback of this technique is that the drillstring vibrations are not sufficiently damped to guarantee an optimal drilling performance.
- $\mathscr{H}_\infty$-*controller* [265]. Results obtained with experimental prototypes have shown that torsional drilling oscillations are reduced by means of an $\mathscr{H}_\infty$ control law which is linear, time-invariant and has robust qualities. However, in order to get a proper control performance, a very accurate model is required. Another disadvantage of this method is that saturation constraints are not well-handled.
- *Active vibration damper* [137]. The basic idea of this method consists in increasing the viscous damping at the bottom end to avoid drillstring vibrations. The damping coefficient is modified via a magnetorheological fluid which allows manipulating the viscous properties of the drilling mud. This strategy allows attenuating the stick-slip vibrations, however an optimal drilling operation requires additional control accions.
- *Sliding mode control* [212]. This control strategy, introduced in [209] and discussed and modified in [212], is based on the bifurcation analysis of a lumped parameter model describing the torsional drilling dynamics developed in [210]. The stick-slip phenomenon can be mathematically seen as a sliding motion, which occurs when the bit velocity is zero. The existence of this sliding motion depends on the weight on the bit and the torque applied by the surface motor. Such a regime is the main cause of bit sticking problems. The sliding mode control consists in introducing another discontinuity surface and forcing the system to evolve along it. On the new surface, the bit speed will follow the top-rotary-system speed after reasonable time, avoiding the bit sticking phenomena. This strategy does not represent an automatic controller, but it should be understood as an off-line safe parameters selection method which helps the driller operators avoiding bit sticking problems.
- *D-OSKIL* [50]. D-OSKIL is a short word for drilling oscillation killer. This method uses the weight on bit as an additional control variable. The control proposal is based on the fact that a large enough weight on bit is required to guarantee a satisfactory rate of penetration, and if it reaches higher values, drilling vibrations may arise. An optimal trade-off between the weight on bit and the rate of penetration has to be found. Experimental implementation of such a mechanism, in a laboratory testbed, is reported in [188]. One disadvantage of this method is that its implementation may require the repetitive addition and removal of drill collar sections to properly adjust the control law, which may result infeasible and could induce axial vibrations.

Despite the development of numerous methods for eliminating drilling vibrations, nowadays such phenomena still greatly affects perforation processes. This is mainly due to the lack of proper understanding of the system's dynamics; in fact, most of the proposed techniques are based on simplified lumped parameter models that disregard the distributed nature of the system, and only consider the torsional drilling behavior.

This monograph compiles research findings and approaches on the modeling and control of vertical rotary oilwell drilling systems. The problem of suppressing drillstring vibrations is addressed within the control theory framework using the wave equation model to describe small amplitude vibrations of the rod. Diverse techniques based on the system stability are proposed and evaluated through numerical analyses. Even though this volume does not presents experimental results, computational analyses are carried out considering parameters that reflect typical operating conditions of a real oil platform. It should be stressed that the description of the actual behavior of the system in all its practical phases of operation is a complex task which can hardly be carried out by models restricted to vibrations of small amplitude motions. Nevertheless, the models presented here give us a general idea of the phenomena occurring during the drilling process and allow us to taking action into this matter.

The aim of this book is threefold. First, the modeling problem is addressed; to fully understand the underlying phenomena giving rise to drilling vibrations, a reliable mathematical model must be conceived. To this end, some theoretical background on the friction laws derived from tribological studies are reviewed. Furthermore, in order to comprehensively characterize the system's dynamics, an infinite-dimensional model, expressed by a set of partial differential equations coupled to nonlinear boundary conditions, which describes the coupled axial and torsional drillstring trajectories is considered. For practical purposes, the proposed model is transformed through the d'Alembert method into a pair of neutral-type time-delay equations. Second, a deeper analysis of the drilling system is provided through a time-delay approach which allows analyzing its stability properties and determining its qualitative dynamic response. Third, the vibration control problem is tackled; several classical control approaches are revisited and some novel techniques based on the dynamic behavior analysis of the system are proposed. The performance of the control proposals is highlighted through simulations of the system which show an effective elimination of coupled drilling vibrations.

This present volume is self-contained and provides operational guidelines and control solutions to deal with drilling vibrations. Since the modeling and control techniques presented here can be generalized to treat diverse engineering problems, it constitutes a useful resource for researchers and specialists working in both control engineering and petroleum engineering area. Furthermore, this monograph can be considered as a complementary tool for teaching the fundamentals of dynamic systems; concepts like modeling, transformation of hyperbolic PDE to delay systems, friction forces between contact surfaces, bifurcation analysis, stability theory, among others are explained in detail and can be easily understood due to the practical application under study.

How to Read the Book?

Three parts compose this contribution.

Part I Modeling

This part provides a summary of the main classical modeling strategies to reproduce the drillstring behavior. Lumped parameter models which have been used to describe torsional and axial drilling dynamics are reviewed. Distributed parameter models allowing a more reliable system representation are next presented. Propagating waves arising from drilling vibrations can be modeled by the standard wave equation; an appropriate choice of boundary conditions allows approximating the phenomena observed at both extremities of the drillstring. The complexity of this modeling strategy is overcome through a direct transformation based on the d'Alembert method which allows deriving a more handleable system representation: a neutral-type time-delay equation.

Since the frictional torque arising from the bit-rock interaction is a crucial aspect of the system, the modeling part of the book includes one chapter entirely devoted to the mathematical description of the friction forces leading to harmful vibrations. Classical friction laws derived from tribological research, including the stiction and the Coulomb, Stribeck, Karnopp, Armstrong, Dahl and LuGre friction models, are briefly discussed.

The first part of the book also presents a set of distributed parameter equations which comprehensively models a vertical oilwell drilling system. The upper extremity modeling includes a description of the actuators' dynamics. The bottom extremity description considers a bit-rock interface model which allows taking into consideration the drilling surface characteristics and the bit geometry.

Part II Analysis

This part of the book provides some basic theoretical tools for analyzing the drilling system dynamics described by time-delay equations of neutral-type. It is worthy of mention that time-delay equations, also known as Delay Differential Equations (DDE), belong to the class of Functional Differential Equations (FDE) which are infinite-dimensional, as opposed to Ordinary Differential Equations (ODE). Time-delay systems are classified into three distinct types: when the rate of change of the state depends on the present and past values of the system state, the model is said to be of *retarded-type*; when the rate of change of the state depends not only on the present and past values of the system state but also on the earlier value of the state rate change, the corresponding equation is of *neutral-type*; when the rate of change of state is determined by future values of the state, the model of *advanced-type*, this last one is rarely encountered in practical applications. This part of the book present an overview on the most important concepts of the general theory of neutral delay differential equations such as the existence and uniqueness of solutions, spectral properties and stability concepts both in frequency-domain and time-domain analysis framework.

A bifurcation analysis of the drilling system is also provided in the second part of the book. By using some tools of the center manifold and normal forms theory, a simplified description of the system is obtained. The reduced model, obtained through spectral projections, allows characterizing the qualitative dynamic response of the system.

Part III Control

Several methodologies are proposed to tackle coupled axial-torsional drilling vibrations; the stick-slip and bit-bounce phenomena are effectively suppressed via stabilizing controllers.

The third part of the book presents a detailed description of the different types of drilling vibrations; detection guidelines, detrimental consequences and empirical control solutions are discussed. Through drilling system simulations, the main practical strategies to suppress the stick-slip phenomenon are evaluated.

Furthermore, a summary of important industrial aspects of the drilling process are presented. These include a review of the different devices absorbing energy to reduce vibrations as well as the different methods used for acquiring, monitoring and transmitting data from the downhole to the surface, and the innovative automated systems that improve the perforation process.

More sophisticated control techniques involving feedback actions are discussed. Low order control schemes to reduce coupled torsional-axial drilling behavior are developed; a pair of delayed proportional and delayed PID feedback controllers are shown to be able to reduce drillstring vibrations.

A different control strategy is designed by exploiting the differential flatness property of the drilling system. This property refers to the ability of a system to be exactly linearized via endogenous feedback. The main attribute of flat systems is that the state and input variables can be directly expressed without integrating any differential equation, in terms of one particular set of variables and a finite number of its derivatives, which helps to tackle, in a simple way, trajectory tracking problems. The design of a pair of nonlinear controllers aimed at steering the drill string trajectories to prescribed paths gives rise to the suppression of coupled drilling vibrations.

Several feedback controllers based on Lyapunov techniques are also presented; asymptotic, exponential and practical stability of the drilling system is achieved. The obtained stability conditions are stated in terms of Linear and Bilinear Matrix Inequalities (LMI, BMI).

The book ends with a comparative/contrastive study that highlights the benefits and vulnerabilities of the different control methods presented here.

It is worthy of mention that the three parts of the book are independent each-other as much as possible. Nevertheless, certain results are required for a proper understanding of the theoretical developments.

Each chapter ends with a section named *Notes and References* which highlights particular aspects of the developments and also includes conclusions, comments and additional reference literature.

Gif-sur-Yvette, December 2014

Martha Belem Saldivar Márquez
Islam Boussaada
Hugues Mounier
Silviu-Iulian Niculescu

Acknowledgments

We would like to thank our collaborators, who significantly contributed to the research results presented in the book. Among them we mention: Sabine Mondié (CINVESTAV-IPN, Mexico), Jean-Jacques Loiseau (IRCCyN, France), Vladimir Răsvan (University of Craiova, Romania), Emilia Fridman (Tel Aviv University, Israel), Alexandre Seuret (LAAS-CNRS, France), Raúl Villafuerte (UAEH, Mexico), Arben Çela (ESIEE Paris, France), Michel Fliess (LIX École Polytechnique and ALIEN, France), Wim Michiels (KU Leuven, Belgium), Pierre Rouchon (Mines ParisTech, France), Joachim Rudolph (ISR, Germany), Torsten Knüppel (TU Dresden-Institute of Control Theory, Germany), Frank Woittennek (TU Dresden-Institute of Control Theory, Germany).

We are grateful to Michael J. Grimble and Michael A. Johnson, Springer (AIC) Series Editors and anonymous reviewers for their constructive feedback and insightful suggestions which greatly helped us in improving the monograph. Next, we wish to thank Oliver Jackson, Editor Engineering at Springer as well as Karin de Bie, Production Editor for their patience and help during the whole process.

The first author (B. Saldivar) would like to thank the Consejo Nacional de Ciencia y Tecnología (CONACyT México) for the financial support (Grant number 204055) during her postdoctoral stay at the Laboratoire des Signaux et Systèmes, L2S (UMR 8506), CNRS—Supélec Université Paris-Sud and for her research position within the *Cátedras CONACyT* program at the Universidad Autónoma del Estado de México.

The second author (I. Boussaada) wishes to thank the Institut Polytechnique des Sciences Avancées, IPSA Paris, for the financial support during the preparation of the manuscript.

The third author (H. Mounier) wishes to thank the Institut Français du Pétrole—Énergies Nouvelles (IFP-EN) for its financial support during his postdoctoral stay in the applied mechanics division.

The fourth author (S.I. Niculescu) would like to thank the French CNRS (National Center for Scientific Research) for its continuous support on developing research in time-delay systems area.

Last but not least, B. Saldivar is greatful to her beloved family, specially to Martinita, Ile, Kary, Liz, Vale, Guz, Dan & Brian, for all the support and the priceless joy moments. I. Boussaada wants to thank his wife, Ouerdia, and his children Rayan, Alissa and Yani for their love and their infinite patience, his gratitude goes also to his parents for their continue support. H. Mounier would like to thank his wife, Amel, for her kind, unswerving, love and support during numerous years of common life. Concerning S.-I. Niculescu, there is a special person in his life, Laura to whom he owes the exceptional support that she gave to Silviu to overcome all the difficulties both professional and extra-professional in the last twenty years. We dedicate this monograph to all of them, in love and and gratitude.

It is important to point out that the matlab packages used for the simulations and analysis purposes represent confidential material and can not be accessible for further use by the readers of the volume.

Gif-sur-Yvette, December 2014

Martha Belem Saldivar Márquez
Islam Boussaada
Hugues Mounier
Silviu-Iulian Niculescu

Contents

Acronyms and Mathematical Symbols

ADV	Active Vibration Damper
AFDE	Advanced Functional Differential Equation
BHA	Bottom Hole Assembly
BMI	Bilinear Matrix Inequality
DDE	Delay Differential Equation
DOF	Degree of Freedom
DVMCS	Drilling Vibration Monitoring and Control System
FDE	Functional Differential Equation
FEM	Finite Element Method
LMI	Linear Matrix Inequality
LWD	Logging While Drilling
MWD	Measurement While Drilling
NDDE	Neutral Delay Differential Equation
NFDE	Neutral Functional Differential Equation
ODE	Ordinary Differential Equation
PDC	Polycrystalline Diamond Compact
PDE	Partial Differential Equation
PID	Proportional-Integral-Derivative
RDS	Robotic Drilling System
RFDE	Retarded Functional Differential Equation
SCC	Stress Corrosion Cracking
SPE	Society of Petroleum Engineers
STRS	Soft Torque Rotary System
ToB	Torque on Bit
WDP	Wired drill pipe
WoB	Weight on Bit
$\mathbb{R}$	Set of real numbers
$\mathbb{R}^n$	n-dimensional Euclidean space
$\mathbb{R}^{n \times m}$	Set of $n \times m$ real matrices

$\mathbb{R}_+$	Set of nonnegative real numbers
$\mathbb{Z}$	Set of integer numbers
$\mathbb{N}$	Set of natural numbers
$\mathbb{C}$	Set of complex numbers
$\mathbb{C}_+$	Open right complex half plane, $\mathbb{C}_+ := \{\lambda : \mathrm{Re}(\lambda) > 0\}$
$\bar{\mathbb{C}}_+$	Closed right complex half plane
$\partial\mathbb{C}_+$	Imaginary axis
$\mathbb{D}$	Open unit disc in the complex plane, $\mathbb{D} := \{\lambda : \lambda \in \mathbb{C}, \lvert\lambda\rvert < 1\}$
$\partial\mathbb{D}$	Unit circle in the complex plane, $\partial\mathbb{D} := \{\lambda : \lambda \in \mathbb{C}, \lvert\lambda\rvert = 1\}$
$\mathbb{D}^c$	Closed exterior of the unit disc, $\mathbb{D}^c := \{\lambda : \lambda \in \mathbb{C}, \lvert\lambda\rvert \geqslant 1\}$
$0_{n\times n}$	Zero $n \times n$ matrix
I	Identity matrix of appropriate dimension
$\mathrm{Re}(\lambda)$	Real part of a complex number $\lambda \in \mathbb{C}$
sgn	Sign function
x_t	Restriction of $x(t)$, $x_t : \theta \to x(t+\theta), \theta \in [-\tau, 0]$
0_τ	$\mathbb{R}^n$-valued trivial function, $0_\tau(\theta) = 0 \in \mathbb{R}^n, \theta \in [-\tau, 0]$
$\sigma(A)$	Spectrum of a given matrix A
$\rho(A)$	Spectral radius of a given matrix A
$\sigma(A, B)$	Set of all generalized eigenvalues of a matrix pair (A, B), $\sigma(A, B) := \{\lambda \in \mathbb{C} : \det(A - \lambda B) = 0\}$
$A \oplus B$	Kronecker sum of a matrix pair (A, B)
$A \otimes B$	Kronecker productof a matrix pair (A, B)
T	Matrix transposition
$\lambda_{\min}(A)$	Smallest eigenvalue of a symmetric matrix A
$\lambda_{\max}(A)$	Largest eigenvalue of a symmetric matrix A
$A > 0$	Symmetric positive definite matrix $A \in \mathbb{R}^{n\times n}$
$A < 0$	Symmetric negative definite matrix $A \in \mathbb{R}^{n\times n}$
Q^T	Transpose of the matrix $Q \in \mathbb{R}^{n\times m}$
$\to$	Mapping of the elements of the domain to the elements of the range
$\Rightarrow$	Implies
$\forall$	For all
$\exists$	Exists
$\in$	Belongs to
$(\cdot)^{-1}$	Inverse operator
$\mathrm{tr}(\cdot)$	Trace operator
$\lVert x\rVert$	Euclidean norm of a vector $x \in \mathbb{R}^n$
$\lVert A\rVert$	Induced norm of a matrix A
$C([-\tau, 0], \mathbb{R}^n)$	Space of $\mathbb{R}^n$-valued continuous functions on $[-\tau, 0]$
$PC([-\tau, 0], \mathbb{R}^n)$	Space of $\mathbb{R}^n$-valued piecewise continuous functions on $[-\tau, 0]$

$C^1([-\tau,0],\mathbb{R}^n)$	Space of $\mathbb{R}^n$-valued continuously differentiable functions on $[-\tau,0]$
$PC^1([-\tau,0],\mathbb{R}^n)$	Space of $\mathbb{R}^n$-valued piecewise continuously differentiable functions on $[-\tau,0]$
$W^{1,2}([a,b],\mathbb{R})$	Sobolev space of absolutely continuous scalar functions $z:[a,b]\to\mathbb{R}$ with square integrable derivatives $z^{(l)}$ of order $l\geq 1$ and with the norm $\|z\|^2_{W^{1,2}} := \int_a^b \left(z^{(1)}\right)^2(\xi)\mathrm{d}\xi$.
$\lvert\varphi\rvert_\tau$	Uniform norm of a function $\varphi\in C^1([-\tau,0],\mathbb{R}^n)$, $\lvert\varphi\rvert_\tau := \sup_{-\tau\leq\theta\leq 0}\lvert\varphi(\theta)\rvert$
$\|\varphi\|_\tau$	Uniform norm of a function $\varphi\in C^1([-\tau,0],\mathbb{R}^n)$, $\varphi_\tau := \max\left(\lvert\varphi\rvert_\tau, \lvert\dot{\varphi}\rvert_\tau\right)$

Chapter 1
Introduction

1.1 Context and Motivation

The presence of drillstring vibrations is the main cause of performance loss in the perforation process for oil and gas. It provokes premature wear and tear of drilling equipment resulting in fatigue and induced failures such as pipe wash-out and twist-off [193]. It also causes a significant wastage of drilling energy [190] and may induce wellbore instabilities reducing the directional control and its overall shape [79]. In the oil industry, the improvement of the drilling performance is a matter of crucial economic interest.

Many studies have been conducted to recognize the different types of vibrations during the drilling operation. These have led to the identification and classification of vibrations into three separate and distinctive categories: *torsional* (stick-slip oscillations), *axial* (bit bouncing phenomenon) and *lateral* (whirl motion due to the out-of-balance of the drillstring).

Currently, drilling of deepwater wells for oil and gas production has opened new horizons for petroleum engineers and experts to try to mitigate the influence of vibrations during the drilling operation. Even though new technology has been deployed, such phenomena still occur, considerably affecting drilling costs and daily operations.

In order to improve perforation processes performance, certain control challenges must be taken into account, e.g., managed pressure control systems design, drilling speed optimization and control, trajectory steering control of directionally controlled drilling systems, supervisory control, development of suitable control architectures, and model inversion. Since 1994, a lot of effort has been focused on developing the so-called "smart-drilling systems", where the system should be "capable of sensing and adapting to conditions around and ahead of the drill bit to reach desired targets," [61, p. 150].

M.B. Saldivar Márquez et al., *Analysis and Control of Oilwell Drilling Vibrations*, Advances in Industrial Control, DOI 10.1007/978-3-319-15747-4_1

The aim of this book is threefold:

(i) Better modeling of the physical phenomena occurring in drilling systems;
(ii) Better understanding of the system dynamics during the drilling process;
(iii) Improvement of the drilling performance through the design of appropriate control laws allowing eliminating undesirable vibrations.

1.2 Drilling System Components

A sketch of a simplified drillstring system is shown in Fig. 1.1.

Roughly speaking, the main components of a drilling system, called a *drilling rig*, consist of drill pipes, steel tubes of typically 10 m length; drill collars, much

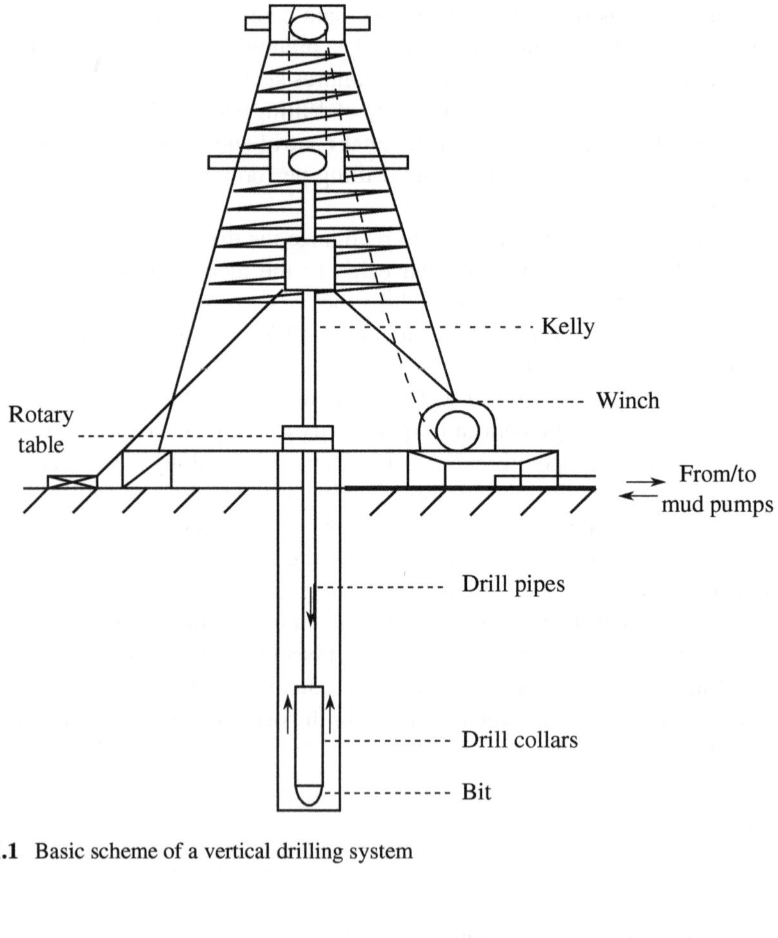

Fig. 1.1 Basic scheme of a vertical drilling system

thicker pipes which provide the necessary weight to perform the perforation; and a rock cutting device, usually known as *bit*. The whole set is called *drillstring*, which is rotated through a motor located at the top. All the pipes of the drillstring are hollow, so that a drilling fluid can be injected by a mud pump in order to evacuate the removed rock.

A more precise description of the system components is presented below [256]:

- **Top extremity**. Its components are: a rotary table, to provide the rotational movement to the drillstring; a traveling block and a drill line going through a crown block, whose function is to hold the drillstring; and a mud pump, for injecting the drilling fluid through the drillstring. There are thus three motors, one for the rotary table, one for the drawworks, and one for the mud pump.
- **Drill pipe string**. Its length varies according to the drilling depth. Typical final hole depth goes from 1000 to 8000 m for ultra-deep drilling.
- **Drill collar string**. Its purpose is to give weight to the drillstring. The drill collars are typically 300–500 m long giving a weight of 40–60 tons at the bottom hole. Some of the drill collars may be thicker; they are called *stabilizers*, and their purpose is to avoid lateral motion of the drill collars in the hole.
- **Drill bit**. This is a cutting tool; the most commonly used are fixed cutter and roller cone. A bit of fixed cutter type has no moving parts; perforation is performed via the rotation of the drillstring. Fixed cutter bits can be either polycrystalline diamond compact (PDC) or grit hot-pressed inserts (GHI). Roller cone bits can be either tungsten carbide inserts (TCI) or milled tooth (MT). Fixed cutter tools are meant to cut the rock, while roller cone ones act more by crushing.

Other components at the bottom hole are the *measurement while drilling* (MWD) device and the *logging while drilling* (LWD) tool. Measurement while drilling tools use accelerometers and magnetometers to measure the inclination and azimuth of the wellbore. Other informations can be measured by MWD tools such as the rotational speed of the drillstring, the downhole temperature, the Torque and Weight on Bit (ToB and WoB respectively), and the mud flow volume.

The measures are transmitted to the surface through one of the three following techniques: mud pulse telemetry, where a downhole valve varies the flow of the drilling fluid, according to the digital information to be transmitted; electromagnetic telemetry, which generates voltage differences between the drillstring sections, and where the data are transmitted through digital modulation; wired drill pipe, which transfers data through electrical wires. Typical mud pulse telemetry technology offers a bandwidth of up to 40 bps, electromagnetic telemetry offers data rates of up to 10 bps, and wired drill pipe yield data rates upwards of 1 Mbps.

Logging while drilling tools are typically used for measuring geological characteristics like density, porosity, resistivity, acoustic-caliper, inclination at the drill bit, magnetic resonance, and formation pressure. The information is not directly transmitted to the surface but is kept in a local memory, and extracted when the LWD tool is brought to the surface.

The set of drill collars, stabilizers, bit and measurement/logging while drilling is called *Bottom Hole Assembly* (BHA).

1.3 Harmful Oscillations

A drillstring is mainly subject to three types of vibrations, each of which can, at the least, cause a premature wear of the various components. Below is a brief description of each of them.

- **Torsional vibration** [43]. Downhole measurements show that applying a constant rotary speed at the surface does not necessarily translate into a steady rotational motion of the bit. In fact, the downhole torsional speed typically exhibits large amplitude fluctuations during a significant fraction of the drilling time. This self-excited rotational motion, also known as *stick-slip*, is induced by the nonlinear relationship between the torque and the angular velocity at the bit [138]. The torsional flexibility of the drilling assembly exacerbates a nonuniform oscillatory behavior causing rotational speeds as high as ten times the nominal rotary table speed or a total standstill of the bit [273]. Torsional vibrations provoke fatigue to drill collar connections, damage the drill bit, and slow down the drilling operation thereby prolonging the overall drilling process. They are detectable at the drillfloor by fluctuations in the power needed to maintain a constant rate of surface rotation.
- **Axial vibration** [286]. This vibration mode consists of irregular movements of the drilling components along its longitudinal axis causing bit bounce and rough drilling behavior that destroys the drill bit, damages the BHA, and increases total drilling time. Additionally, due to downhole coupling mechanisms, it also excites lateral displacements of the string [270]. The bit bounce pattern may be detected at the surface; it is likely to develop when drilling with a bit of roller-cone type, also called tricone, consisting of multiple lobes which leads to an erratic interaction of the bit with the bottom of the well making the bit to lose contact with the rock formation.
- **Lateral vibration** [291]. One of the most destructive drillstring oscillations is the whirling phenomenon, since it may be unleashed with no indication at the surface. Deep in the hole, the rotating BHA interacts with the borehole wall generating shocks from lateral vibrations. The collisions with the borehole wall will produce eccentric hole and the shocks can damage components of the BHA [197]. The lateral oscillations of the drillstring cause severe damage to the borehole wall and affect the overall drilling direction [137]. Drill collars whirling are simply the centrifugally induced bowing of the drill collar resulting from rotation. If the center of gravity of the drill collar is not initially located precisely on the centerline of the hole, then as the collar rotates, a centrifugal force acts at the center of gravity causing the collar to bend [291]. Forward and backward whirling behaviors can further intensify due to the combined effect of fluid damping, stabilizer clearance, and friction of the drilling assembly against the borehole wall [289].

1.4 Book Contents

This book is organized into three parts; the main ideas can be resumed as follows:

- **Part I. Modeling**. An overview of the existing finite and infinite-dimensional representations of torsional and axial drillstring vibrations is provided. Next, the d'Alembert procedure to transform the wave equation drilling model into a delay system of neutral type is detailed. Classical friction models allowing characterizing the bit-rock interface are reviewed. The main empirical strategies to tackle the problem of torsional oscillations in drilling platforms are evaluated; operational guidelines to avoid the stick-slip phenomenon are discussed. Finally, a comprehensive drilling model leading to a complete characterization of the system dynamics is derived from the fundamental laws of mechanics.
- **Part II. Analysis**. As mentioned before, the drilling dynamics can be described by a system of delay-equations; such systems are infinite-dimensional and can be represented by functional differential equations (FDE). The time-delay system under consideration is said to be of neutral-type since the state time derivative depends not only on the delayed states but also on its delayed derivative. In this part of the book, a theoretical background on the stability of this class of systems is provided. The qualitative dynamic response of the drilling system is characterized through a bifurcation analysis. Based on the center manifold theorem and normal forms theory, the Neutral Functional Differential Equation (NFDE) model is reduced to a finite-dimensional system described by an Ordinary Differential Equation (ODE) which simplifies the analysis task. Algebraic tools for the design of a state estimator to approximate the angular and axial bit speeds are provided. The flatness property of the distributed model describing coupled drilling vibrations is established. Ultimate bounds on the response of the distributed parameter drilling model are determined through Lyapunov techniques.
- **Part III. Control**. Based on the simplified ODE drilling model, a delayed Proportional-Integral-Derivative (PID) controller and a delayed proportional feedback control strategy ensuring locally stable system trajectories are determined. A pair of flatness-based controllers allowing the exponential convergence of the system trajectories is shown to be effective in eliminating torsional and axial drillstring oscillations. Within the framework of Lyapunov theory and based on the NFDE drilling model several control methodologies to tackle axial and torsional vibrations are developed; stabilizing conditions in terms of Linear and Bilinear Matrix Inequalities (LMI, BMI) lead to the synthesis of linear and nonlinear feedback controllers. Simulation results highlight the performance of the proposed strategies and allow a comparative analysis of the control solutions.

Part I
Modeling

Chapter 2
An Overview of Drillstring Models

Modeling of drillstring dynamics constitutes the basis for system analysis and control of harmful vibrations. Over the last half-century, extensive research effort has been conducted to mathematically describe the physical phenomena occurring in real wells.

Existing drilling models can be classified into the following general categories:

- **Lumped parameter models**. The drillstring is regarded as a mass-spring-damper system which can be described by an ordinary differential equation. This finite-dimensional system representation provides a rough description of the dynamics taking place at different levels of the string; it can be of one to several degrees of freedom.
- **Distributed parameter models**. The drillstring is considered as a beam subject to axial and/or torsional efforts. A system of partial differential equations provides a characterization of the drilling variables in an infinite-dimensional setting. The price paid for the model accuracy is the complexity involved in its analysis and simulations.
- **Neutral-type time-delay models**. These models, which are directly derived from the distributed parameter ones (when damping is considered negligible), provide an input–output system description. The involved time delays (which are dependent on the string length) are related to the speed of the oscillatory waves traveling throughout the rod. This type of model provides a good trade-off between system representation accuracy and complexity of the description. Furthermore, it offers some good matching with data transmission models that use delays in their mathematical representation.

This chapter presents a brief compilation of the most popular modeling strategies allowing the analysis and control of a vertical oilwell drilling system.

M.B. Saldivar Márquez et al., *Analysis and Control of Oilwell Drilling Vibrations*, Advances in Industrial Control, DOI 10.1007/978-3-319-15747-4_2

2.1 Lumped Parameter Models

2.1.1 Torsional Dynamics

The use of reduced models for vibration analysis is motivated by the need to define a simple description of drilling dynamics. Roughly speaking, the continuous system, consisting of drillpipes and of the bottom hole assembly, is regarded as a torsional pendulum described by a lumped parameter model with one or multiple degrees-of-freedom (DOF).

Figure 2.1 shows the simplified two-degree-of-freedom torsional model of a conventional vertical drillstring proposed in [209]. The inertial masses I_p and I_b, locally damped by d_p and d_b, are connected one to each other by a linear spring with torsional stiffness k and torsional damping c. The equations of motion can be represented as follows:

$$\begin{cases} I_p\ddot{\Phi}_p + c(\dot{\Phi}_p - \dot{\Phi}_b) + k(\Phi_p - \Phi_b) + d_p\dot{\Phi}_p = u_T, \\ I_b\ddot{\Phi}_b - c(\dot{\Phi}_p - \dot{\Phi}_b) - k(\Phi_p - \Phi_b) + d_b\dot{\Phi}_b \ = -T(\dot{\Phi}_b), \end{cases} \tag{2.1}$$

where Φ_p and Φ_b are the angular displacements of the rotary table and of the BHA, respectively. The control signal u_T is the drive torque coming from the rotary table transmission box used to regulate the rotary angular velocity $\dot{\Phi}_b$. The frictional torque T represents the torque on bit and the nonlinear frictional forces along the drill collars.

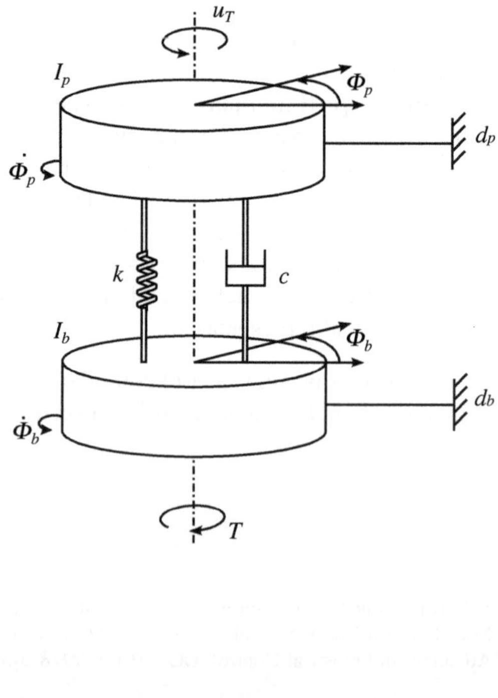

Fig. 2.1 Lumped parameter model of a rotary drilling rig

2.1.2 Axial Dynamics

A simplified model described by an ordinary differential equation is presented in [53]; the modeling strategy is inspired by the fact that any mass subject to a force in stable equilibrium acts as a harmonic oscillator for small vibrations. More precisely, the damped harmonic oscillator model describing the longitudinal drillstring motion is:

$$m_0 \ddot{v} + c_0 (\dot{v} + \rho(t)) + k_0 v = -\mu_1 T(\dot{\Phi}_b), \tag{2.2}$$

where the variable v is defined as: $v = U_b - \rho_0 t$.

Notations m_0, c_0, and k_0 represent the mass, damping, and spring constant; U_b, $\dot{U}_b$ and $\ddot{U}_b$ denote the bottom hole axial variables: position, velocity, and acceleration, respectively. The system is controlled through the rate of penetration $\rho(t)$, which is an axial speed imposed at the surface, ρ_0 is a constant nominal value. The torque on bit function T depending on the bit angular velocity couples the axial model with the torsional one.

The coefficient μ_1 is given by the following relation [72]:

$$\mu_1 = 2 \left(R_b \mu_{bit} c_{bit}\right)^{-1},$$

where R_b is the bit radius, μ_{bit} is the friction coefficient at the bit-rock contact, and c_{bit} is the so-called bit coefficient. For a bladed bit, c_{bit} is equal to the dimensionless length of the cutting edge (and independent of the number of blades). For a flat bit, c_{bit} is computed as follows:

$$c_{bit} = \frac{6 + 4\rho_{bit}}{6 + 3\rho_{bit}}, \tag{2.3}$$

where ρ_{bit} is the radial rate of increase of cutter density. Notice that in (2.3), c_{bit} varies between 1 and 4/3 and may be considered as a constant.

2.1.3 Torsional-Axial Dynamics

A discrete model that takes into consideration the axial and torsional vibration modes of a rotary drilling system with drag bit[1] is presented in [243]. Figure 2.2 shows the physical model as a two degrees-of-freedom system (one-axial and one-torsional) to describe the axial and torsional vibrations of the drillstring. The torsional part of

[1] A drag bit is a specific type of cutting device consisting of n identical blades symmetrically distributed around the axis of revolution.

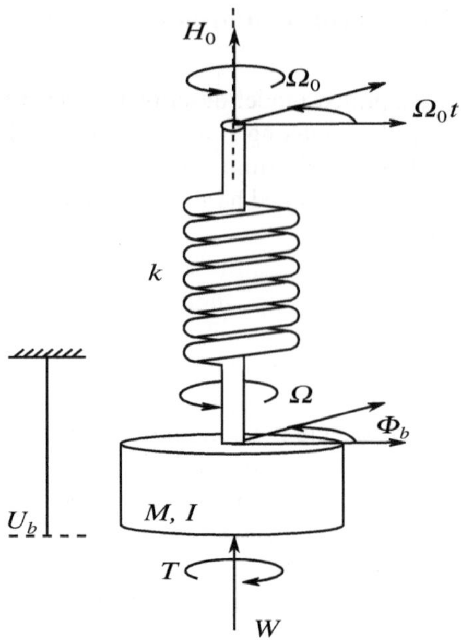

Fig. 2.2 Simplified model of the drilling system proposed in [243]

the model idealizes the drillstring as a torsional pendulum, and the axial part of the model idealizes the BHA and the drillstring as a lumped mass M.

The governing equations of motion are given by:

$$\begin{cases} I\frac{d^2\Phi_b}{dt^2} + k(\Phi_b - \Phi_{b0}) = T_0 - T \\ M\frac{d^2U_b}{dt^2} = W_0 - W, \end{cases} \tag{2.4}$$

where the variables U_b and Φ_b stand for the vertical and angular positions of the bit, respectively. The mechanical elements representing the BHA are: M, the point mass; I, the moment of inertia; and k, the spring stiffness representing the torsional rigidity of the drill pipe. The torque and the weight on bit are denoted by T and W, respectively. The stationary variables Φ_{b0}, T_0 and W_0, associated with the trivial solution of (2.4), satisfy:

$$\Phi_{b0} = \Omega_0 t - \frac{T_0}{k},$$

$$W_0 = W_s - H_0$$

where Ω_0 is a prescribed angular velocity, H_0 is a constant upward tension and W_s is the submerged weight of the drillstring.

A shortcoming of model (2.4) is that the dissipation in axial and torsional components is omitted, which may result unrealistic. Furthermore, the axial compliance of the drillpipes is not considered.

Some of these deficiencies were accounted for in the model proposed in [31], given by:

$$\begin{cases} I\frac{d^2\Phi_b}{dt^2} + k_t(\Phi_b - \Omega_0 t) = T_0 - T \\ M\frac{d^2U_b}{dt^2} + d\frac{dU_b}{dt} + k_a\,(U_b - \rho_0 t) = W_0 - W, \end{cases} \tag{2.5}$$

where k_t, k_a denote the torsional and axial spring stiffness and the friction parameter d characterizes viscous friction along the BHA.

An improved version of the lumped parameter model describing the coupled torsional-axial drilling dynamics is proposed in [207]. The suggested governing equations are given by:

$$\begin{cases} I\frac{d^2\Phi_b}{dt^2} + d_t\frac{d\Phi_b}{dt} + k_t(\Phi_b - \Omega_0 t) = \quad T_0 - T \\ M\frac{d^2U_b}{dt^2} + d_a\frac{dU_b}{dt} + k_a\,(U_b - \rho_0 t) = W_0 - W, \end{cases} \tag{2.6}$$

where d_t and d_a represent the viscous damping coefficients accounting for dissipation in the torsional and axial directions, respectively.

Model (2.6) is used for developing a local stability analysis to determine the stable and unstable operating regimes in the WoB-rotary speed parameter plane yielding some operating guidelines for practical purposes [207].

2.1.4 Torsional-Axial-Lateral Dynamics

A discrete system model for analytical and numerical studies of the coupled axial-torsional-lateral motions of the drillstring is developed in [181]. The discretized model of the drilling system considers two disks connected together over one constant length span, each disk with different translational inertia and rotary inertia properties. The inertia properties of the drill pipes are lumped into the first disk, while the inertial properties for the BHA are lumped into the second one. The dynamics of each disk is described collectively by a total of four DOF, with one DOF for axial motions, one DOF for torsional motions, and two DOF for lateral motions. The discretized system is driven by a motor with a constant angular speed Ω_0 and the two discrete inertial elements are connected through two identical massless, elastic elements which have associated axial, torsional, and bending potential energy contributions that reflect the elastic properties of the drill pipes.

The positions of each disk are described by the lateral, longitudinal, and torsional coordinates $(x_1, y_1, z_1, \varphi_1)$ and $(x_2, y_2, z_2, \varphi_2)$.

The governing equations of motion for the eight degree-of-freedom system are given by:

$$
\begin{cases}
m_1\ddot{x}_1 + c_l\dot{x}_1 + 2k_l\left(x_1 - \frac{x_2}{2}\right) = m_1 e\dot{\varphi}_1^2\cos\varphi_1 + F_{1x}\\
m_1\ddot{y}_1 + c_l\dot{y}_1 + 2k_l\left(y_1 - \frac{y_2}{2}\right) = m_1 e\dot{\varphi}_1^2\sin\varphi_1 + F_{1y}\\
m_1\ddot{z}_1 + c_a\dot{z}_1 + k_a\left(z_1 - z_2\right) = m_1 g - H_0\\
J_1\ddot{\varphi}_1 + c_t\dot{\varphi}_1 + k_t\left(2\varphi_1 - \varphi_2\right) = -T_1 + k_t\Omega_0 t\\
m_2\ddot{x}_2 + c_l\dot{x}_2 + k_l\left(\frac{x_2}{2} - x_1\right) = m_2 e\dot{\varphi}_2^2\cos\varphi_2 + F_{2x}\\
m_2\ddot{y}_2 + c_l\dot{y}_2 + k_l\left(\frac{y_2}{2} - y_1\right) = m_2 e\dot{\varphi}_2^2\sin\varphi_1 + F_{2y}\\
m_2\ddot{z}_2 + c_a\dot{z}_2 + k_a\left(z_2 - z_1\right) = m_2 g - F_{2z}\\
J_2\ddot{\varphi}_2 + c_t\dot{\varphi}_2 + k_t\left(\varphi_2 - \varphi_1\right) = -T_2
\end{cases}
$$

where m and J are the mass and the rotary inertia of each disk, e is the eccentricity which is assumed to be the same, and the influence of torsional acceleration on the lateral centrifugal force has been neglected. The stiffness and damping are denoted by k and c, respectively. The subscripts a, t, and l correspond to the direction in which these properties have influence: axial, torsional, and lateral. The quantities F_{1x}, F_{1y}, and T_1 are the respective force components and the torque that arises due to interactions between the drill pipe and wellbore. Similarly, the quantities F_{2x}, F_{2y}, F_{2z}, and T_2 are the respective force components and torque that arise due to interactions between the drill bit and rock. The axial, torsional, and lateral vibrations of the discrete system are coupled through these interaction forces and torques, see [181].

2.2 Distributed Parameter Models

To the best of the author's knowledge, the very first model of drilling vibrations was developed in the 1960s at the Shell Development Company, by Bailey [20] and Finnie [89]. They proposed a wave equation model to analytically treat the longitudinal and torsional drilling vibrations. Due to the accuracy of this modeling approach in reproducing the drilling behavior, regarding its distributed nature, it constitutes the basis of several recent contributions on drilling analysis and control.

In order to have a better understanding of the system' dynamics, the following section explains the physical aspects of a flexible bar giving rise to the wave equation model.

2.2.1 Model Derivation

Let us analyze the axial dynamics of a flexible metal bar of length L and of cross-section σ_0. Let $q(x,t)$ be the displacement of a point x of the bar with respect to

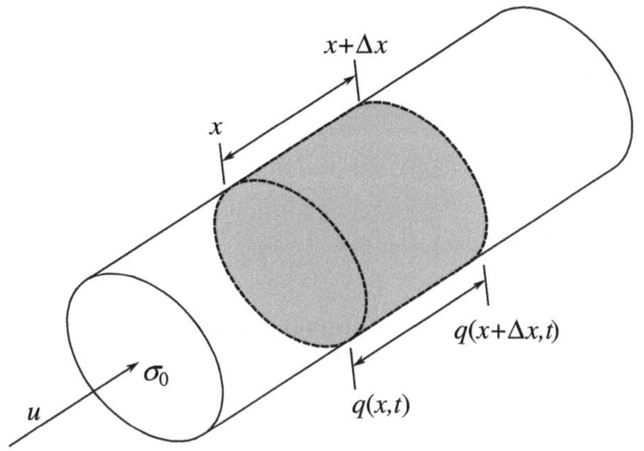

Fig. 2.3 Flexible bar

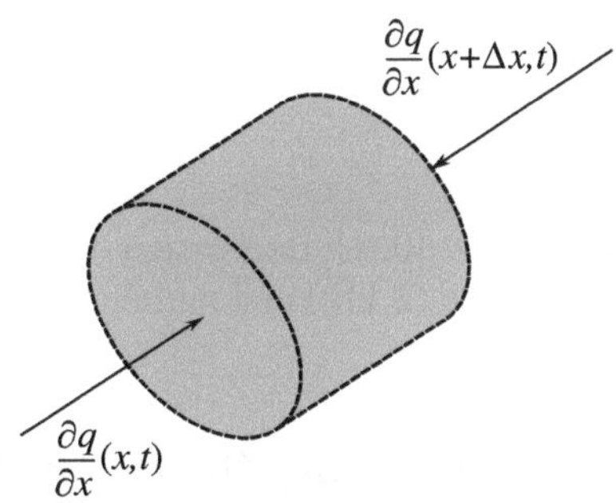

Fig. 2.4 Tension applied to a short segment of the bar

its equilibrium position, and $T(x,t)$, the tension applied at the point x at time "t" (Figs. 2.3 and 2.4).

Consider an element of length l_0 under the mean tension T_0.

The *fundamental elasticity law* establishes the following relation between the elongation $dl := l - l_0$ and the infinitesimal tension $dT := T - T_0$:

$$\frac{dT}{\sigma_0} = E_0 \frac{dl}{l_0}, \tag{2.7}$$

where E_0 is the Young modulus, or elasticity factor under the tension T_0. This law only applies for a sufficiently small relative elongation dl/l_0. At time "t", the segment $(x, x+\Delta x)$ has a static length of l_0 and takes the position $(x+q(x,t),\ x+\Delta x+q(x+\Delta x,t))$. Under a tension force, the segment length increases from $l_0 = \Delta x$ to $l = l_0+dl = \Delta x+(\partial q/\partial x)\Delta x$, we then have $dl/l_0 = \partial q/\partial x$. The elasticity law implies:

$$T - T_0 = E_0\sigma_0 \frac{\partial q}{\partial x}. \tag{2.8}$$

Let ρ_0 be the linear density (mass per unit length) of the bar.

The fundamental principle of dynamics states:

$$\rho_0 \frac{\partial^2 q}{\partial t^2} \Delta x = \frac{\partial T}{\partial x} dx, \tag{2.9}$$

by introducing (2.8), we obtain:

$$\rho_0 \frac{\partial^2 q}{\partial t^2} = E_0 \sigma_0 \frac{\partial^2 q}{\partial x^2}, \tag{2.10}$$

which constitutes the wave equation with propagation speed: $c = \sqrt{E_0 \sigma_0 / \rho_0}$.

The normalized model can be written as:

$$\frac{\partial^2 q}{\partial t^2}(x,t) = c^2 \frac{\partial^2 q}{\partial x^2} \tag{2.11a}$$

$$\frac{\partial q}{\partial x}(0,t) = -u(t) \qquad \frac{\partial q}{\partial x}(1,t) = 0 \tag{2.11b}$$

$$q(x,0) = q_0(x) \qquad \frac{\partial q}{\partial t}(x,0) = q_{t0}(x) \tag{2.11c}$$

where $x \in [0, 1]$. Expression (2.11a) is known as the wave equation with boundary conditions (2.11b) and initial conditions (2.11c).

2.2.2 Drillstring Oscillatory Behavior

The propagation of torsional waves along a drillstring of length L can be modeled by the following hyperbolic partial differential equation [53]:

$$GJ \frac{\partial^2 \Phi}{\partial s^2}(s,t) - I \frac{\partial^2 \Phi}{\partial t^2}(s,t) - \gamma \frac{\partial \Phi}{\partial t}(s,t) = 0, \qquad s \in (0, L), \tag{2.12}$$

where the twist angle Φ depends on the length coordinate "s" and time "t." The shear modulus and the second moment of area (also known as geometric moment of inertia) are denoted by G and J, respectively. The inertia I is such that $I = \rho_a J$, where ρ_a is the area density. A viscous damping $\gamma \geqslant 0$ is assumed along the structure.

Since most of the energy dissipation in drilling systems is taking place at the bit-rock interface, we may consider that the damping γ is null.

Thus, the distributed parameter model (2.12) reduces to the one-dimensional wave equation:

$$\frac{\partial^2 \Phi}{\partial s^2}(s,t) = \tilde{c}^2 \frac{\partial^2 \Phi}{\partial t^2}(s,t), \quad s \in (0, L), \tag{2.13}$$

where $\tilde{c} = \sqrt{I/GJ} = \sqrt{\rho_a/G}$.

An appropriate choice of boundary conditions allows characterizing the propagating torsional waves along the drillstring. Several equations have been proposed to describe the dynamics taking place at the upper and lower ends of the drillstring; the main ones are presented in the sequel.

2.2.2.1 Kinematic Boundary Conditions

In [53], the following boundary conditions are considered:

$$\Phi(0,t) = \Omega t, \tag{2.14a}$$

$$GJ\frac{\partial \Phi}{\partial s}(L,t) + I_B\frac{\partial^2 \Phi}{\partial t^2}(L,t) = -T\left(\frac{\partial \Phi}{\partial t}(L,t)\right). \tag{2.14b}$$

A lumped inertia I_B is chosen to represent the assembly at the bottom hole. It is assumed that the speed at the surface ($s = 0$) is restricted to a constant value Ω, the other extremity ($s = L$), is subject to a torque T, which is a function of the bit speed. The model of T includes a linear term $c_b \partial\Phi/\partial t(L,t)$ representing the viscous damping torque which approximates the influence of the mud drilling and a nonlinear term $F\,(\partial\Phi/\partial t(L,t))$ representing the dry friction torque which models the bit-rock contact, i.e.,

$$T\left(\frac{\partial \Phi}{\partial t}(L,t)\right) = c_b\frac{\partial \Phi}{\partial t}(L,t) + F\left(\frac{\partial \Phi}{\partial t}(L,t)\right). \tag{2.15}$$

2.2.2.2 Boundary Conditions Considering a Speed Difference at the Top Extremity

The boundary condition (2.14b) satisfactorily reproduces the behavior at the ground level; however, (2.14a) is only a pure kinematic description of the drillstring upper part. The angular velocity coming from the rotor Ω does not match the rotational speed of the load $\partial\Phi/\partial t(0,t)$, this sliding speed results in the local torsion of the drillstring. In order to take into account this phenomenon, the following boundary conditions are introduced in [253]:

$$GJ\frac{\partial \Phi}{\partial s}(0,t) = \beta\left(\frac{\partial \Phi}{\partial t}(0,t) - \Omega(t)\right) = \beta\frac{\partial \Phi}{\partial t}(0,t) - u_T(t) \tag{2.16a}$$

$$GJ\frac{\partial \Phi}{\partial s}(L,t) = -I_B\frac{\partial^2 \Phi}{\partial t^2}(L,t) - T\left(\frac{\partial \Phi}{\partial t}(L,t)\right), \tag{2.16b}$$

where β denotes the angular momentum at the top extremity.

2.2.2.3 Newtonian Boundary Conditions

The following equation constitutes an alternative top boundary condition inspired by the Newton's second law of motion, see for instance [64]:

$$I_T \frac{\partial^2 \Phi}{\partial t^2}(0,t) = -T_T(t) + u_T(t).$$

The effective moment of inertia of the top-drive is denoted by I_T, u_T corresponds to the external torque delivered by the rotary table taken as a control input, and the function $T_T(t)$, describing the transmitted torque and damping due to viscous effects, is given by:

$$T_T(t) = -GJ\frac{\partial \Phi}{\partial s}(0,t) + \beta \frac{\partial \Phi}{\partial t}(0,t).$$

The boundary conditions are then written as:

$$GJ\frac{\partial \Phi}{\partial s}(0,t) = I_T \frac{\partial^2 \Phi}{\partial t^2}(0,t) + \beta \frac{\partial \Phi}{\partial t}(0,t) - u_T(t) \tag{2.17a}$$

$$GJ\frac{\partial \Phi}{\partial s}(L,t) = -I_B \frac{\partial^2 \Phi}{\partial t^2}(L,t) - T\left(\frac{\partial \Phi}{\partial t}(L,t)\right). \tag{2.17b}$$

2.2.3 *Coupled Axial-Torsional Vibrations*

It is well known that torsional vibrations contribute to the excitation of axial oscillations, in fact, these self-excited oscillations are intimately coupled together and may occur simultaneously.

Coupled axial-torsional excitations of a drillstring of length L, described by the rotary angle $\Phi(s,t)$ and the longitudinal position $U(s,t)$ can be modeled by the wave equations [39]:

$$\frac{\partial^2 \Phi}{\partial s^2}(s,t) = \tilde{c}^2 \frac{\partial^2 \Phi}{\partial t^2}(s,t), \tag{2.18a}$$

$$\frac{\partial^2 U}{\partial s^2}(s,t) = c^2 \frac{\partial^2 U}{\partial t^2}(s,t), \tag{2.18b}$$

with boundary conditions

$$GJ\frac{\partial \Phi}{\partial s}(0,t) = \beta \frac{\partial \Phi}{\partial t}(0,t) - u_T(t) \tag{2.19a}$$

$$GJ\frac{\partial \Phi}{\partial s}(L,t) = -I_B \frac{\partial^2 \Phi}{\partial t^2}(L,t) - T\left(\frac{\partial \Phi}{\partial t}(L,t)\right), \tag{2.19b}$$

and

$$E\Gamma\frac{\partial U}{\partial s}(0,t) = \alpha\frac{\partial U}{\partial t}(0,t) - u_H(t) \tag{2.20a}$$

$$E\Gamma\frac{\partial U}{\partial s}(L,t) = -M_B\frac{\partial^2 U}{\partial t^2}(L,t) - T\left(\frac{\partial \Phi}{\partial t}(L,t)\right). \tag{2.20b}$$

The spatial variable "s" is chosen such that $s = 0$ denotes the top of the drillstring and $s = L$ its bottom. The propagation speeds of the axial and torsional waves v_U, v_Φ, defined as: $v_U = c^{-1}$ and $v_\Phi = \tilde{c}^{-1}$ can be computed from material parameters, namely Young modulus E, the shear modulus G and the density ρ_a, by means of

$$c = \sqrt{\frac{\rho_a}{E}} \quad \text{and} \quad \tilde{c} = \sqrt{\frac{\rho_a}{G}}. \tag{2.21}$$

In the boundary condition (2.20a), u_H is the brake motor control (upward hook force) and $\alpha\partial U/\partial t(0,t)$ represents a friction force of viscous type (where α is the viscous friction coefficient). In (2.19a), u_T represents the torque produced by the rotary table motor and $\beta\partial\Phi/\partial t(0,t) - u_T(t)$ designates the difference between the motor speed and rotational speed of the first pipe. It is assumed that the drilling system can be controlled by the boundary force u_H and the boundary torque u_T.

The model contains geometrical parameters of the drill string, that are assumed to be spatially and timely constant. These comprise the drillstring's cross-section Γ and its second moment of area J, as well as the mass M_B and the inertia moment[2] of the drill bit I_B. The function T considered in the bottom boundary conditions accounts for the frictional torque resulting from the interaction between the drill bit and the rock.

Notice that the boundary conditions (2.14b), (2.19b) and (2.20b), corresponding to the bottom of the rod, involve a frictional torque arising from the bit-rock interaction. The modeling of the torque on the bit constitutes a crucial aspect of the system description since it allows to reproduce the vibrational phenomena; this subject will be discussed in Chap. 3.

2.3 Neutral-Type Time-Delay Models

The wave equation model provides a realistic description of the distributed system variables; however, under certain circumstances, it is convenient to deal with a relatively simpler model involving just the primary interest variables. This section presents a direct procedure to derive, from the wave equations, equivalent input-output

[2] The inertia moment is such that $I_B = M_B r^2$, where r is taken as the averaged radius of drillpipe.

models described by neutral-type time-delay equations relating the variables at both ends of the drilling rod.

Integration along characteristics of the hyperbolic PDE allows the association of a certain system of functional differential equations to the mixed problem, more precisely, a one-to-one correspondence may be established and proved between the solutions of the mixed problem for hyperbolic PDE and the initial value problem for the associated system of functional equations [238].

By reducing a boundary value problem to a neutral-type time-delay equation we are able to exploit techniques from delay systems theory to gain insight into the complexity involved in the analysis and simulation of PDE models.

The *method of d'Alembert* provides a solution to the one-dimensional wave equation. Introducing the variables $\gamma = t + \tilde{c}s$ and $\eta = t - \tilde{c}s$, the general solution of the undamped wave equation (2.13), describing the torsional drilling behavior, is given by:

$$\Phi(s,t) = \varphi(\gamma) + \psi(\eta), \tag{2.22}$$

where φ and ψ are arbitrary continuously differentiable real-valued functions, with φ representing an arbitrary up-traveling wave and ψ an arbitrary down-traveling wave.

The boundary conditions (2.19a, 2.19b) can be rewritten as:

$$\tilde{c}\frac{\partial\varphi}{\partial\gamma}(t) - \tilde{c}\frac{\partial\psi}{\partial\eta}(t) = \frac{\beta}{GJ}\left(\frac{\partial\varphi}{\partial\gamma}(t) + \frac{\partial\psi}{\partial\eta}(t)\right) - \frac{1}{GJ}u_T(t), \tag{2.23}$$

$$\begin{aligned}\tilde{c}\frac{\partial\varphi}{\partial\gamma}(t+\tau) - \tilde{c}\frac{\partial\psi}{\partial\eta}(t-\tau) = &-\frac{I_B}{GJ}\left(\frac{\partial^2\varphi}{\partial\gamma^2}(t+\tau) + \frac{\partial^2\psi}{\partial\eta^2}(t-\tau)\right) \\ &-\frac{1}{GJ}T\left(\frac{\partial\varphi}{\partial\gamma}(t+\tau) + \frac{\partial\psi}{\partial\eta}(t-\tau)\right),\end{aligned} \tag{2.24}$$

where $\tau = \tilde{c}L$.

We define $\dot{\Phi}_b(t)$ as the angular velocity at the bottom extremity of the rod:

$$\dot{\Phi}_b(t) = \frac{\partial\Phi}{\partial t}(L,t) = \dot{\varphi}(t+\tau) + \dot{\psi}(t-\tau). \tag{2.25}$$

Equations (2.23) and (2.24) can be rewritten as:

$$\tilde{c}\dot{\varphi}(t) - \tilde{c}\dot{\psi}(t) = \frac{\beta}{GJ}\left(\dot{\varphi}(t) + \dot{\psi}(t)\right) - \frac{1}{GJ}u_T(t), \tag{2.26}$$

$$\tilde{c}\dot{\varphi}(t+\tau) - \tilde{c}\dot{\psi}(t-\tau) = -\frac{I_B}{GJ}\ddot{\Phi}_b(t) - \frac{1}{GJ}T\left(\dot{\Phi}_b(t)\right). \tag{2.27}$$

Equation (2.25) gives

$$\dot{\varphi}(t) = -\dot{\psi}(t-2\tau) + \dot{\Phi}_b(t-\tau), \tag{2.28}$$

substituting (2.28) into (2.27) yields

$$\dot{\psi}(t-\tau) = \frac{1}{2}\dot{\Phi}_b(t) + \frac{I_B}{2\tilde{c}GJ}\ddot{\Phi}_b(t) + \frac{1}{2\tilde{c}GJ}T\left(\dot{\Phi}_b(t)\right). \tag{2.29}$$

Substituting (2.29) into (2.28) gives

$$\dot{\varphi}(t) = \frac{1}{2}\dot{\Phi}_b(t-\tau) - \frac{I_B}{2\tilde{c}GJ}\ddot{\Phi}_b(t-\tau) - \frac{1}{2\tilde{c}GJ}T\left(\dot{\Phi}_b(t-\tau)\right). \tag{2.30}$$

We can write (2.26) as

$$\left(\tilde{c} - \frac{\beta}{GJ}\right)\dot{\varphi}(t) - \left(\tilde{c} + \frac{\beta}{GJ}\right)\dot{\psi}(t) = -\frac{1}{GJ}u_T(t). \tag{2.31}$$

Substituting the expressions for $\dot{\psi}$ and $\dot{\varphi}$ given in (2.29) and (2.30) into (2.31) yields:

$$\left(\tilde{c} - \frac{\beta}{GJ}\right)\left(\frac{1}{2}\dot{\Phi}_b(t-2\tau) - \frac{I_B}{2\tilde{c}GJ}\ddot{\Phi}_b(t-2\tau) - \frac{1}{2\tilde{c}GJ}T\left(\dot{\Phi}_b(t-2\tau)\right)\right)$$
$$-\left(\tilde{c} + \frac{\beta}{GJ}\right)\left(\frac{1}{2}\dot{\Phi}_b(t) + \frac{I_B}{2\tilde{c}GJ}\ddot{\Phi}_b(t) + \frac{1}{2\tilde{c}GJ}T\left(\dot{\Phi}_b(t)\right)\right) = -\frac{1}{GJ}u_T(t-\tau).$$

Simplifying, we get a torsional drilling model described by the neutral-type time-delay equation:

$$\begin{aligned}\ddot{\Phi}_b(t) - \Upsilon\ddot{\Phi}_b(t-2\tau) &= -\Psi\dot{\Phi}_b(t) - \Upsilon\Psi\dot{\Phi}_b(t-2\tau) - \frac{1}{I_B}T\left(\dot{\Phi}_b(t)\right)\\ &\quad + \frac{1}{I_B}\Upsilon T\left(\dot{\Phi}_b(t-2\tau)\right) + \Pi_\Omega u_T(t-\tau),\end{aligned} \tag{2.32}$$

where $\dot{\Phi}_b(t)$ is the angular velocity at the bottom extremity, and

$$\Pi_\Omega = \frac{2\Psi}{\beta + \tilde{c}GJ}, \quad \Upsilon = \frac{\beta - \tilde{c}GJ}{\beta + \tilde{c}GJ}, \quad \Psi = \frac{\tilde{c}GJ}{I_B}, \quad \tau = \tilde{c}L. \tag{2.33}$$

Similarly, the wave equation (2.18b) with boundary conditions (2.20a, 2.20b), corresponding to the axial drilling dynamics is transformed into the following neutral-type equation:

$$\begin{aligned}\ddot{U}_b(t) - \tilde{\Upsilon}\ddot{U}_b(t-2\tilde{\tau}) &= -\tilde{\Psi}\dot{U}_b(t) - \tilde{\Upsilon}\tilde{\Psi}\dot{U}_b(t-2\tilde{\tau}) - \frac{1}{M_B}T\left(\dot{\Phi}_b(t)\right)\\ &\quad + \frac{1}{M_B}\tilde{\Upsilon}T\left(\dot{\Phi}_b(t-2\tilde{\tau})\right) + \tilde{\Pi}_\rho u_H(t-\tilde{\tau}),\end{aligned} \tag{2.34}$$

where $\dot{U}_b(t)$ is the axial velocity at the bottom extremity, and

$$\tilde{\Pi}_\rho = \frac{2\tilde{\Psi}}{\alpha + cE\Gamma}, \quad \tilde{\Upsilon} = \frac{\alpha - cE\Gamma}{\alpha + cE\Gamma}, \quad \tilde{\Psi} = \frac{cE\Gamma}{M_B}, \quad \tilde{\tau} = cL. \tag{2.35}$$

It is important to point out that the delays τ and $\tilde{\tau}$ represent the time that the torsional and the axial waves take to travel from one to the other extremity of the drillstring. They are computed as follows: $\tau = L/v_\Phi$ and $\tilde{\tau} = L/v_U$, where L is the length of the drilling rod and v_Φ, v_U are the speeds of propagation of the rotational and longitudinal waves.

2.4 Notes and References

The most commonly used models for describing the drillstring behavior are the lumped parameter ones. The model accuracy is related to the involved DOF number. In [178], a one DOF model is used to investigate the effects of viscous damping, rotary speed, and natural frequency on the stick-slip phenomenon. A similar model is considered in [128] to design a control strategy allowing a smooth bit rotation to reduce axial and lateral drilling vibrations. A two DOF lumped parameter model is used in [43] to analyze torsional vibrations resulting from the perforation characteristics when drilling with a polycrystalline diamond compact (PDC) bit. This model is also used in [138, 266] to design control strategies (active damping and $\mathcal{H}_\infty$ control, respectively) to tackle torsional drilling oscillations. The empirical methods studied in [209] and the D-OSKILL control technique proposed in [51] are tested through a similar two DOF model. The sliding-mode control proposed in [210] is based on a discontinuous lumped parameter torsional model of four DOF. In [181], a discrete model of a drillstring system of eight DOF is proposed and used to study their nonlinear motions. Coupled axial-torsional dynamics of a drill string are studied in [182] through a discrete 32-segment model with 128 states that considers nonlinearities such as dry friction, loss of contact, and state-dependent time delays; the normal strain contours of the spatial-temporal system demonstrate the existence of strain wave propagation along the drill string. Different reduced-order models to describe the drilling behavior can be found in [177].

It is well known that lumped parameter models constitute a simplified representation of the physical mechanism neglecting the distributed nature of the system. The wave equation model provides higher accuracy in reproducing the rod oscillatory behavior. The very first analytic studies on the subject were carried out using the classical wave equation to describe the torsional behavior of drillstring assemblies [20]. Several recent contributions have adopted this modeling approach. In [53], a distributed parameter model subject to mixed boundary conditions is used to investigate the drilling system stability. A quite similar mathematical description is considered in [284] to design drilling vibration controllers. It is worth mentioning that the wave equation model plays a key role in the analysis and control of torsional drilling oscillations developed in the forthcoming chapters.

As explained in the previous paragraphs, in our opinion, the complexity involved in the analysis and simulations of a nonlinear partial differential equation can be avoided by transforming it into a delay system of neutral type. To the best of the author's knowledge, the transformation process, developed via the d'Alembert method, was presented for the first time in [3] and investigated in [62]. Furthermore, it was employed in [91] to study the controllability and motion planning of a flexible rod, and in [19] to analyze the drilling oscillatory dynamics.

A similar transformation of PDE boundary value problems to time-delay equations plays a significant role in a wide range of other biological, physical, and engineering problems. These include cardiovascular system dynamics [136], homeostatic mechanisms for maintaining blood pressure (the system model is presented in [223] and a bifurcation analysis in [265]), laser optical fibers, power transmission line networks [42], sonar/radar ranging technologies [25], and many other applications [134].

A different modeling method for describing structural waves propagating in a drill system is proposed in [129]. The proposed strategy is based on a vibration transfer matrix approach in which a drill pipe section is modeled through an analytical vibration transfer matrix between two sets of structural wave variables at the two ends of the pipe section. This model requires significantly less computational resources compared to the conventional Finite Element Method (FEM) [34], which requires a large number of meshes, complicating the numerical analysis of the system.

There are some other modeling techniques to describe drilling vibrations. For instance, in [114], the drillstring vibrations are investigated using both a coupled nonlinear elastodynamic mathematical model and a dynamic FEM model. In [85], a mechanics model has been developed to analyze axial and torsional vibrations in the frequency domain and provide vibration indices indicative of dysfunction in these modes. The model makes use of transfer matrices solve harmonic perturbations around a baseline solution obtained from a torque-and-drag type analysis.

A drilling system model is not complete if the frictional interface between the bit and the rock formation is not considered. In order to reproduce the nonlinearities giving rise to the drillstring vibration phenomena, an appropriate torque on bit model must be built. The following chapter addresses this important subject. It provides first the basic concepts of tribology that allows describing the force resisting the motion between two surfaces in contact; then, a compilation of the most popular modeling techniques which approximate the frictional force between the cutting device and the drilling surface is presented.

Chapter 3
Bit-Rock Frictional Interface

It is usually assumed that the growth of instabilities eventually leading to stick-slip and bit-bounce oscillations arises from the friction model, which empirically captures the interaction between the cutting device and the rock. An appropriate model of friction allows gaining insight into vibrational drillstring phenomena thus characterizing the dynamic behavior at bit level and making possible the development of appropriate control strategies to tackle this problem.

This chapter pursues two objectives: first, to compile the main classical friction models and second, to describe the main modeling strategies used to approximate the frictional torque on bit.

First, a summary of the most popular models of friction is presented. The most simple and widely used one is the Coulomb friction model, which helps predicting the direction and magnitude of the friction force between two bodies with dry surfaces in contact. Based on the structure of the Coulomb model, more elaborated friction models have been constructed. Most mechanisms involve complicate friction dynamics which are derived, for instance, from the presence of lubricants and adhesion characteristics. In order to take into account these physical features, several modeling approaches have been proposed. An important phenomenon studied in this chapter is the so-called *Stribeck effect* which occurs due to the irregular geometry of the contacting surfaces and from which the stick-slip phenomenon arises.

The second part of the chapter is devoted to the modeling of the friction due to the contact between the cutting device and the drilling surface. Most of the approaches include the Coulomb friction model and the model proposed by Karnopp. They consider the stiction and the Stribeck effects which entail the inclusion of decaying terms. A brief description of the different structures of the decaying terms is also presented.

It is worth mentioning that to properly reproduce the drilling vibrations, the general system model must be subject to a suitable friction law; the present chapter provides the basic ideas to choose an accurate frictional torque model.

M.B. Saldivar Márquez et al., *Analysis and Control of Oilwell Drilling Vibrations*,
Advances in Industrial Control, DOI 10.1007/978-3-319-15747-4_3

3.1 Friction Modeling

Friction is a natural phenomenon, occurring in all mechanical systems, described as the resistance to motion when two surfaces slide against each other. Friction is a useful property making ordinary things like walking and the brake in a car possible; however it can cause undesirable effects like steady-state errors, limit cycles, and poor performance.

Friction can also be described as the tangential reaction force between two surfaces in contact. These reaction forces are the results of different mechanisms which depend on physical aspects such as the contact geometry and topology, the material composition of the contact bodies, their relative displacement and velocity, and the presence of lubrication.

The friction phenomenon has been studied for hundreds of years. To the best of the authors' knowledge, the first experimental investigations were developed by Leonardo Da Vinci (1452–1519) in 1495. He observed the following friction properties:

1. The frictional force is proportional to the normal force, or load.
2. The frictional force is independent of the contact surface area.

These friction properties (currently known as the *Fundamental Laws of Friction*) were rediscovered by Guillaume Amontons (1663–1705) in 1699 and further developed by Charles-Augustin de Coulomb (1736–1806) in 1781.

This section is devoted to summarize the main classical models of friction allowing the characterization of the resisting force between surfaces in relative motion.

3.1.1 Coulomb Friction Model

Through detailed experimental investigations, Coulomb determined that the frictional force between two bodies which are pressed together with a normal force F_N, exhibit the following simple properties [233]:

Static Friction. In order to set in motion a body lying on an even surface in a state of rest, a critical force F_s (static friction force), must be overcome. This force is roughly proportional to the normal force F_N ($F_s = \mu_s F_N$). The coefficient μ_s, called static friction coefficient, depends on the pairing of the contacting materials, however, it is not dependent on contact area or roughness.

Kinetic Friction. This friction corresponds to the resisting force acting on a body after the static friction force has been overcome. Coulomb experimentally determined the following properties of kinetic friction F_R:

- Kinetic friction is proportional to the normal force $F_R = \mu_k F_N$.
- It shows no considerable dependence on the contact area or roughness of the surface.

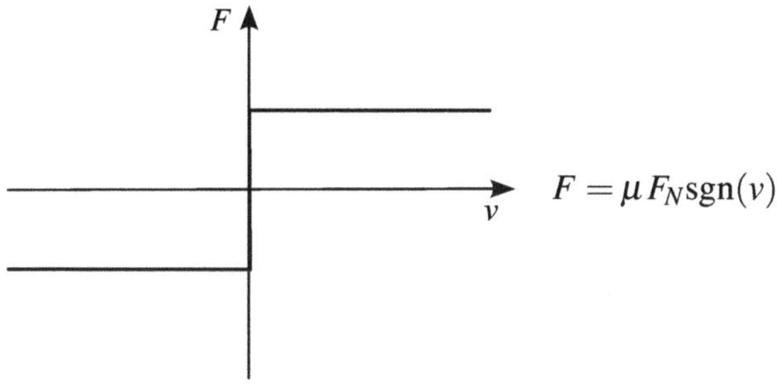

Fig. 3.1 Coulomb friction

- The coefficient of kinetic friction is approximately equal to the coefficient of static friction ($\mu_k \approx \mu_s$).
- The kinetic friction is independent of, or rather very weakly dependent on, the sliding velocity.

The most basic classical friction model, known as Coulomb model, is mathematically stated as follows:

$$F = F_C \mathrm{sgn}(v), \qquad F_C = \mu F_N, \tag{3.1}$$

where v is the relative velocity of the surfaces. Figure 3.1 shows the Coulomb friction curve. Notice that the Coulomb friction model does not specify the friction force for zero velocity. It may be zero or it can take on any value in the interval between $-F_C$ and F_C, depending on how the "sgn" function is defined [222].

3.1.2 Viscous Friction

It has been well known since ancient times that lubrication reduces frictional forces and wear, inhibiting direct contact between two bodies and, thereby, replacing dry friction with fluid friction. However, it was not until around 1880, that a pragmatic study of lubrication was developed. Nikolai Pavlovich Petrov (1836–1920) and Osborne Reynolds (1842–1912) recognized the hydrodynamic nature of lubrication, and introduced a theory of fluid film lubrication. Still today, Reynolds' steady-state equation[1] is valid for hydrodynamic lubrication of thick films (>μm).

The viscous friction force is normally described as:

$$F = F_v v,$$

where F_v is the viscous friction coefficient.

[1] The steady-state Reynolds' equation states that the frictional force, F, is proportional to both the sliding velocity, v, and the bulk fluid viscosity η, and is inversely proportional to the film lubricant thickness, D, i.e., $F \propto v\eta/D$.

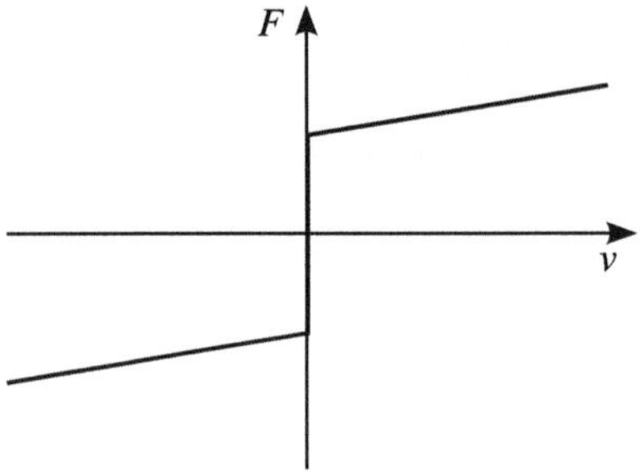

Fig. 3.2 Coulomb plus viscous friction

In several applications, viscous friction is combined with Coulomb friction, see Fig. 3.2.

A better approximation to experimental observations is obtained by a nonlinear dependence on velocity, e.g. [222],

$$F = F_v\,|v|^{\delta_v}\,\text{sgn}(v),$$

where δ_v depends on the geometry of the application.

3.1.3 Stiction

It is well known that surfaces in contact, moving relative to each other, experience a frictional force which is contributed by adhesion and deformation of the contacting asperities. In most cases, the primary friction contributor is the adhesion component, thus, the static friction force required to initiate the movement is larger than the kinetic friction force necessary to maintain the motion. Furthermore, in the presence of lubricants, high static friction may occur due to viscous effects; this is referred to as stiction [33].

The term stiction was coined at IBM General Products Division labs in San Jose, CA around 1980 when they encountered that head slider getting stuck to the disk surface while resting at high humidity due to liquid-mediated adhesion [32]. High lateral force had to be applied to initiate sliding to overcome high static friction or sticking.

Stiction counteracts external forces below a certain level and thus enable relative movement of stationary surfaces in contact. It is obvious that stiction cannot be described as a function of the velocity, but it must be dependent on the external force, as in the following model:

$$F = \begin{cases} F_e & \text{if } v = 0 \text{ and } |F_e| < F_s \\ F_s\text{sgn}(F_e) & \text{if } v = 0 \text{ and } |F_e| \geqslant F_s \end{cases}$$

The classical friction components can be combined in different ways, see Figs. 3.2 and 3.3; any such combination is referred to as a classical model [222].

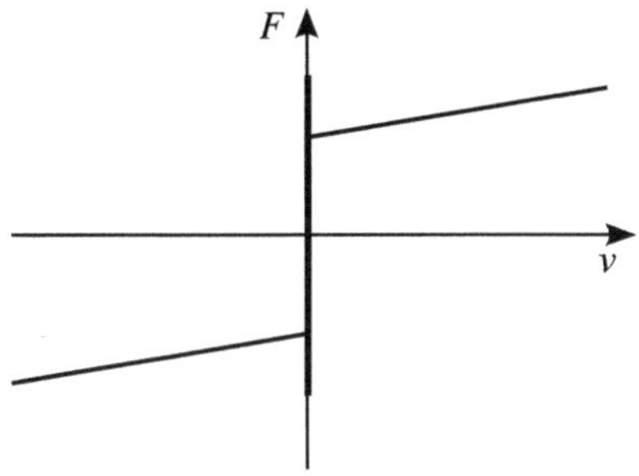

$$F = \begin{cases} F_e & \text{if } v = 0 \text{ and } |F_e| < F_s \\ F_s \mathrm{sgn}(F_e) & \text{if } v = 0 \text{ and } |F_e| \geqslant F_s \\ F_{C,v} & \text{if } v \neq 0 \end{cases}$$

Fig. 3.3 Stiction plus Coulomb and viscous friction

3.1.4 Stribeck Friction

An important contribution to the study of friction was developed in 1902 by Richard Stribeck (1861–1950) at the Royal Prussian Technical Testing Institute in Berlin. His research led to the characterization of the friction coefficient in lubricated surfaces [278].

Figure 3.4 shows the Stribeck curve which describes the friction behavior regarding the relative sliding speed.

The Stribeck curve comprises four phases [13]:

I. No sliding (zero velocity).
II. Boundary lubrication, the velocity is not enough to entrain fluid lubricant into the junction, then the sliding occurs with solid-to-solid contact.
III. Partial fluid lubrication, the velocity is adequate to entrain some fluid into the junction, but not enough to completely separate the surfaces.
IV. Full fluid lubrication, the surfaces are completely separated by a fluid film.

Fig. 3.4 Stribeck curve

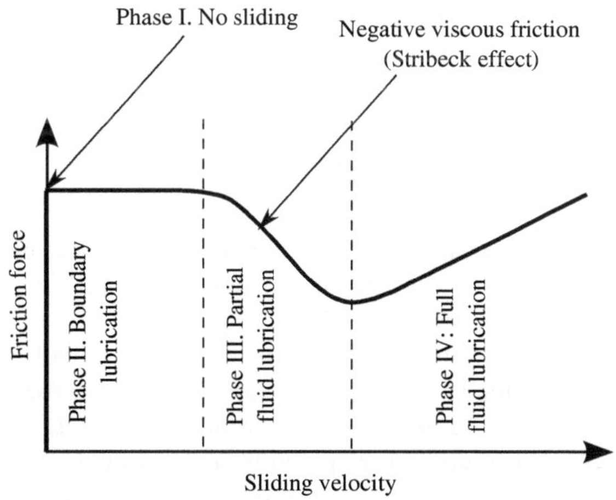

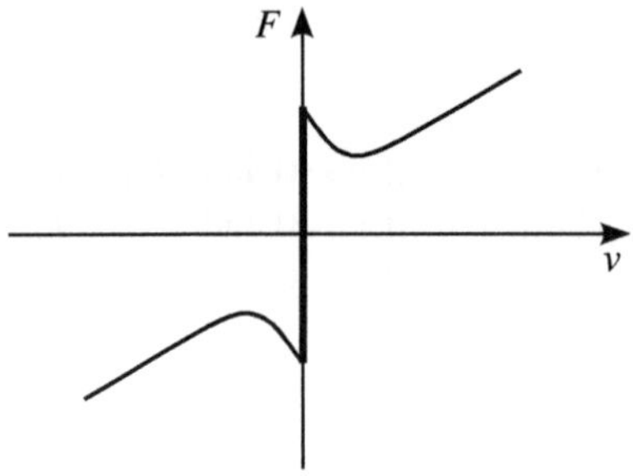

$$F = \begin{cases} F_e & \text{if } |v| < \varepsilon \text{ and } |F_e| < F_s \\ F_s \text{sgn}(F_e) & \text{if } |v| < \varepsilon \text{ and } |F_e| \geqslant F_s \\ F(v) & \text{if } |v| \geqslant \varepsilon \end{cases}$$

Fig. 3.5 Karnopp friction model

The negative going portion of the curve arises from the contact riding up on a lubricant film in the regime of partial fluid lubrication: as the lubricant film grows thicker with increasing velocity, the friction decreases. This portion of the curve gives a substantial destabilizing effect and it is usually considered as the source of the stick-slip phenomenon in several mechanisms [11].

According to the Stribeck curve, the friction force, in practical applications, does not decrease discontinuously (as in Fig. 3.3), but is continuously dependent on the sliding velocity (see for example, the Karnopp friction model in Fig. 3.5).

The Stribeck friction can be modeled as follows [222]:

$$F = \begin{cases} F(v) & \text{if } v \neq 0 \\ F_e & \text{if } v = 0 \text{ and } |F_e| < F_s \\ F_s \text{sgn}(F_e) & \text{otherwise} \end{cases}$$

where $F(v)$ is an arbitrary function which may take the form of the curve in Fig. 3.5. A common form of the nonlinearity is

$$F(v) = F_C + (F_s - F_C)\mathrm{e}^{-|v/v_\sigma|^{\delta_\sigma}} + F_v v$$

where v_σ is called the Stribeck velocity.

3.1.5 Karnopp Friction Model

In 1985, Dean Karnopp proposed a friction model to overcome the problems with zero velocity detection and to avoid switching between different state equations for sticking and sliding [145].

The friction Karnopp model considers a small velocity window near zero speed. Outside the velocity window (slip region), the friction force is an arbitrary function of velocity. Inside the velocity window (stick region), the friction force is determined by other forces in the system in such a way that the velocity remains constant until the breakaway force (stiction) value is reached, i.e., the friction force takes on the force required to keep zero acceleration. The system enters the sticking region when the speed is in the velocity window and the net external force is less than or equal to the limiting static friction force. The system leaves the sticking region when the net external force exceeds the breakaway value of fiction force.

The Karnopp friction model is widely used to describe the friction in practical actuators and mechanical problems [281].

3.1.6 Armstrong Friction Model

In 1991, Brian Armstrong-Hélouvry developed a seven-parameter friction model for control applications [12]. The proposed model includes Coulomb, viscous, and Stribeck friction plus frictional memory, time-dependent sticking friction, and preslip displacement. The Armstrong friction model effectively describes relevant behavior in several sticking or slipping scenarios, however, the main shortcoming is the necessity of identifying seven different friction parameters.

The proposed model consists of two sub models; the first one corresponds to the stiction phase, and the second to the sliding phase.

The friction force in the stiction phase is described by:

$$F(x) = \varsigma_0 x$$

where ς_0 is the tangential stiffness of the static contact (microstiffness) and x is the displacement of the friction surface. In the sliding phase, the friction force is modeled as:

$$\begin{cases} F(v,t) = \left(F_C + F_S(\gamma, t_d)\dfrac{1}{1 + (v(t-\tau_e)/v_S)^2}\right)\mathrm{sgn}(v) + F_v v \\ F_S(\gamma, t_d) = F_{S,a} + \left(F_{S,\infty} - F_{S,a}\dfrac{t_d}{t_d + \gamma}\right) \end{cases}$$

where F_S is the magnitude of the Stribeck friction, $F_{S,a}$ is the Stribeck friction at the end of the previous sliding period, $F_{S,\infty}$ is the Stribeck friction after a long time at rest (with a slow application of force), v_S is the characteristic velocity of the Stribeck friction, τ_e is the time constant of frictional memory, t_d is the time since becoming stuck (dwell time), and γ is a the temporal parameter of the rising static friction. The seven-parameter model captures the dynamic friction force by introducing a time delay in the sliding phase [290].

The model presented below is commonly used to numerically simulate the friction force between moving bodies [15]:

$$F = \begin{cases} \left(F_{v_{th}} + F_C + (F_s - F_C)\mathrm{e}^{-c_v v_{th}}\right) v/v_{th} & \text{if } |v| < v_{th} \\ (F_C + (F_s - F_C)\mathrm{e}^{-c_v|v|})\mathrm{sgn}(v) + F_v v & \text{if } |v| \geqslant v_{th} \end{cases} \tag{3.2}$$

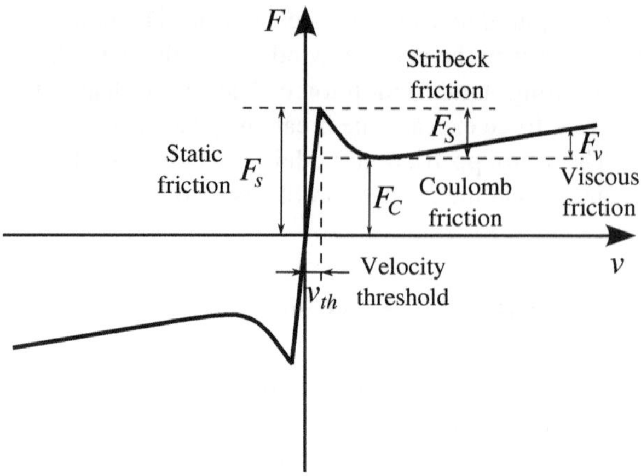

Fig. 3.6 Friction force between moving contact surfaces proposed in [15]

This friction model involves Stribeck, Coulomb, and viscous components. In order to avoid a discontinuity at $v = 0$, it considers a small region around zero velocity vicinity (velocity threshold v_{th}). In this region the friction force is assumed to be linearly proportional to the relative velocity v, see Fig. 3.6. Notice that when an acting force drops below the stiction level, the friction force described in (3.2) does not stop the relative motion, but the contact surfaces creep relative to each other at a small velocity proportional to the acting force.

3.1.7 Dahl Friction Model

In 1968, P.R. Dahl developed a simple dynamic model to simulate control systems with friction [65]. The experimental study, performed with ball bearings in servo systems, allowed to conclude that bearing friction behaves similar to solid friction.

The Dahl's model is based on the stress–strain curve in classical solid mechanics (see for instance [262]). It is well known that an elastic object subject to a small displacement returns to its original position; besides, the maximum stress decreases with increasing strain until rupture occurs. This property is reproduced by the Coulomb friction in the Dahl's model. Conversely, a large displacement results in a plastic deformation leading to a permanent displacement. The maximum stress that can be attained in the stress–strain characteristic is approximated by the stiction force.

The Dahl's model is given by the following differential equation:

$$\frac{dF}{dx} = \varsigma \left(1 - \frac{F}{F_C} \operatorname{sgn}(v) \right)^{\alpha},$$

where x is the relative displacement, ς is the stiffness coefficient, and the parameter α defines the shape of the curve. The case $\alpha = 1$ is frequently considered. Notice that if $|F(0)| < F_C$, then $|F|$ is always smaller than F_C, see Fig. 3.7.

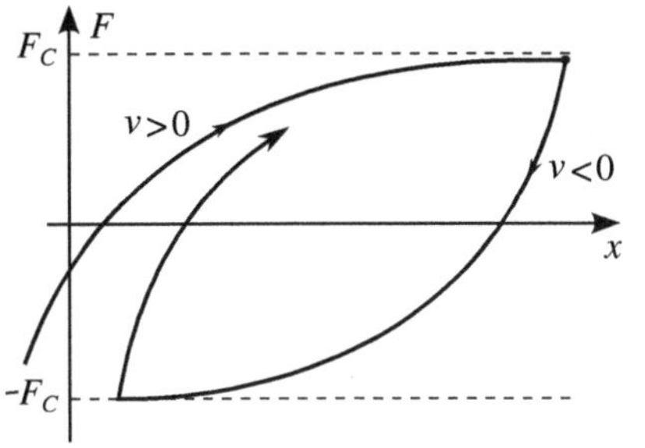

$$\frac{dF}{dt} = \varsigma v - \frac{F}{F_C}\varsigma\,|v|$$

Fig. 3.7 Dahl friction model

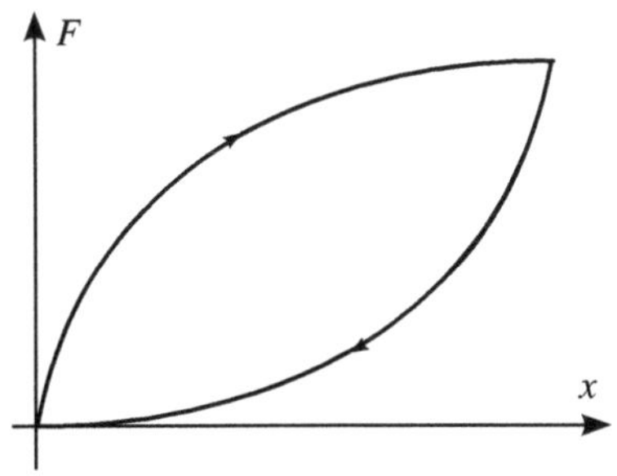

$$F = \varsigma_0 z + \varsigma_1 \dot{z} + \varsigma_2 v$$

Fig. 3.8 LuGre friction model

The above expression can be written as:

$$\frac{dF}{dt} = \frac{dF}{dx}\frac{dx}{dt} = \frac{dF}{dx}v = \varsigma\left(1 - \frac{F}{F_C}\operatorname{sgn}(v)\right)^{\alpha} v.$$

For $\alpha = 1$, it follows that

$$\frac{dF}{dt} = \varsigma v - \frac{F}{F_C}\varsigma\,|v|\,. \tag{3.3}$$

This expression effectively models predisplacements and hysteresis in dynamical systems, but it is unable to reproduce other phenomena like the Stribeck effect. However, the Dahl's model constitutes the basis of more sophisticated friction models such as the LuGre model reviewed below.

3.1.8 LuGre Friction Model

The LuGre model, introduced in [49], is derived from the Dahl friction model. In view of (3.3), the introduction of the state variable $z = F/\varsigma_0$ yields:

$$\frac{dz}{dt} = \frac{1}{\varsigma_0}\frac{dF}{dx}\frac{dx}{dt} = \frac{1}{\varsigma_0}\frac{dF}{dx}v = v - \varsigma_0\frac{|v|}{F_C}z$$

where v is the relative velocity and z is the internal friction state. Instead of F_C, the LuGre model considers a positive function $g(v)$ which is chosen according to the contact surfaces characteristics, such as temperature and lubrication:

$$\dot{z} = v - \varsigma_0 \frac{|v|}{g(v)} z.$$

When v is constant, the deflection z approaches the value

$$z_{ss} = \frac{v}{|v|} g(v) = g(v)\text{sgn}(v).$$

The LuGre model includes two additional terms to account for damping and viscous friction, as follows:

$$F = \varsigma_0 z + \varsigma_1 \dot{z} + \varsigma_2 v,$$

where ς_0 is the stiffness and ς_1 is a damping coefficient (see Fig. 3.8).

According to [49], the function $\varsigma_0 g(v) + \varsigma_2 v$ is determined by the steady-state friction force at constant velocity. The Stribeck effect can be reproduced by taking:

$$\varsigma_0 g(v) = F_C + (F_s - F_C)\,\mathrm{e}^{-(v/v_\sigma)^2},$$

where F_C, F_s, and v_σ are the Coulomb friction, the stiction force, and the Stribeck velocity, respectively. For steady-state motion, the relation between velocity and friction is defined by:

$$\begin{aligned} F_{ss}(v) &= \varsigma_0 g(v)\text{sgn}(v) + \varsigma_2 v \\ & F_C\text{sgn}(v) + (F_s - F_C)\, e^{-(v/v_\sigma)^2}\text{sgn}(v) + \varsigma_2 v. \end{aligned}$$

The LuGre friction model reproduces different phenomena such as the presliding displacement, the frictional lag, and the stick-slip motion.

3.2 Bit-Rock Interaction Laws

The description of the bit-rock interaction is a crucial aspect of drilling vibrations modeling since it is well known that the oscillation mechanism arises from the friction force at the bottom extremity. Classical friction models lead the characterization of the frictional torque on bit occurring during the drilling process. Several modeling approaches can be found in the specialized literature; this section presents the main bit-rock interaction laws approximating the physical phenomena at the bit level.

3.2.1 *Velocity Weakening Law*

In [53], the model:

$$T(\dot{U}_b, \dot{\Phi}_b) = \zeta \dot{U}_b \mathrm{e}^{-\bar{\alpha}\dot{\Phi}_b}$$

is chosen to represent the bit-rock interaction, $\dot{U}$ and $\dot{\Phi}_b$ stand for the axial and angular velocities at the bottom extremity and ζ denotes the ability of the rock to be cut.

3.2.2 Stiction Plus Coulomb Friction

In [138, 266], the torque on the bit is modeled through the expression:

$$T(\dot{\Phi}_b(t)) = c_b\dot{\Phi}_b(t) + T_{fb}(\dot{\Phi}_b(t))$$

where c_b is the damping viscous coefficient at the bit level and T_{fb} is the classical Coulomb plus static friction (dry friction) model, that is,

$$T_{fb}(\dot{\Phi}_b(t)) = \begin{cases} (T_{sb} - T_{cb})\text{sgn}(\dot{\Phi}_b(t)) & \text{if } \dot{\Phi}_b(t) \neq 0 \\ T_{sb} & \text{if } \dot{\Phi}_b(t) = 0 \end{cases} \tag{3.4}$$

with $T_{sb} = \mu_{sb}W_{ob}R_b$ and $T_{cb} = \mu_{cb}W_{ob}R_b$ the static and Coulomb friction torques, μ_{sb}, $\mu_{cb} \in (0, 1)$ the static and Coulomb friction coefficients, W_{ob} is the weight on the bit, and R_b is the bit radius.

The use of this modeling strategy is explained in [266] as follows: the maximum torque T_{sb} clamping the bit to zero speed is substantially larger than the Coulomb friction T_{cb} experienced when the bit is rotating. If $\dot{\Phi}_b = 0$, the friction torque will adjust to the torque in the drillstring maintaining a static equilibrium of the bit (see, for instance Fig. 3.9).

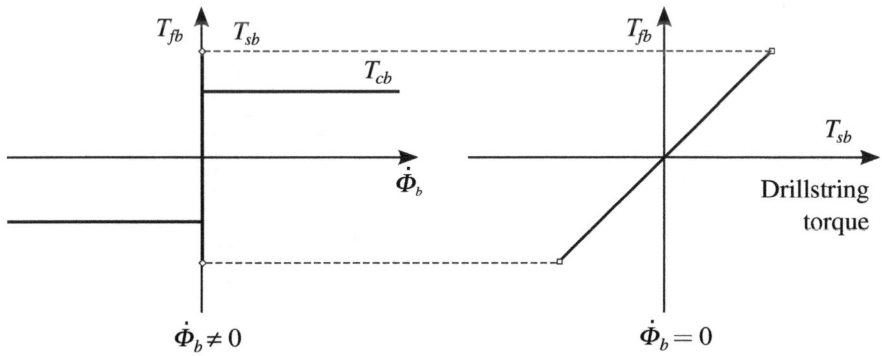

Fig. 3.9 Coulomb and static friction

3.2.3 Dry Friction Plus Karnopp's Model

Another common model for $T_{f_b}(\dot{\Phi}_b(t))$ is defined below

$$T_{f_b}(\dot{\Phi}_b(t)) = \begin{cases} T_{eb} & \text{if } \left|\dot{\Phi}_b(t)\right| < D_v,\ |T_{eb}| \leqslant T_{sb} \\ & \text{(stick)} \\ T_{sb}\mathrm{sgn}(T_{eb}) & \text{if } \left|\dot{\Phi}_b(t)\right| < D_v,\ |T_{eb}| > T_{sb} \\ & \text{(stick-to-slip transition)} \\ T_{cb}\mathrm{sgn}(\dot{\Phi}_b(t)) & \text{if } \left|\dot{\Phi}_b(t)\right| \geqslant D_v, \\ & \text{(slip)} \end{cases}$$

where $D_v > 0$ specifies a small enough neighborhood of $\dot{\Phi}_b(t) = 0$ and T_{eb} is the applied external torque that must overcome the static friction torque T_{sb} to make the bit move. For a lumped parameter model of the form (2.1), T_{eb} is modeled as follows:

$$T_{eb} = c\left(\dot{\Phi}_p - \dot{\Phi}_b\right) + k\left(\Phi_p - \Phi_b\right) - c_b\dot{\Phi}_b. \tag{3.5}$$

For a PDE model, the external torque T_{eb} can be described by:

$$T_{eb} = c\frac{\partial^2 \Phi}{\partial t \partial x}(L,t) + k\frac{\partial \Phi}{\partial x}(L,t) - c_b\frac{\partial \Phi}{\partial t}(L,t). \tag{3.6}$$

This model combines the dry friction model (3.4) with the Karnopp's model proposed in [145] and introduces a zero velocity band (see Fig. 3.10).

Fig. 3.10 Dry friction plus Karnopp's model

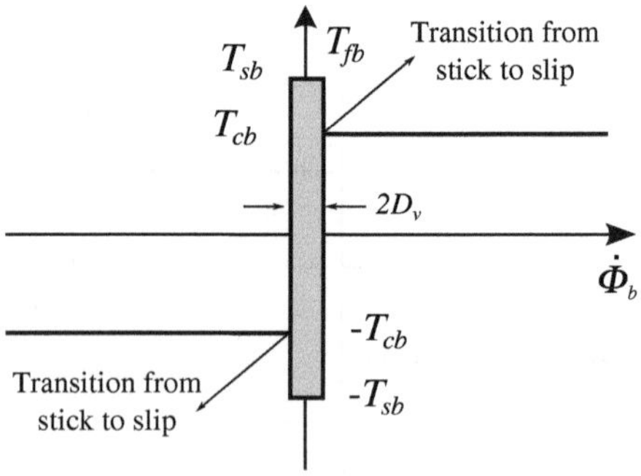

3.2.4 Karnopp's Model with a Decaying Friction Term

The function governing friction in the slip phase is chosen as a decaying function inspired by the experimental results given in [43]:

$$T_{f_b}(\dot{\Phi}_b(t)) = \begin{cases} \min\{|T_{eb}, T_{sb}|\}\mathrm{sgn}(T_{eb}) & \text{if } \left|\dot{\Phi}_b(t)\right| < D_v \\ f_b(\dot{\Phi}_b(t))\mathrm{sgn}(\dot{\Phi}_b(t)) & \text{if } \left|\dot{\Phi}_b(t)\right| \geqslant D_v \end{cases} \tag{3.7}$$

with

$$f_b(\dot{\Phi}_b(t)) = W_{ob}R_b\mu_b(\dot{\Phi}_b(t))$$
$$\mu_b(\dot{\Phi}_b(t)) = \frac{\mu_{sb} - \mu_{cb}}{1 + \gamma_b\left|\dot{\Phi}_b(t)\right|} + \mu_{cb},$$

T_{eb} is defined in (3.5) for an ODE model, and in (3.6) for a PDE model. The static friction torque T_{sb} is given by $T_{sb} = \mu_{sb}W_{ob}R_b$, where μ_b is the dry friction coefficient at the bit and γ_b is a positive constant defining the decaying velocity of T_{f_b} (see Fig. 3.11).

3.2.5 Karnopp's Model with an Exponential Decaying Friction Term

An alternative model for the torque on the bit defines an exponential decaying term in the slip phase [210]. It considers the expression given in (3.7) with:

$$f_b(\dot{\Phi}_b(t)) = W_{ob}R_b\mu_b(\dot{\Phi}_b(t)) \tag{3.8}$$
$$\mu_b(\dot{\Phi}_b(t)) = \mu_{cb} + (\mu_{sb} - \mu_{cb})\mathrm{e}^{-\gamma_b\left|\dot{\Phi}_b(t)\right|}. \tag{3.9}$$

Simulation results presented in [252] validate the proposed model.

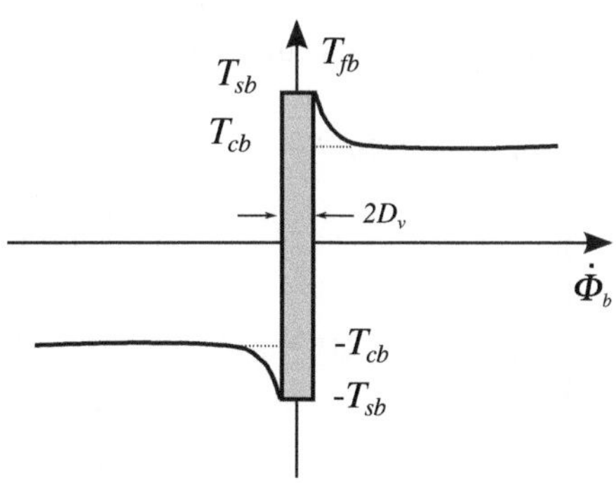

Fig. 3.11 Karnopp's model with a decaying friction term

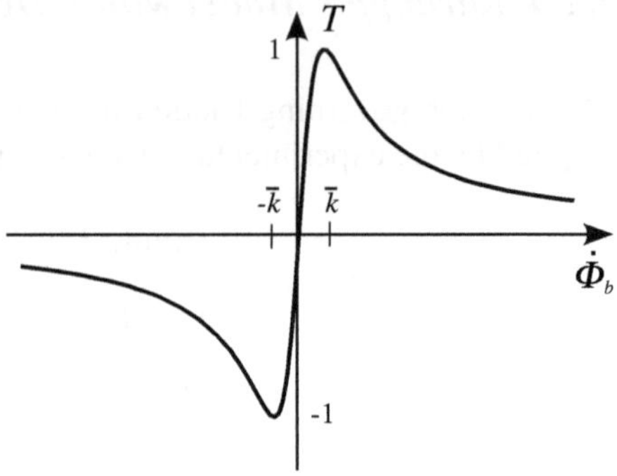

Fig. 3.12 Simplified torque on bit model proposed in [155]

3.2.6 Simplified Torque on Bit Model

Is worth mentioning that the models presented above may be complicated for practical purposes. For this reason, the construction of a simplified model which captures their essential dynamic properties is of interest.

In [155] the following model is introduced:

$$T(\dot{\Phi}_b(t)) = \frac{2\bar{k}\dot{\Phi}_b(t)}{\dot{\Phi}_b^2(t) + \bar{k}^2} \tag{3.10}$$

where $\bar{k}$ is a positive parameter. This model is simpler than those presented above and effectively reproduces the behavior of the friction at the bit level (see Fig. 3.12).

3.3 Friction-Driven Drilling Vibrations

It is well known that the proper choice of a friction model guarantees an effective reproduction of the stick-slip and bit-bounce phenomena. In this section, this fact is highlighted through simulations of the drilling system under different rock-bit frictional interface models.

Consider the pair of coupled neutral-type time-delay equations describing the torsional and axial drilling dynamics:

$$\begin{cases} \ddot{\Phi}_b(t) - \Upsilon\ddot{\Phi}_b(t-2\tau) = -\Psi\dot{\Phi}_b(t) - \Upsilon\Psi\dot{\Phi}_b(t-2\tau) - \frac{1}{I_B}T\big(\dot{\Phi}_b(t)\big) \\ \qquad + \frac{1}{I_B}\Upsilon T\big(\dot{\Phi}_b(t-2\tau)\big) + \Pi\Omega_0 \\ \ddot{U}_b(t) - \tilde{\Upsilon}\ddot{U}_b(t-2\tilde{\tau}) = -\tilde{\Psi}\dot{U}_b(t) - \tilde{\Upsilon}\tilde{\Psi}\dot{U}_b(t-2\tilde{\tau}) - \frac{1}{M_B}T\big(\dot{\Phi}_b(t)\big) \\ \qquad + \frac{1}{M_B}\tilde{\Upsilon}T\big(\dot{\Phi}_b(t-2\tilde{\tau})\big) + \tilde{\Pi}\rho_0, \end{cases} \tag{3.11}$$

where $\dot{\Phi}_b(t)$ and $\dot{U}_b(t)$ are the torsional and axial velocities at the bottom extremity of the drillstring, and

$$\Pi = \frac{2\Psi\beta}{\beta + \tilde{c}GJ}, \quad \tilde{\Pi} = \frac{2\tilde{\Psi}\alpha}{\alpha + cE\Gamma}.$$

Notations Υ, $\tilde{\Upsilon}$, Ψ, and $\tilde{\Psi}$ are defined in (2.33) and (2.35). The system is assumed to be subject to constant angular and axial velocities Ω_0, ρ_0 imposed at the surface.

The simulations presented below are developed by considering $\Omega_0 = 10\,\text{rad s}^{-1}$, $\rho_0 = 0.1\,\text{m s}^{-1}$, and the numerical parameters given in Table C.1 of Appendix C.

Firstly, let us consider a linear model of the frictional torque $T(\dot{\Phi}_b(t))$:

$$T(\dot{\Phi}_b(t)) = T_0\dot{\Phi}_b(t). \tag{3.12}$$

Figure 3.13 shows the axial and torsional trajectories of the drill bit when the linear ToB friction model (3.12) is considered. Notice that although torsional vibrations can be observed, the stick-slip phenomenon is not reproduced.

Consider now the velocity weakening law:

$$T(\dot{U}_b, \dot{\Phi}_b) = \zeta\,\dot{U}_b \mathrm{e}^{-\bar{\alpha}\dot{\Phi}_b}, \tag{3.13}$$

where ζ is a constant related to the rock hardness.

Using the friction model (3.13), which considers an exponential decaying term, the stick-slip behavior can be observed within the first few seconds of the simulation, see Fig. 3.14.

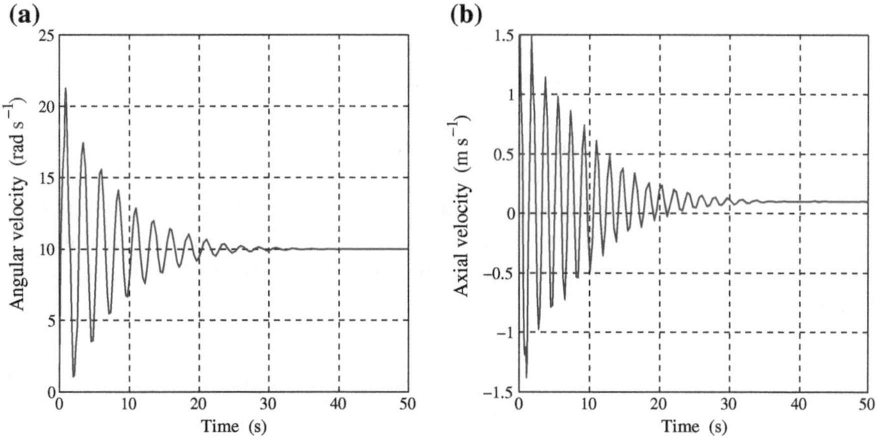

Fig. 3.13 Simulation of model (3.11) under the linear model of the frictional torque on bit (3.12) for $T_0 = 25\,\text{Nms rad}^{-1}$. **a** Bit angular velocity. **b** Bit axial velocity

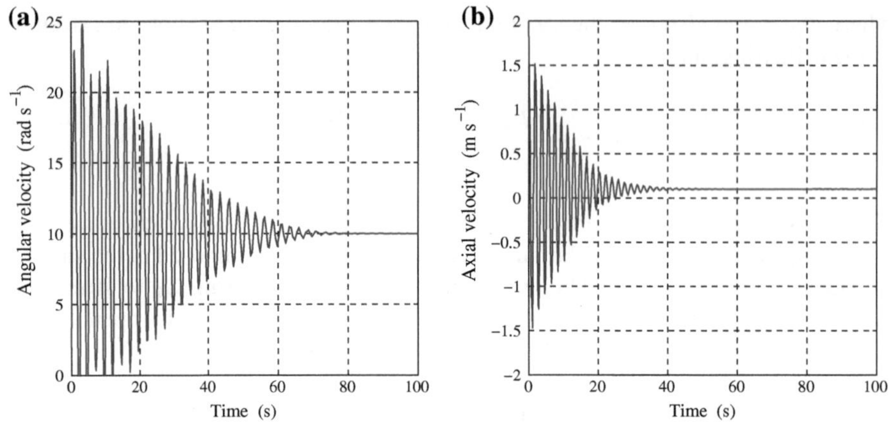

Fig. 3.14 Simulation of model (3.11) under the velocity weakening law (3.13) describing the frictional ToB for $\zeta = 650\,\text{Ns}$ and $\bar{\alpha} = 0.1$. **a** Bit angular velocity. **b** Bit axial velocity

A more sophisticated model to describe the rock-bit interaction is given by the following nonlinear function, derived from the decaying friction model (3.7)–(3.9):

$$T\left(\dot{\Phi}_b(t)\right) = c_b\dot{\Phi}_b(t) + W_{ob}R_b\mu_b\left(\dot{\Phi}_b(t)\right)\text{sgn}\left(\dot{\Phi}_b(t)\right). \qquad (3.14)$$

The term $c_b\dot{\Phi}_b(t)$ represents the viscous damping torque at the bottom end and the expression $W_{ob}R_b\mu_b\left(\dot{\Phi}_b(t)\right)\text{sgn}\left(\dot{\Phi}_b(t)\right)$ approximates the dry friction torque. Notations R_b and W_{ob} stand for the bit radius and the weight on bit, respectively. The friction coefficient $\mu_b(\dot{\Phi}_b(t))$ is given by:

$$\mu_b\left(\dot{\Phi}_b(t)\right) = \mu_{cb} + (\mu_{sb} - \mu_{cb})\text{e}^{-\gamma_b|\dot{\Phi}_b(t)|}, \qquad (3.15)$$

where μ_{cb}, μ_{sb} denote the Coulomb and static friction coefficients, the constant $0 < \gamma_b < 1$ defines the velocity decrease rate.

A simulation of the drilling system model (3.11) subject to the friction torque model (3.14) with the friction coefficient given in (3.15) is shown in Fig. 3.15. Observe that the angular velocity varies between zero and twice the reference velocity $\Omega_0 = 10\,\text{rad s}^{-1}$, which clearly characterizes the stick-slip phenomenon.

As reviewed before, the frictional torque on bit can be approximated by the function (3.10); a simulation of the drilling system allows assessing the accuracy of this simplified model. Figure 3.16 shows the angular and axial bit trajectories for a frictional model of the form:

$$T(\dot{\Phi}_b(t)) = \frac{2p\bar{k}\dot{\Phi}_b(t)}{\dot{\Phi}_b^2(t) + \bar{k}^2}, \quad p > 0,\ \bar{k} > 0. \qquad (3.16)$$

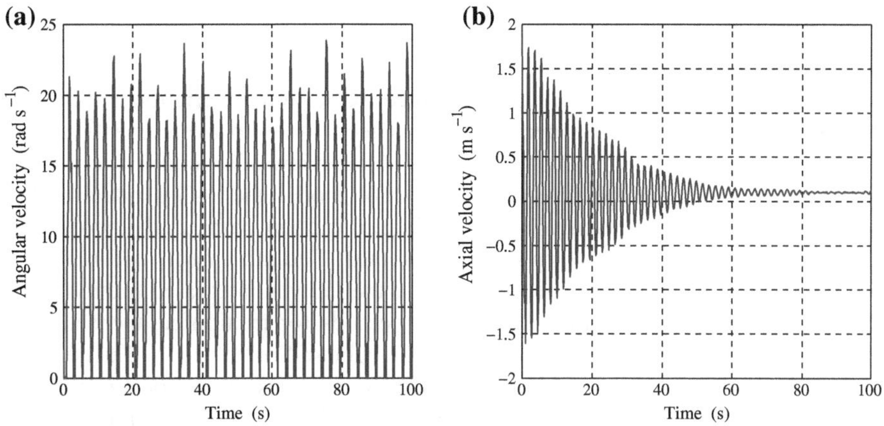

Fig. 3.15 Simulation of model (3.11) subject to the Karnopp-type friction function (3.14) with the friction coefficient given in (3.15). **a** Bit angular velocity. **b** Bit axial velocity

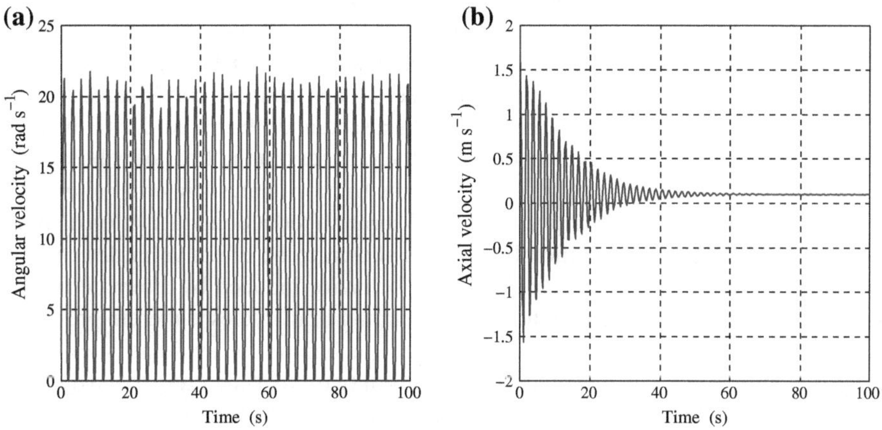

Fig. 3.16 Simulation of model (3.11) under the simplified frictional torque on bit model (3.16) for $p = 1{,}000$. **a** Bit angular velocity. **b** Bit axial velocity

In this context, it is important to point out that the similarity between the bit trajectories depicted in Figs. 3.15 and 3.16 is remarkable, showing that the simplified friction function accurately approximates the Karnopp-type friction model.

3.4 Notes and References

There is a wide range of contributions devoted to the study of friction and its compensation. Several modeling approaches and techniques to overcome friction in mechanical systems have been proposed. In [16] a review of several friction models is presented. It includes the classical Dahl friction model, the LuGre model, and the Bliman-Sorine model. An analysis on the effects of friction, particularly the stick-slip phenomenon, in control systems is also presented. The aforementioned models of friction and some other classical ones, such as the Leuven integrated friction model can be found in [290]. Furthermore, some methods for friction compensation are therein reviewed. In [222], a comparison of the Bliman-Sorine and the LuGre model is presented. The considered aspects include the analysis of rate dependency, the oscillatory behavior at low velocities, damping, and dissipativity. Control systems applications are also discussed. In [49], a dynamic model for friction is proposed. It includes the Stribeck effect, hysteresis, spring-like characteristics for stiction, and varying breakaway force. A survey of models, analysis tools, and compensation methods for the control of machines with friction can be found in [14]. In [27], a review of the literature on friction modeling for dynamical systems across engineering and science disciplines is presented. Another survey of friction laws is presented in [17]. It is focused on dry friction in engineering applications; advantages and disadvantages of frictional effects are therein discussed and illustrated. More than 300 references cited these survey papers evince the progress in the field of tribology.

Numerous research papers are devoted to the friction analysis in order to treat the oscillatory behavior of mechanical systems. Vibration reduction via friction dampers is analyzed in [187]; the Coulomb model is shown to be effective in determining the maximum energy dissipation. Stick-slip behavior induced by alternate friction models is studied in [171]; a method for calculating limit cycles is therein presented. In [80], a proportional-derivative (PD) control technique for avoiding stick-slip is proposed; the control design entails the analysis of two experimentally based dynamic friction models. In [299], the friction forces giving rise to torsional oscillations of a rotor with continuous stator contact are studied; a reduced-order model has been used to develop analytical and numerical studies of the system.

Much attention has been paid to the study of the frictional torque occurring at the rock-bit interface in drilling systems giving rise to harmful oscillations. Several modeling approaches have been proposed; see for instance [64], where two different representations of Coulomb friction in the context of a dynamic simulation of torsional drilling vibrations are analyzed. Different friction modeling approaches and an analysis of stick-slip behavior in a drilling system can be found in [208]; the proposed models are oriented to avoid simulation problems due to the discontinuities originated by the presence of dry friction. A frictional torque model derived from field data measured by a sensing device known as *Télévigile* was introduced in [226]. This experimentally obtained model is validated in [8]; a specially designed test rig with a braking device was constructed for this purpose.

Besides the friction force arising from the contact between the cutting surface and the bit, the regenerative phenomena have influence on the occurrence of vibrations. In drilling operations, the depth of cut is determined by both the current and previous position of the bit. The regenerative cutting effect, which refers to this dependence on the current and previous states, is analyzed in [183] through a PDE model with switching boundary conditions.

The models presented in Chap. 2, coupled to the friction laws reviewed in this chapter allow an acceptable representation of the vibrations occurring when drilling an oil well. The wave equation model describing the torsional and axial propagating waves along the drilling rod constitutes a reliable description of the undesirable drilling vibrations since it considers the infinite-dimensional nature of the system. The whole drillstring is considered as a flexible pendulum subject to external forces acting at their both extremities. An improvement in the model accuracy is achieved by considering the different characteristics of each of the drillstring components, for example, a distinction between the drill pipes and the drill collars should be taken into account. The next chapter presents a comprehensive model that considers such differences; furthermore, a frictional ToB model that comprises the bit geometry and the drilling surface characteristics is reviewed.

Chapter 4
Comprehensive Modeling of a Vertical Oilwell Drilling System

Classical models of the drillstring and different modeling approaches to approximate the frictional interface at the bit level were presented in preceding chapters. As previously explained, series pipe sections and the BHA (consisting of drill collars and the bit) constituting the drillstring are commonly regarded as a single flexible beam under the action of external forces at both extremities. A precise representation of the entire oilwell drilling system requires an independent modeling of each set of components sharing consistent physical characteristics.

This chapter provides a comprehensive mathematical description of the system which embeds individual sub models corresponding to different drillstring sections. The proposed distributed parameter modeling scheme considers the drilling dynamics at different levels of the string:

- **Top extremity**. To properly describe the string behavior at the upper end, it is essential to consider the actuators' dynamics; induction and DC motor models are then recalled. Furthermore, top boundary conditions must be defined to characterize axial-torsional drilling trajectories at this level.
- **Drillstring sections**. To accurately model the drillstring dynamics, it is important to make a distinction between the different sections composing the drilling rod, i.e., the series of drillpipes of length L_p and the BHA of length L_b, so that $L = L_p + L_b$, where L is the total length of the rod, as illustrated in Fig. 4.1.
- **Bottom extremity**. The system trajectories at the bit level are characterized by the bottom boundary conditions involving a nonlinear term to represent the bit-rock contact from which undesirable vibrations arise. This chapter presents an accurate experience-based model of the frictional torque which considers the drilling surface characteristics and the bit geometry.

For the sake of clarity, the model parameters used in this chapter are defined in Table C.2 of Appendix C.

M.B. Saldivar Márquez et al., *Analysis and Control of Oilwell Drilling Vibrations*, Advances in Industrial Control, DOI 10.1007/978-3-319-15747-4_4

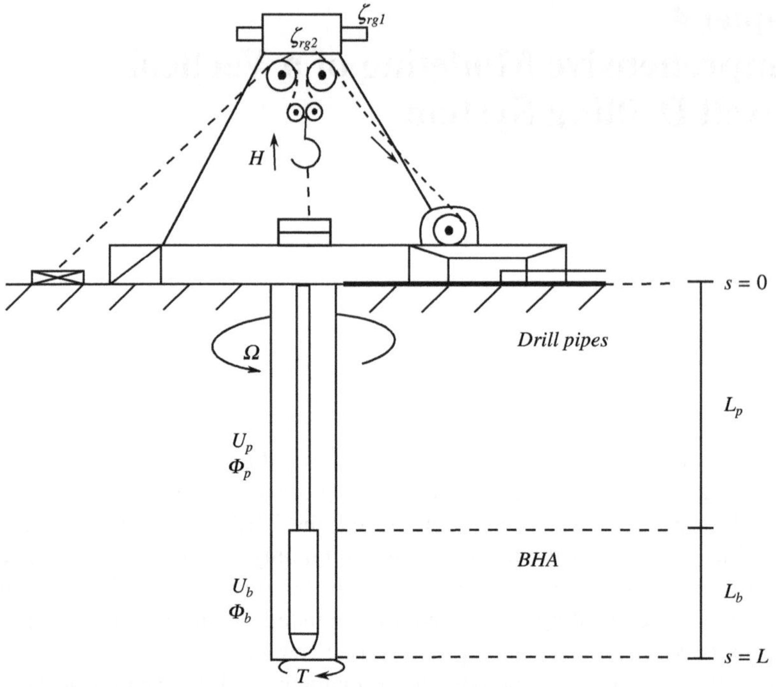

Fig. 4.1 Simplified scheme depicting an oilwell rotary drilling system

4.1 Top Drilling Dynamics

4.1.1 Actuators

The drilling system includes essentially three motors converting electrical energy into mechanical energy: one of them produces the rotational movement of the string (rotary table), the second one provides a means of raising and lowering the traveling blocks allowing the drill string to be moved up and down (drawworks), and the third one adds additional power to the bit while drilling (mud pump) and serves to drain crushed rock. Each engine is modeled by a system of mechatronic equations as outlined below.

DC Motor

A common type of motor is the direct current armature control motor. The torque developed by this motor is proportional to the stator's flux and the current in the armature:

$$\Gamma_t = k_f \, \psi \, K_a \, I, \tag{4.1}$$

where Γ_t is the shaft torque, ψ is the magnetic flux in the stator field, which is assumed to be constant and I is the current in the motor armature. Since the flux is maintained constant, we can also write

$$\Gamma_t = k_T\, I \quad \text{where} \quad k_T = k_f\, \psi\, K_a. \tag{4.2}$$

When a current carrying conductor passes through a magnetic field, a voltage V_b corresponding to the so-called back electromagnetic force, appears:

$$V_b = k_e\, \omega, \tag{4.3}$$

where ω is the rotation speed of the motor shaft. The constants k_T and k_e have the same value. Kirchhoff's law yields the electric equation of the motor:

$$V - V_{res} - V_{coil} - V_b = 0, \tag{4.4}$$

where V is the input voltage, $V_{res} = -R\,I$ is the armature resistor voltage (R being the armature resistor), and $V_{coil} = L\,\dot{I}$ is the armature inductance voltage (L being the armature inductance). The motor's electrical equation is then

$$L\,\dot{I} = -k_e\, \omega - R\,I + V. \tag{4.5}$$

Induction Motor

Another common type of actuator is the induction motor. When AC current is applied to an induction engine, a rotating magnetic field is set up in the stator. This rotating field moves with respect to the rotor windings, inducing a current flow in the rotor. The current flowing in the rotor windings sets up its own magnetic field. In the stationary reference frame ($\omega_S = 0$), the stator voltage vector is expressed as:

$$v_s{}^S = R_s i_s{}^S + \dot{\psi}_s^S, \tag{4.6}$$

where i_S^S and ψ_S^S are the stator current and rotor flux vectors. In the same way, the rotor voltage vector is expressed in terms of a rotating frame (ω_R) fixed with respect to the rotor:

$$v_r{}^R = R_r i_r{}^R + \dot{\psi}_r^R, \tag{4.7}$$

where i_r^R and ψ_r^R are the rotor current and rotor flux vectors. A transformation in an arbitrary rotating reference frame (ω_k) yields:

$$\begin{cases} v_s^k = R_s i_s^k + \dot{\psi}_s^k + jp\omega_k \psi_s^k \\ v_r^k = R_r i_r^k + \dot{\psi}_r^k + jp(\omega_k - \omega_r)\psi_r^k. \end{cases} \tag{4.8}$$

The flux vectors are expressed as:

$$\begin{cases} \psi_s^k = L_s i_s^k + L_m i_r^k \\ \psi_r^k = L_r i_r^k + L_m i_s^k. \end{cases} \tag{4.9}$$

Hereinafter, the so-called d/q-reference frame, which is aligned to the rotor flux vector, will be used. The superscripts will be omitted, i.e.,

$$\psi_r = \psi_{rd} + j\psi_{rq} = \psi_r{}^S e^{-j\varsigma}, \tag{4.10}$$

in the above expression, ς stands for the rotor flux angle in the stationary reference frame.

A nonlinear differential equation model of the induction motor consisting of a four-dimensional electric representation and of a two-dimensional mechanical description is obtained by substituting the stator flux and rotor current vectors in (4.8), using (4.9) and introducing the variables $\eta = 1-(L_m^2/L_sL_r)$, $\chi = L_m^2 R_r/\sigma L_s L_r^2 + R_s/\sigma L_s$, $\zeta = R_r/L_r$, and $\xi = L_m/\sigma L_s L_r$.

To further simplify the notations, we shall set for the components of the rotor flux and the current and voltage of the stator in the d/q reference frame: $\psi_{rd} = \psi_d$, $I_{sd} = I_d$, $I_{sq} = I_q$, $v_{sd} = v_d$, and $v_{sq} = v_q$.

The current/flux equations are given by:

$$\begin{cases} \dot{\psi}_d = -\zeta(\psi_d - L_m I_d) \\ \dot{\varsigma} = p\omega_R + \zeta L_m \dfrac{I_d}{\psi_d} \\ \dot{I}_d = -\chi I_d + \zeta\xi\psi_d + p\omega_R I_q + \zeta L_m \dfrac{I_q^2}{\psi_d} + \dfrac{v_d}{\eta L_s} \\ \dot{I}_q = -\chi I_q - p\xi\omega_R I_d - \zeta L_m \dfrac{I_q I_d}{\psi_d} + \dfrac{v_q}{\eta L_s}, \end{cases}$$

and the mechanical model is defined by

$$\begin{cases} J\dot{\omega}_R = \mu\psi_{rd} I_{sq} - T_l \\ \dot{\theta}_R = \omega_R, \end{cases} \tag{4.11}$$

where $\mu = 3pM/2L_r$, θ_R is the rotor angle, and T_l is the load torque.

4.1.2 Top Boundary Conditions

In view of the physical aspects of the system at the upper extremity ($s = 0$), the top boundary condition, corresponding to the torsional drilling dynamics, is given by:

$$GJ_p \frac{\partial \Phi_p}{\partial s}(0,t) = -I_T \frac{\partial^2 \Phi_p}{\partial t^2}(0,t) + \beta \frac{\partial \Phi_p}{\partial t}(0,t) - u_T(t), \tag{4.12}$$

where Φ_p is the angular position of the drill pipes, I_T is the top-drive inertia, G is the shear modulus of the string, J_p the polar moment of inertia (second area moment) of one pipe section, and u_T is the torque produced by the rotary table motor, taken as a control input.

Note that the above expression can be regarded as the mechanical part of an induction motor model and for the sake of uniformity, it could be considered instead of (4.11).

The boundary condition corresponding to the axial motion at the top of the string is given by:

$$E\Gamma_p \frac{\partial U_p}{\partial s}(0,t) = -M_T \frac{\partial^2 U_p}{\partial t^2}(0,t) + \alpha \frac{\partial U_p}{\partial t}(0,t) - u_H(t), \tag{4.13}$$

where U_p is the longitudinal drill pipes position, M_T the top-drive mass, E is the Young modulus of the drillstring steel, Γ_p is the pipes cross-section, and u_H is the control input corresponding to the upward hook force.

The variables Γ_p and J_p are given by:

$$\Gamma_p = \pi(r_{po}^2 - r_{pi}^2), \qquad J_p = \frac{\pi}{2}(r_{po}^4 - r_{pi}^4),$$

with r_{po} and r_{pi} the outer and inner pipe radius.

The initial conditions are taken such that Φ_p, $\frac{\partial \Phi_p}{\partial t}$, $\frac{\partial \Phi_p}{\partial s}$, U_p, $\frac{\partial U_p}{\partial t}$, $\frac{\partial U_p}{\partial s}$ vanish at $t = 0$.

Following [227], the dynamics taking place at the upper part of the drilling structure (comprising the derrick, the crown block and the traveling block) are modeled by a coupled spring-mass system:

$$\begin{cases} M_{rg_1}\ddot{\zeta}_{rg_1}(t) + \gamma_{rg_1}\dot{\zeta}_{rg_1}(t) + M_{rg_1}g = u_F(t) + k_{rg_{12}}(\zeta_{rg_2}(t) - \zeta_{rg_1}(t)) - k_{rg_{01}}\zeta_{rg_{ini}}(t) \\ M_{rg_2}\ddot{\zeta}_{rg_2}(t) + \gamma_{rg_2}\dot{\zeta}_{rg_2}(t) + M_{rg_2}g = -u_H(t) - k_{rg_{12}}(\zeta_{rg_2}(t) - \zeta_{rg_1}(t)) \end{cases} \tag{4.14}$$

where ζ_{rg_2} accounts for vibrations in cables, crown, and traveling blocks, ζ_{rg_1} approximates the vibrations of all other components of the upper drilling rig mechanism, the effort $k_{rg_{01}}\zeta_{rg_{ini}}$ represents the ground reaction force, and $u_F(t) = k_{rg_{01}}(\zeta_{rg_1}(t) - \zeta_{rg_0}(t))$ is a tension force in the cable at the drawworks level considered as an additional control input (it is directly related to the drawworks rotation motor). The parameters M, γ, and k (with corresponding subscripts) stand for equivalent masses, damping, and stiffness coefficients.

4.2 Drillstring Dynamics

4.2.1 Drilling Pipe

The drilling pipe deformation is modeled through a pair of wave equations involving viscous and viscoelastic Kelvin-Voigt internal damping:

$$\begin{cases} \rho_a J_p \dfrac{\partial^2 \Phi_p}{\partial t^2}(s,t) = G J_p \dfrac{\partial^2 \Phi_p}{\partial s^2}(s,t) + \varepsilon^i_{\Phi_p} \dfrac{\partial^3 \Phi_p}{\partial t \partial s^2}(s,t) - \gamma^v_{\Phi_p} \dfrac{\partial \Phi_p}{\partial t}(s,t) \\ \rho_a \Gamma_p \dfrac{\partial^2 U_p}{\partial t^2}(s,t) = E \Gamma_p \dfrac{\partial^2 U_p}{\partial s^2}(s,t) + \varepsilon^i_{U_p} \dfrac{\partial^3 U_p}{\partial t \partial s^2}(s,t) - \gamma^v_{U_p} \dfrac{\partial U_p}{\partial t}(s,t), \end{cases} \tag{4.15}$$

where $0 < s < L_p$ and ρ_a represents the steel density. The internal damping coefficients are denoted by $\varepsilon^i_{U_p}$, $\varepsilon^i_{\Phi_p}$, and the viscous damping coefficients by $\gamma^v_{U_p}$, $\gamma^v_{\Phi_p}$.

4.2.2 Bottom Hole Assembly (BHA)

Following the previous modeling approach, the governing equations of axial and torsional movement of the BHA are defined as:

$$\begin{cases} \rho_a J_b \dfrac{\partial^2 \Phi_b}{\partial t^2}(s,t) = G J_b \dfrac{\partial^2 \Phi_b}{\partial s^2}(s,t) + \varepsilon^i_{\Phi_b} \dfrac{\partial^3 \Phi_b}{\partial t \partial s^2}(s,t) - \gamma^v_{\Phi_b} \dfrac{\partial \Phi_b}{\partial t}(s,t) \\ \rho_a \Gamma_b \dfrac{\partial^2 U_b}{\partial t^2}(s,t) = E \Gamma_b \dfrac{\partial^2 U_b}{\partial s^2}(s,t) + \varepsilon^i_{U_b} \dfrac{\partial^3 U_b}{\partial t \partial s^2}(s,t) - \gamma^v_{U_b} \dfrac{\partial U_b}{\partial t}(s,t), \end{cases} \tag{4.16}$$

where $L_p < s < L$, and Γ_b, J_b are the cross-section and the polar inertia moment of one BHA section, given by

$$\Gamma_b = \pi(r_{bo}^2 - r_{bi}^2), \qquad J_b = \frac{\pi}{2}(r_{bo}^4 - r_{bi}^4),$$

with r_{bo} and r_{bi} the outer and inner drill collar radius.

To achieve continuity in speed and effort, Φ_p, Φ_b, U_b, and U_p must satisfy the following connection conditions:

$$\begin{cases} \dfrac{\partial \Phi_b}{\partial t}(L_p,t) = \dfrac{\partial \Phi_p}{\partial t}(L_p,t) \\ \dfrac{\partial \Phi_b}{\partial s}(L_p,t) = \dfrac{J_p}{J_b}\dfrac{\partial \Phi_p}{\partial s}(L_p,t) \\ \dfrac{\partial U_b}{\partial t}(L_p,t) = \dfrac{\partial U_p}{\partial t}(L_p,t) \\ \dfrac{\partial U_b}{\partial s}(L_p,t) = \dfrac{\Gamma_p}{\Gamma_b}\dfrac{\partial U_p}{\partial s}(L_p,t), \end{cases} \tag{4.17}$$

where J_* and Γ_* are given by $J_* = \pi(r_{*o}^4 - r_{*i}^4)/2$ and $\Gamma_* = \pi(r_{*o}^2 - r_{*i}^2)$, with $* \in \{p, b\}$.

4.3 Downhole Drilling Dynamics

4.3.1 Bottom Boundary Conditions

Torsional dynamics occurring at the lower end of the string ($s = L$) are approximated by the following equation:

$$GJ_b \frac{\partial \Phi_b}{\partial s}(L,t) = -I_B \frac{\partial^2 \Phi_b}{\partial t^2}(L,t) - T(t), \tag{4.18}$$

where Φ_b stands for the rotational movement of the bit and T represents the reaction torque at the lower end.

The bottom boundary condition corresponding to the longitudinal string motion is given by:

$$E\Gamma_b \frac{\partial U_b}{\partial s}(L,t) = -M_B \frac{\partial^2 U_b}{\partial t^2}(L,t) - W(t) \tag{4.19}$$

where U_b represents the bit axial position, M_B the mass of the bit and $W(t)$ a reaction force due to the so-called dynamic weight on bit.

4.3.2 Bit-Rock Interface

As explained in Chap. 3, there are several modeling approaches to approximate the interaction between the cutting tool and the rock formation; most of them consider the bit-rock interface as an equivalent frictional contact. However, when considering a drag-type bit, it is convenient to characterize the action of rock cutting by individual cutters.

A refined frictional interface representation is proposed in [243]. It considers a drag bit composed by n identical radial blades regularly spaced by an angle equal to $2\pi/n$. When such a bit is drilling rock, the depth of cut per blade d_n (i.e., the thickness of the rock ridge in front of the blade) is constant along the blade and identical for each blade. Furthermore, d_n is related to the axial position of the bit U according to

$$d_n(t) = U_b(L,t) - U_b(L, t - t_n),$$

where t_n is the time required for the bit to rotate by an angle $2\pi/n$ to its current position at time "t", as schematically illustrated in Fig. 4.2. The delay $t_n(t)$ is a solution of

$$\Phi_b(L,t) - \Phi_b(L, t - t_n) = 2\pi/n.$$

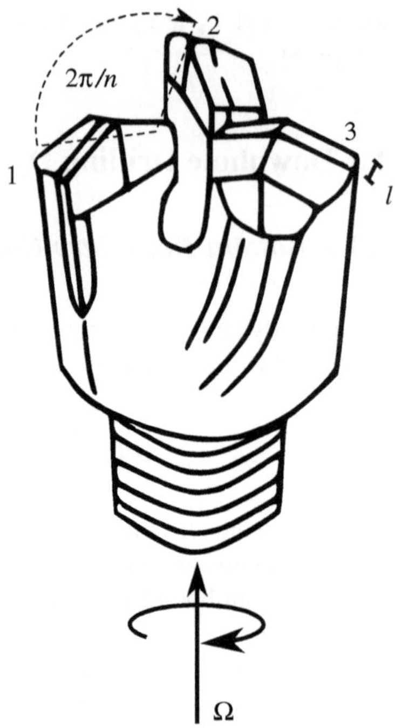

Fig. 4.2 Three blade drag bit

The range of t_n is given by $t_{n0} = 2\pi/(n\Omega_0)$, with Ω_0 a nominal rotating speed at the top. The combined bit depth of cut is simply $d(t) = nd_n(t)$ or

$$d(t) = n\left[U_b(L,t) - U_b(L,t-t_n)\right]. \tag{4.20}$$

The torque T and the weight on bit W, in the boundary conditions (4.18), (4.19) and in the drilling model given in (2.4), are functions of the history of Φ_b and of the history of U_b, indeed their cutting components are proportional to the depth of cut $d(t)$, as discussed below.

The modeling approach proposed in [243] decomposes T and W into a contribution associated to the forces transmitted by the pure cutting process of each cutter (denoted by the subscript c) and another term corresponding to the forces arising from the frictional contact (denoted by the subscript f):

$$T = T_c + T_f, \qquad W = W_c + W_f.$$

Next, the cutting and friction elements of the above expression are characterized.

Friction Force/Moment. The frictional elements T_f and W_f are related according to

$$2T_f(t) = \mu R_b \varkappa W_f(t).$$

In [287], these components are modeled as follows:

$$T_f(t) = \frac{R_b^2}{2}\varkappa\mu\varsigma l\,\mathscr{F}(\|V_b(L,t)\|), \qquad W_f(t) = R_b l\varsigma\,\mathscr{F}(\|V_b(L,t)\|),$$

where R_b is the bit radius, l the length of the wearflat, ς the contact stress, $\varkappa$ is a bit geometry number, greater than 1, that characterizes the orientation and spatial distribution of the frictional contact surfaces associated with the cutters, μ is a coefficient of friction related to the internal friction angle of the rock [4] (ratio between the horizontal and the vertical components of the frictional force), $V_b = (\frac{\partial U_b}{\partial t}, \frac{\partial \Phi_b}{\partial t})$, and $\mathrm{sgn}(V_b)$ designate the orientation of V_b with respect to the horizontal plane, and $\mathscr{F}$ is an dimensionless friction function which can be approximated by the following model, proposed in [287]:

$$\mathscr{F}(x) = \frac{\alpha_0 x}{\sqrt{x^2+\varepsilon^2}}, \quad \alpha_0 > 0.$$

Alternatively, in [286], the following model is considered

$$\mathscr{F}(x) = \alpha_0\left(\tanh(x) + \frac{x}{1+\alpha_1 x^2}\right), \quad \alpha_0 > 0, \alpha_1 > 0.$$

Cutting Force/Moment. In [110], the cutting components of the torque T_c and weight W_c are modeled as:

$$T_c(t) = \frac{R_b^2}{2}\varepsilon d(t), \qquad W_c(t) = R_b\zeta\varepsilon d(t),$$

where ε is the intrinsic specific energy (a parameter related to the rock strength under certain conditions [71]) and ζ is a number characterizing the orientation of the cutting force (the ratio of the vertical to the horizontal force for a sharp cutter). In [287], T_c is approximated by:

$$T_c(t) = -a_4(\mathscr{F}(\|V_b(L,t)\|))^2 d(t)$$

where a_4 is a constant related to the Young modulus. Notice that, according to the considered frictional interface, the set of equations given in (2.4) constitutes a retarded dynamic model characterized by a state-dependent delay t_n.

4.4 Notes and References

A quasi-complete model of a vertical oilwell drilling system consisting in a set of PDEs to describe the torsional and axial drilling rod behavior at different levels of the string was presented. The distributed parameter system is inspired by the dynamic model proposed in [200, 201], where the trajectories of the drilling system are characterized by the wave equation involving viscous and viscoelastic Kelvin-Voigt internal damping. Based on similar models, torsional drilling dynamics and the stick-slip are investigated in [45], and the bit-bounce phenomenon related to axial drilling displacements is studied in [46].

A more complete description of the drilling dynamics should take into account three interconnected systems:

(i) Mechanical system.
(ii) Mechatronic system.
(iii) Transmission system.

The mechanical system comprises the dynamics of the drill pipes, the drill collars, and the bit. This book is focused on the analysis and control of the mechanical part of the system which reflects the occurrence of detrimental vibrations in the drillstring.

The mechatronic system is composed of the drilling motors providing the rotational and axial string motion. As explained above, the drilling system includes DC and induction motors which transform electrical energy into mechanical energy, giving rise to the drilling operation.

The transmission system transfers the drilling parameters measured at the downhole end to the drill operator; there are several data transmission methods used in oilwell platforms, the main ones will be reviewed in Sect. 8.5.

The comprehensive model presented in this chapter provides an accurate description of the dynamics occurring along the drillstring during the perforation process; however, for analysis purposes, it is preferable to handle a simpler model that preserves the infinite-dimensional characteristics of the system. With this in mind, subsequent chapters address the analysis of a vertical oilwell drilling system modeled through the wave equation models studied in Sect. 2.2. As previously explained, these models can be transformed into time-delay systems of neutral-type (see Sect. 2.3) providing the same information about the system dynamics but simplifying the simulation and analysis tasks. In order to analyze the system dynamics and design control laws for the suppression of drilling vibrations in a successful manner, the basic notions of the theory of neutral-type time-delay systems are required. The following chapter provides the most important tools for analyzing the stability of such systems, both in frequency and time domain; on that basis, in Chaps. 6 and 7, the drilling system dynamics are studied.

Part II
Analysis

Chapter 5
Neutral-Type Time-Delay Systems: Theoretical Background

Time-delay systems are known as systems with aftereffect or dead time, equations with deviating argument or differential-difference equations. They belong to the class of functional differential equations (FDE) which are infinite dimensional, as opposed to the "standard" ODE. Time-delay systems are classified into three distinct types: retarded, neutral, and advanced. When the rate of change of the state depends on the present and past values of the system state, the model corresponds to a retarded functional differential equation (RFDE). When the rate of change of the state depends not only on the present and past values of the system state but also on the earlier value of the state rate change (under appropriate constraints on the corresponding delay difference operator), the corresponding equation is of neutral-type, also known as neutral functional differential equation (NFDE). When the rate of change of state is determined by future values of the state, the system is described by an advanced functional differential equation (AFDE). It should be mentioned that AFDE is rarely used in practical applications.

Numerous motivational issues give rise to the continuous interest and development on time-delay systems research; some of them are listed below [242].

- It is well known that most processes include aftereffect phenomena in their dynamics. There are examples of that in economics, physiology, population dynamics, biology, chemistry, viscoelasticity, physics, information technologies, mechanics, engineering sciences, etc. Research work on functional differential equations continuously grows in all scientific areas, especially in control engineering.
- In general, classical controllers are not necessarily able to stabilize delay systems. A common technique to avoid the complexity of delay systems consists in approximating them by finite-dimensional representations; however, neglecting aftereffects is not a "good" solution: in the best case scenario (constant and known delays), it leads to the same degree of complexity in the control design, in worst situations (time-varying delays, for example), it is potentially harmful for the system's stability.
- Delay characteristics are surprising since several studies have shown that voluntary introduction of delays can also benefit the control, see for instance [195].

M.B. Saldivar Márquez et al., *Analysis and Control of Oilwell Drilling Vibrations*, Advances in Industrial Control, DOI 10.1007/978-3-319-15747-4_5

- In spite of their complexity, time-delay systems, however, often appear as simple infinite-dimensional models in the area of PDEs. As mentioned in [158], "it is usually not difficult to show that the appearance of delay in a differential equation results of some essential simplification of the model." For instance, hyperbolic PDEs can be locally understood as functional differential equations of neutral type [157] and, conversely, any time delay can be represented by a classical transport equation [163].

As explained in the first part of the book, a system of neutral-type time-delay equations provides a reliable representation of the coupled axial-torsional drilling dynamics. This representation allows us to take advantage of the wide range of existing tools for the stability analysis of time-delay systems.

This chapter provides the main theoretical elements to address the stability analysis of the drilling system in the framework of neutral functional differential equations.

5.1 Preliminaries

This section presents some basic notions of the general theory of NFDE.

Several mathematical representations have been established to describe a time-delay system of neutral type. One of the most frequently used was proposed by Jack K. Hale (1928–2009) in the 1970s; see for instance [126], where a NFDE is defined as:

$$\frac{d}{dt}\mathscr{D}(t, x_t) = f(t, x_t),$$

where $f : \Pi \to \mathbb{R}^n, \mathscr{D} : \Pi \to \mathbb{R}^n$ (for an open set $\Pi \subseteq \mathbb{R}\times C$) are given continuous functions. The function $\mathscr{D}$ is called *difference operator*.

We are interested in a particular class of NFDE having a linear difference operator. Thus, in the sequel, we consider the following representation:

$$\frac{d}{dt}\left[x(t) - \sum_{i=1}^{m} D_i x(t - \tau_i)\right] = f(t, x_t), \tag{5.1}$$

where D_i are given $n \times n$ matrices and τ_i $(i = 1, \dots, m)$ are constant delays. Special attention will be paid to the case when $f(t, x_t)$ rewrites as $f\,(t, x(t), x(t - \tau_1), \dots, x(t - \tau_m))$.

The information required to characterize a particular solution of the system consists of an initial time instant $t_0 \geqslant 0$ and an initial function $\varphi \in PC^1\left([-\bar{\tau}, 0], \mathbb{R}^n\right)$ such that $\varphi : [-\bar{\tau}, 0] \to \mathbb{R}^n$, $\bar{\tau} = \max\{\tau_1, \dots, \tau_m\}$ and

$$x(t_0 + \theta) = \varphi(\theta), \quad \theta \in [-\bar{\tau}, 0]. \tag{5.2}$$

It is assumed that $f(t,\varphi)$ is continuous in both arguments and is defined for all $t \in [0,\infty)$,

$$f : [0,\infty) \times PC^1\left([-\bar{\tau},0],\mathbb{R}^n\right) \to \mathbb{R}^n.$$

The system state denoted as $x(t,t_0,\varphi)$ or $x_t(t_0,\varphi)$ at $t \geqslant t_0$ is defined as:

$$x_t : \theta \to x(t+\theta), \quad \theta \in [-\bar{\tau},0],$$

and, as suggested by Nikolai Nikolaevich Krasovskii (1924–2012), represents an appropriate "piece-of-trajectory" of the corresponding system.

In the sequel, we consider that the following assumptions hold.

1. The difference operator $x(t,t_0,\varphi) - \sum_{i=1}^{m} D_i x(t-\tau_i,t_0,\varphi)$ acting on the system's trajectories is continuous and differentiable for $t \geqslant t_0$ almost everywhere (excepting only a countable number of points).
2. The right-hand side derivative of $x(t,t_0,\varphi) - \sum_{i=1}^{m} D_i x(t-\tau_i,t_0,\varphi)$ is assumed at $t = t_0$.

The integral form of the initial value problem (5.1) with the initial function (5.2) is given by:

$$x(t) = \sum_{i=1}^{m} D_i x(t-\tau_i) + \left[\varphi(0) - \sum_{i=1}^{m} D_i \varphi(-\tau_i)\right] + \int_{t_0}^{t} f\left(s,x_s\right) ds,$$

where $x(t)$ is a solution of (5.1) with the initial function (5.2).

The following theorems provide the conditions under which the existence and uniqueness of the system solution is guaranteed; for the sake of brevity, we will not detail their proofs (see for instance [122]).

Theorem 5.1.1 *Let a functional*

$$f : [0,\infty) \times PC^1\left([-\bar{\tau},0],\mathbb{R}^n\right) \to \mathbb{R}^n$$

satisfying the following conditions:

1. *For any* $H > 0$ *there exists* $M(H) > 0$ *such that*

$$\|f(t,\varphi)\| \leqslant M(H), \quad (t,\varphi) \in [0,\infty) \times PC^1\left([-\bar{\tau},0],\mathbb{R}^n\right), \quad \|\varphi\|_\tau \leqslant H.$$

2. *The functional* $f(t,\varphi)$ *is continuous with respect to both arguments.*
3. *The functional* $f(t,\varphi)$ *is Lipschitz with respect to the second argument, i.e., for any* $H > 0$, *there exists a Lipschitz constant* $L(H)$ *such that the inequality*

$$\left\| f(t,\varphi^{(1)}) - f(t,\varphi^{(2)}) \right\| \leqslant L(H) \left\| \varphi^{(1)} - \varphi^{(2)} \right\|_\tau$$

holds for $t \geqslant 0$, $\varphi^{(k)} \in PC^1\left([-\bar{\tau},0],\mathbb{R}^n\right)$, *and* $\left\|\varphi^{(k)}\right\|_\tau \leqslant H$, $k = 1,2$.

Then, for $t_0 \geqslant 0$ *and* $\varphi \in PC^1([-\bar{\tau}, 0], \mathbb{R}^n)$ *there exists a positive* $\bar{\tau^*}$ *such that the initial value problem* (5.1) *with the initial function* (5.2) *admits a unique solution defined on the segment* $\left[t_0 - \bar{\tau}, t_0 + \bar{\tau^*}\right]$.

Theorem 5.1.2 *If system* (5.1) *satisfies the conditions of Theorem* 5.1.1 *and additionally* $f(t, \varphi)$ *satisfies the inequality*

$$\|f(t, \varphi)\| \leqslant \eta\left(\|\varphi\|_\tau\right), \quad t \geqslant 0, \ \varphi \in PC^1\left([-\bar{\tau}, 0], \mathbb{R}^n\right),$$

where the function $\eta(r), r \in [0, \infty)$, *is continuous, nondecreasing and the following condition holds for any* $r_0 \geqslant 0$:

$$\lim_{R \to \infty} \int_{r_0}^{R} \frac{dr}{\eta(r)} = \infty,$$

then, any solution $x(t, t_0, \varphi)$ *of the system is defined on* $[0, \infty)$.

5.2 Stability of Neutral Systems

Roughly speaking, an equilibrium point is said to be "stable" if all solutions starting at "nearby" points stay "nearby"; otherwise, it is unstable. It is asymptotically stable if all solutions starting at nearby points not only stay nearby, but also tend to the equilibrium point as time approaches infinity. The exponential stability is a stronger property since it guarantees that the solutions remain bounded by a decaying exponential function. These concepts are formally defined below.

Definition 5.2.1 The trivial solution of system (5.1) is said to be stable if for any $\varepsilon > 0$ and $t_0 \geqslant 0$ there exists $\delta(\varepsilon, t_0) > 0$ such that for every initial function $\varphi \in PC^1([-\bar{\tau}, 0], \mathbb{R}^n)$, with $\|\varphi\|_\tau < \delta(\varepsilon, t_0)$, the following inequality is satisfied:

$$\|x(t, t_0, \varphi)\| < \varepsilon, \quad t \geqslant t_0.$$

If $\delta(\varepsilon, t_0)$ can be chosen independently of t_0, then the trivial solution is said to be uniformly stable.

Definition 5.2.2 The trivial solution of system (5.1) is said to be asymptotically stable if for any $\varepsilon > 0$ and $t_0 \geqslant 0$ there exists $\delta_1(\varepsilon, t_0) > 0$ such that for every initial function $\varphi \in PC^1([-\bar{\tau}, 0], \mathbb{R}^n)$, with $\|\varphi\|_\tau < \delta_1(\varepsilon, t_0)$, the following conditions hold.

1. $\|x(t, t_0, \varphi)\| < \varepsilon$, for $t \geqslant t_0$.
2. $x(t, t_0, \varphi) \to 0$, as $t - t_0 \to \infty$.

If $\delta_1(\varepsilon, t_0)$ can be chosen independently of t_0 and there exists $H_1 > 0$ such that $x(t, t_0, \varphi) \to 0$, because $t - t_0 \to \infty$, uniformly with respect to $t_0 \geqslant 0$, and

$\varphi \in PC^1([-\bar{\tau}, 0], \mathbb{R}^n)$, with $\|\varphi\|_\tau < H_1$, then the trivial solution is said to be uniformly asymptotically stable.

Definition 5.2.3 The trivial solution of system (5.1) is said to be exponentially stable if there exist $\delta_2 > 0$, $\sigma > 0$, and $\gamma \geqslant 1$ such that for every $t_0 \geqslant 0$ and $\varphi \in PC^1([-\bar{\tau}, 0], \mathbb{R}^n)$, with $\|\varphi\|_\tau < \delta_2$, the following inequality holds:

$$\|x(t, t_0, \varphi)\| < \gamma \mathrm{e}^{-\sigma(t-t_0)} \|\varphi\|_\tau, \quad t \geqslant t_0.$$

In what follows, we present two different approaches to analyze the stability of NFDE: the frequency-domain approach for linear time-invariant systems (zero locations of analytic functions) and the time-domain approach (Lyapunov-based method).

5.2.1 Frequency-Domain Approach

This analysis framework, frequently referred to as Laplace domain approach, is a powerful tool to investigate the stability of the solution of a given linear NFDE. As in the finite-dimensional case, the steady-state solution is said to be asymptotically stable if and only if the roots of the corresponding characteristic equation have negative real part.

The characteristic function of a neutral-type time-delay system of the form

$$\dot{x}(t) - \sum_{i=1}^{m} D_i \dot{x}(t - \tau_i) = \sum_{i=0}^{m} A_i x(t - \tau_i), \tag{5.3}$$

is given by $\Delta(\lambda) : \mathbb{C} \to \mathbb{C}$,

$$\Delta(\lambda) = \lambda I - \sum_{i=1}^{m} D_i \lambda \mathrm{e}^{-\tau_i \lambda} - \sum_{i=0}^{m} A_i \mathrm{e}^{-\tau_i \lambda}, \tag{5.4}$$

where $\tau_0 = 0$. Then, system (5.3) is stable for the given constant delays $\tau_i \geqslant 0$ $(i = 0, \ldots, m)$ if the characteristic quasipolynomial, defined as:

$$p(\lambda, \mathrm{e}^{-\tau_1 \lambda}, \ldots, \mathrm{e}^{-\tau_m \lambda}) = \det \Delta(\lambda) \tag{5.5}$$

is such that for some $\varepsilon < 0$,

$$p(\lambda, \mathrm{e}^{-\tau_1 \lambda}, \ldots, \mathrm{e}^{-\tau_m \lambda}) \neq 0, \quad \forall \lambda \in \mathbb{C}_{\varepsilon+},$$

where $\mathbb{C}_{\varepsilon+} = \{\lambda : \mathrm{Re}(\lambda) > \varepsilon\}$. If this condition is satisfied for any τ_i, then system (5.3) is said to be *delay-independent stable*.

As discussed in the previous chapters, the drilling model under consideration is given by the following pair of coupled partial differential equations:

$$\begin{cases} \dfrac{\partial^2 \Phi}{\partial s^2}(s,t) = \tilde{c}^2 \dfrac{\partial^2 \Phi}{\partial t^2}(s,t), \quad \tilde{c} = \sqrt{\dfrac{\rho_a}{G}} \\ GJ\dfrac{\partial \Phi}{\partial s}(0,t) = \beta \dfrac{\partial \Phi}{\partial t}(0,t) - u_T(t) \\ GJ\dfrac{\partial \Phi}{\partial s}(L,t) = -I_B \dfrac{\partial^2 \Phi}{\partial t^2}(L,t) - T\left(\dfrac{\partial \Phi}{\partial t}(L,t)\right) \end{cases}$$

$$\begin{cases} \dfrac{\partial^2 U}{\partial s^2}(s,t) = c^2 \dfrac{\partial^2 U}{\partial t^2}(s,t), \quad c = \sqrt{\dfrac{\rho_a}{E}} \\ E\Gamma\dfrac{\partial U}{\partial s}(0,t) = \alpha \dfrac{\partial U}{\partial t}(0,t) - u_H(t) \\ E\Gamma\dfrac{\partial U}{\partial s}(L,t) = -M_B \dfrac{\partial^2 U}{\partial t^2}(L,t) - T\left(\dfrac{\partial \Phi}{\partial t}(L,t)\right) \end{cases}$$

where $\Phi(s,t)$ and $U(s,t)$ denote the rotary angle and the longitudinal position along the drillstring. Recall that the above model is transformed, through the d'Alembert method, into the following pair of neutral-type time delay equations:

$$\begin{cases} \ddot{\Phi}_b(t) - \Upsilon \ddot{\Phi}_b(t-2\tau) = -\Psi \dot{\Phi}_b(t) - \Upsilon\Psi \dot{\Phi}_b(t-2\tau) - \dfrac{1}{I_B} T\left(\dot{\Phi}_b(t)\right) \\ \qquad + \dfrac{1}{I_B}\Upsilon T\left(\dot{\Phi}_b(t-2\tau)\right) + \Pi\Omega(t-\tau) \\ \ddot{U}_b(t) - \tilde{\Upsilon}\ddot{U}_b(t-2\tilde{\tau}) = -\tilde{\Psi}\dot{U}_b(t) - \tilde{\Upsilon}\tilde{\Psi}\dot{U}_b(t-2\tilde{\tau}) - \dfrac{1}{M_B} T\left(\dot{\Phi}_b(t)\right) \\ \qquad + \dfrac{1}{M_B}\tilde{\Upsilon} T\left(\dot{\Phi}_b(t-2\tilde{\tau})\right) + \tilde{\Pi}\rho(t-\tilde{\tau}), \end{cases} \tag{5.6}$$

where $\dot{\Phi}_b(t)$ and $\dot{U}_b(t)$ are the angular and axial bit velocities. Notice that setting, $\tau_1 = 2\tau$, $\tau_2 = 2\tilde{\tau}$, $x = \left[\dot{\Phi}_b \; \dot{U}_b\right]^T$, the linearized form of system (5.6) can be rewritten as:

$$\dot{x}(t) - D_1\dot{x}(t-\tau_1) - D_2\dot{x}(t-\tau_2) = A_0x(t) + A_1x(t-\tau_1) + A_2x(t-\tau_2), \tag{5.7}$$

where the matrices D_1, D_2, A_0, A_1 and A_2 take the form:

$$D_1 = \begin{bmatrix} d_{11}^1 & 0 \\ 0 & 0 \end{bmatrix}, \quad D_2 = \begin{bmatrix} 0 & 0 \\ 0 & d_{22}^2 \end{bmatrix}, \quad A_0 = \begin{bmatrix} a_{11}^0 & 0 \\ a_{12}^0 & a_{22}^0 \end{bmatrix}, \tag{5.8}$$

$$A_1 = \begin{bmatrix} a_{11}^1 & 0 \\ 0 & 0 \end{bmatrix}, \quad A_2 = \begin{bmatrix} 0 & 0 \\ a_{21}^2 & a_{22}^2 \end{bmatrix}.$$

The particular form of the characteristic quasipolynomial of the system under consideration is stated in the following proposition.

Proposition 5.2.4 *The characteristic quasipolynomial of the drilling model described by a NFDE of the form* (5.7) *is given by:*

$$p(\lambda, e^{-\tau_1\lambda}, e^{-\tau_2\lambda}) = \prod_{i=1}^{2}\left(\lambda - d_{ii}^{i}\lambda e^{-\tau_i\lambda} - a_{ii}^{0} - a_{ii}^{i} e^{-\tau_i\lambda}\right). \tag{5.9}$$

Proof According to Eq. (5.4), the characteristic function of system (5.7) is given by:

$$\Delta(\lambda) = \lambda I - D_1\lambda \mathrm{e}^{-\tau_1\lambda} - D_2\lambda \mathrm{e}^{-\tau_2\lambda} - A_0 - A_1\mathrm{e}^{-\tau_1\lambda} - A_2\mathrm{e}^{-\tau_2\lambda},$$

thus, in view of (5.5) and through direct computations, expression (5.9) is obtained.

Remark 5.2.5 For further discussions and a deeper analysis of the spectrum of linear neutral-type time-delay systems, we refer to [196].

In system (5.7), $x(t) - D_1x(t - \tau_1) - D_2x(t - \tau_2)$ represents the corresponding difference operator.

It is well known that the stability of the difference operator is a necessary condition for the stability and the stabilization of neutral-type time-delay systems [186]. This property, known as *formal stability*, was introduced and discussed in [47]. According to [232], it simply means that a NFDE has only a finite number of zeros in the right-half complex plane.

Definition 5.2.6 ([186]) System (5.7) is said to be formally stable if

$$\text{Rank}\left(I - \hat{D}(\lambda)\right) = n, \qquad \forall\lambda \text{ such that } \mathrm{Re}\,(\lambda) \geqslant 0,$$

where

$$\hat{D}(\lambda) = \sum_{i=1}^{2} D_i \mathrm{e}^{-\tau_i\lambda} \tag{5.10}$$

and $\mathrm{Re}(\lambda)$ denotes the real part of the complex number λ.

The following proposition establishes the formal stability of the drilling model under consideration.

Proposition 5.2.7 *The drilling model described by a NFDE of the form* (5.7) *is formally stable.*

Proof According to Definition 5.2.6, and in view of matrices D_1 and D_2 defined in (5.8), $\hat{D}(\lambda)$ is given by

$$\hat{D}(\lambda) = \begin{bmatrix} d_{11}^1 e^{-\tau_1 \lambda} & 0 \\ 0 & d_{22}^2 e^{-\tau_2 \lambda} \end{bmatrix}.$$

By using the following property:

$$\text{Rank}\left(I - \hat{D}(\lambda)\right) = n \Leftrightarrow \det\left(I - \hat{D}(\lambda)\right) \neq 0,$$

the formal stability condition of Definition 5.2.6 can be rewritten as:

$$\left(1 - d_{11}^1 e^{-\tau_1 \lambda}\right)\left(1 - d_{22}^2 e^{-\tau_2 \lambda}\right) \neq 0 \quad \forall \lambda \text{ such that } \mathrm{Re}\,(\lambda) \geqslant 0. \tag{5.11}$$

Observe that, since

$$e^{-\tau_1 \lambda} \leqslant 1, \quad \forall \lambda \text{ such that } \mathrm{Re}\,(\lambda) \geqslant 0,$$

and

$$\left|d_{11}^1\right| = \left|\frac{\beta - \tilde{c}GJ}{\beta + \tilde{c}GJ}\right| < 1, \quad \text{for any } \beta > 0,$$

it is clear that $d_{11}^1 e^{-\tau_1 \lambda} \neq 1$. Similarly, as

$$e^{-\tau_2 \lambda} \leqslant 1, \quad \forall \lambda \text{ such that } \mathrm{Re}\,(\lambda) \geqslant 0,$$

and

$$\left|d_{22}^2\right| = \left|\frac{\alpha - cE\Gamma}{\alpha + cE\Gamma}\right| < 1, \quad \text{for any } \alpha > 0,$$

then, $d_{22}^2 e^{-\tau_2 \lambda} \neq 1$ and the condition (5.11) is satisfied. We conclude that the neutral-type time-delay system describing the coupled torsional-axial drilling dynamics is formally stable.

5.2.1.1 Illustrative Example: Scalar Case

A useful proposition to determine the stability of a single-delay scalar neutral-type equation is presented below.

Consider a time-delay equation of the form:

$$\dot{x} - d\dot{x}(t - \tau) = a_0 x(t) + a_1 x(t - \tau) \tag{5.12}$$

for which the characteristic function is given by

$$\Delta(\lambda) = \lambda \left(1 - d\,\mathrm{e}^{-\lambda\tau}\right) - a_0 - a_1\,\mathrm{e}^{-\lambda\tau}. \tag{5.13}$$

It is well known that if $|d| > 1$ then the trivial solution of (5.12) is unstable. However, when $|d| < 1$, there is a finite number of roots of (5.13) with positive real part.

The following proposition gives the conditions under which the steady-state solution of system (5.12) with $|d| < 1$ is asymptotically stable.

Proposition 5.2.8 ([305]) *The roots of the characteristic equation $\Delta(\lambda) = 0$ lie in the half complex plane $\mathbb{C}_-$ if and only if*

$$a_0\tau < 1 - d, \quad a_0 + a_1 < 0 \quad \textit{and} \quad a_1 > -\frac{\delta}{\tau}\,\sin\delta - a_0\,\cos\delta,$$

where δ is a root of the equation:

$$\frac{1}{\tau}\,\cos\delta - \frac{a_0}{\delta}\,\sin\delta - \frac{d}{\tau} = 0.$$

The proof of Proposition 5.2.8, presented below, is based on several results due to Lev Semenovich Pontryagin (1908–1988) [232] which provides the conditions under which transcendental equations have only zeros with negative real part (see Appendix D).

Proof The characteristic equation associated with system (5.12) ($\Delta(\lambda) = 0$, where $\Delta(\lambda)$ is defined in (5.13)) can be written as:

$$\Delta(\lambda) = (\lambda - a_0)\,\mathrm{e}^{\lambda\tau} - d\lambda - a_1 = 0.$$

Defining $\bar{\lambda} = \lambda\tau$, we have that

$$\Delta(\bar{\lambda}) = \left(\bar{\lambda} - \bar{a}_0\right)\mathrm{e}^{\bar{\lambda}} - d\bar{\lambda} - \bar{a}_1 = 0,$$

where $\bar{a}_0 = a_0\tau$ and $\bar{a}_1 = a_1\tau$. Since $\mathrm{Re}(\bar{\lambda}) < 0$ if and only if $\mathrm{Re}(\lambda) < 0$, it suffices to consider $\Delta(\lambda) = 0$ with $\tau = 1$. Consider now Theorems D.1.1 and D.1.2 given in Appendix D.

If one takes $\lambda = j\delta$, $\Delta(j\delta)$ can be written as $\Delta(j\delta) = L(\delta) + jM(\delta)$, where the real and the imaginary parts of $\Delta(j\delta)$ are given by:

$$\begin{aligned} L(\delta) &= -a_0\cos\delta - \delta\sin\delta - a_1 \\ M(\delta) &= \delta\cos\delta - a_0\sin\delta - d\delta. \end{aligned}$$

The function $M(\delta) = m(\delta, \cos\delta, \sin\delta)$ has the principal term $\delta\cos\delta$ and, according Theorem D.1.2, we have that $\phi_*(1)(\lambda) = \cos\lambda$, and we may take $\varepsilon = 0$.

Sufficiency. If all zeros of $\Delta(\lambda) = 0$ are in the left half-plane, the zeros of $L(\delta)$ and $M(\delta)$ are real, simple and alternate, and condition (D.1) of Theorem D.1.1 holds for all $\delta \in \mathbb{R}$. δ is a zero of M only if $\delta = m(\delta)$ with

$$m(\delta) = \frac{a_0 \sin\delta}{\cos\delta - d}.$$

Note that $\delta = 0$ is a zero of M. From condition (D.1), $M'(0) = 1 - d - a_0$ cannot be zero, thus $a_0 \neq 1 - d$. M has one root in each interval $[k\pi, (k+1)\pi]$, $k \in \mathbb{Z}$. Theorem D.1.2 requires that M has exactly $4k+1$ roots in $[-2k\pi, 2k\pi]$ for all $k \in \mathbb{N}$. This is only possible if M has exactly 3 roots in $[-\pi, \pi]$, which is always the case if $a_0 < 0$. If $a_0 > 0$, the slope of δ/a_0 at $\delta = 0$ must be greater than that of $m(\delta)$ to ensure three roots in $[-\pi, \pi]$. The slope of $m(\delta)$ at $\delta = 0$ is $1/(1-d)$. Hence, we must have $a_0 < 1 - d$. Also, given that

$$M'(0)L(0) = (1 - d - a_0)(-a_0 - a_1),$$

condition (D.1) holds for $\delta = 0$ only if $a_0 + a_1 > 0$.

Now, let us establish the last of the three inequalities of Theorem D.1.1. Let δ_0 be the root of M in $(0, \pi)$. First, observe that the function

$$h(d) = -\delta_0 + \frac{1}{2}\sin(2\delta_0) + d(\delta_0 \cos\delta_0 - \sin\delta_0) = \sin\delta_0 M'(\delta_0)$$

reaches its maximum at $d = 1$ and $h(1) < 0$ for all $\delta_0 \in (0, \pi)$. This implies that $M'(\delta_0) < 0$ and condition (D.1) gives

$$M'(\delta_0)L(\delta_0) = M'(\delta_0)(-a_0\cos\delta_0 - \delta_0\sin\delta_0 - a_1)$$

or

$$a_1 > -a_0\cos\delta_0 - \delta_0\sin\delta_0.$$

Necessity. As before, if $a_0 < 1 - d$, M has exactly $4k + 1$ roots in $[-2k\pi, 2k\pi]$ for all $k \in \mathbb{N}$ and all roots must be real. Thus, we must verify that (D.1) holds for all roots of M. Clearly, (D.1) holds for $\delta = 0$ and $\delta = \delta_0$. Since both L and M are odd functions, $M'(\delta)L(\delta) - L'(\delta)M(\delta)$ is even in δ and we only need to establish (D.1) for all positive roots of M. Let δ_n be the root of M in $(n\pi, (n+1)\pi)$. Then $L(\delta_n) = -a_0 q(\delta_n) - a_1$, where the function

$$q(\delta) = \frac{1 - d\cos\delta}{\cos\delta - d}$$

is increasing on $(2n\pi, (2n+1)\pi)$ with singularities at $\cos^{-1} d + 2n\pi$, decreasing on $((2n+1)\pi, (2n+2)\pi)$ with singularities at $-\cos^{-1} d + (2n+2)\pi$ and $|q(\delta)| \geqslant 1$.

Consider $\delta_{2n} \neq \cos^{-1} d + 2n\pi$. Since $\sin\delta_{2n} > 0$, we must show that $L(\delta_{2n}) < 0$.

If $a_0 > 0$, then the sequence $\{\delta_{2n} - 2n\pi\} \subset (0, \pi)$ is increasing and $\delta_0 < \delta_{2n} - 2n\pi$. Thus, $q(\delta_0) < q(\delta_{2n} - 2n\pi) = q(\delta_{2n})$. Therefore, $L(\delta_0) - L(\delta_{2n}) > 0$ or $L(\delta_{2n}) < L(\delta_0)$.

If $a_0 < 0$, then the sequence $\{\delta_{2n} - 2n\pi\} \subset (0, \pi)$ is decreasing and $\delta_0 > \delta_{2n} - 2n\pi$. Thus, $q(\delta_0) > q(\delta_{2n} - 2n\pi) = q(\delta_{2n})$. Therefore, $L(\delta_0) - L(\delta_{2n}) > 0$ or $L(\delta_{2n}) < L(\delta_0) < 0$.

Let us now turn to $\delta = \delta_{2n+1}$. Here we must show that $L(\delta_{2n+1}) > 0$.

If $a_0 > 0$, $\delta_{2n+1} \in \left((2n+1)\pi, (2n+2)\pi - \cos^{-1} d\right)$. In this interval $q(\delta)$ is negative and $-a_0 q(\delta_{2n+1}) > 0$. The assumption $a_0 + a_1 < 0$ amounts to requiring $-a_1 > 0$ and we obtain $L(\delta_{2n+1}) > 0$.

If $a_0 < 0$, $\delta_{2n+1} \in \left((2n+1)\pi - \cos^{-1} d, (2n+2)\pi\right)$. In this interval $q(\delta) > 1$, thus, $-a_0 q(\delta_{2n+1}) > -a_0$, which together with the condition $-a_1 > a_0$ gives $L(\delta_{2n+1}) > 0$.

It is worth mentioning that, as suggested by Nussbaum in [219], the same result can be derived by using the Rouché's Theorem.

5.2.2 *Time-Domain Approach*

This section present some basic stability results within the framework of time-domain techniques (Lyapunov-Krasovskii approach) for a neutral-type system of the form (5.1).

In the sequel, the following assumptions are considered:

1. System (5.1) satisfies the conditions stated in Theorem 5.1.1 about the uniqueness of the solution.
2. System (5.1) admits the trivial solution:

$$f(t, 0_\tau) \equiv 0, \quad \text{for } t \geqslant 0.$$

3. Matrix D is Schur stable, i.e., the spectrum of the matrix lies in the open unit disc of the complex plane.

The following definition about the positive definiteness of a given functional is used in subsequent stability theorems.

Definition 5.2.9 A functional $V(t, \varphi)$ is positive definite if the following conditions hold for some $H > 0$,

1. $V(t, 0_\tau) = 0, t \geqslant 0$.
2. $V(t, \varphi)$ is defined for $t \geqslant 0$ and $\varphi \in PC^1([-\bar{\tau}, 0], \mathbb{R}^n)$, with $\|\varphi\|_\tau \leqslant H$.
3. There exists a positive definite function $V_1(x)$ satisfying

$$V_1(\varphi(0) - D\varphi(-\tau)) \leqslant V(t, \varphi), \quad t \geqslant 0,$$
$$\varphi \in PC^1\left([-\bar{\tau}, 0], \mathbb{R}^n\right), \text{ with } \|\varphi\|_\tau \leqslant H.$$

4. For a given $t_0 \geqslant 0$, $V(t_0, \varphi)$ is continuous in φ at 0_τ, i.e., for $\varepsilon > 0$ there exists some $\delta > 0$ such that $\|\varphi\|_\tau < \delta$ implies

$$|V(t_0, \varphi) - V(t_0, 0_\tau)| = V(t_0, \varphi) < \varepsilon.$$

The following theorems provide stability conditions for a neutral-type time-delay system of the form (5.1); their proofs can be found, for instance, in [150].

Theorem 5.2.10 *The trivial solution of a system of the form* (5.1) *is said to be stable if and only if there exists a positive definite functional* $V(t, \varphi)$ *such that* $V(t_0, x_t)$ *as a function of "t" does not increase.*

Theorem 5.2.11 *The trivial solution of system* (5.1) *is uniformly stable if and only if it is stable (according to Theorem* 5.2.10*) and additionally there exists a positive definite functional* $V(t, \varphi)$ *which is continuous in* φ *at the point* 0_τ*, uniformly for* $t \geqslant 0$.

Theorem 5.2.12 *The trivial solution of system* (5.1) *is asymptotically stable if and only if it is stable (according to Theorem* 5.2.10*) and for any* $t_0 \geqslant 0$ *there exists a positive value* $\mu(t_0)$ *such that if* $\varphi \in PC^1([-\bar{\tau}, 0], \mathbb{R}^n)$ *and* $\|\varphi\|_\tau < \mu(t_0)$*, then a positive definite functional* $V(t, x_t(t_0, \varphi))$ *decreases monotonically to zero as* $t - t_0 \to \infty$.

The following theorems give sufficient conditions under which the trivial solution of system (5.1) is asymptotically and exponentially stable.

Theorem 5.2.13 *Asymptotic stability of the trivial solution of system* (5.1) *is achieved if there exists a positive definite function* $W(x)$ *and a positive definite functional* $V(t, \varphi)$ *such that* $V(t, \varphi)$ *is differentiable and the inequality*

$$\frac{dV}{dt}(t, x_t) \leqslant -W(x(t) - Dx(t - \tau))$$

is satisfied along the system solutions.

Theorem 5.2.14 *Exponential stability of the trivial solution of system* (5.1) *is achieved if there exists a positive definite functional* $V(t, \varphi)$ *satisfying:*

1. *The inequality*

$$\alpha_1 \|\varphi(0) - D\varphi(-\tau)\|^2 \leqslant V(t, \varphi) \leqslant \alpha_2 \|\varphi\|_\tau^2, \qquad t \geqslant 0$$

   is satisfied for some constants $\alpha_1 > 0$, $\alpha_2 > 0$ *and* $\varphi \in PC^1([-\bar{\tau}, 0], \mathbb{R}^n)$*, with* $\|\varphi\|_\tau \leqslant H$.
2. $V(t, \varphi)$ *is differentiable and the inequality*

$$\frac{dV}{dt}(t, x_t) + \sigma_1 V(t, x_t) \leqslant 0$$

   is satisfied along the solutions for a constant $\sigma_1 > 0$.

The following result provides asymptotic stability conditions derived from Theorem 5.2.13.

Theorem 5.2.15 ([293]) *A neutral system of the form*

$$\dot{x} - D\dot{x}(t-\tau) = A_0x(t) + A_1x(t-\tau) \tag{5.14}$$

is delay-independent asymptotically stable if:

(i) A_0 *is Hurwitz stable matrix, i.e., each eigenvalue of* A_0 *has strictly negative real part;*
(ii) D *is Schur-Cohn stable matrix, i.e., all the eigenvalues of* D *lie within the unit circle;*
(iii) *there are two symmetric and positive definite matrices* R *and* Q *such that the Ricatti equation:*

$$A_0^TP+PA_0+S+Q+[P(A_0D+A_1)+SD]R^{-1}[D^TS+(A_1^T+D^TA_0^T)P]=0 \tag{5.15}$$

has a symmetric and positive definite solution P, *and* S *is a symmetric and positive definite solution of the Lyapunov equation:*

$$D^TSD - S + R = 0. \tag{5.16}$$

Proof Consider the Lyapunov-Krasovskii functional candidate:

$$V(x_t) = (x(t) - Dx(t-\tau))^T\, P\,(x(t) - Dx(t-\tau)) + \int_{-\tau}^{0} x^T(t+\theta)\, Sx\,(t+\theta)\, d\theta$$

where P and S are the solutions of the Riccati equation (5.15) and of the Lyapunov equation (5.16), respectively, and consider the difference operator $\mathscr{D} : \Pi \to \mathbb{R}^n$:

$$\mathscr{D}(\varphi) = \varphi(0) - D\varphi(-\tau), \qquad \varphi \in PC^1\left([-\tau,0], \mathbb{R}^n\right). \tag{5.17}$$

Notice that the functional $V(x_t)$ satisfies the condition:

$$u\left(|\mathscr{D}(\varphi)|\right) \leqslant V(\varphi) \leqslant v\left(\|\varphi\|_\tau\right), \tag{5.18}$$

where $u(s) = \lambda_{\min}(P)s^2$ and $v(s) = [\lambda_{\max}(P) + \tau\lambda_{\max}(S)]\,s^2$. The derivative of $V(x_t)$ along the trajectory of the neutral system (5.14) is given by:

$$\begin{aligned}\dot{V}(x_t) = {} & (A_0x(t) + A_1x(t-\tau))^T\, P\,(x(t) - Dx(t-\tau)) \\ & + (x(t) - Dx(t-\tau))^T\, P\,(A_0x(t) + A_1x(t-\tau)) \\ & + x^T(t)Sx(t) - x^T(t-\tau)Sx(t-\tau).\end{aligned} \tag{5.19}$$

Simple computation allows to rewrite Eq. (5.19) as follows:

$$\begin{aligned}\dot{V}(x_t) = &(x(t) - Dx(t-\tau))^T \left(A_0^T P + PA_0 + S\right)(x(t) - Dx(t-\tau)) \\ &+ (x(t) - Dx(t-\tau))^T PA_0 Dx(t-\tau) \\ &+ x^T(t-\tau) D^T A_0^T P (x(t) - Dx(t-\tau)) \\ &+ (x(t) - Dx(t-\tau))^T SDx(t-\tau) + x^T(t-\tau) D^T S (x(t) - Dx(t-\tau)) \\ &+ x^T(t-\tau) D^T SDx(t-\tau) - x^T(t-\tau) Sx(t-\tau) \\ &+ (x(t) - Dx(t-\tau))^T PA_1 x(t-\tau) \\ &+ x^T(t-\tau) A_1^T P (x(t) - Dx(t-\tau)).\end{aligned}$$

Since S is the positive definite solution of the Lyapunov equation (5.16) and using the difference operator (5.17), the above relation is rewritten as:

$$\begin{aligned}\dot{V}(x_t) = &\mathscr{D}^T(x_t)\left(A_0^T P + PA_0 + S\right)\mathscr{D}(x_t) \\ &+ \mathscr{D}^T(x_t)(PA_0 D + PA_1 + SD)\, x(t-\tau) \\ &+ x^T(t-\tau)\left(D^T S + D^T A_0^T P + A_1^T P\right)\mathscr{D}(x_t) - x^T(t-\tau) Rx(t-\tau).\end{aligned}$$

Since P is the symmetric and positive definite solution of the Riccati equation (5.15) and using the Schur complement property, we have:

$$\begin{aligned}\dot{V}(x_t) \leqslant &- \mathscr{D}^T(x_t) Q \mathscr{D}(x_t) \\ &- \left[\left(D^T A_0^T P + A_1^T + D^T S\right)\mathscr{D}(x_t) - Rx(t-\tau)\right]^T R^{-1} \\ &\left[\left(D^T A_0^T P + A_1^T + D^T S\right)\mathscr{D}(x_t) - Rx(t-\tau)\right] \\ \leqslant &- \mathscr{D}^T(x_t) Q \mathscr{D}(x_t).\end{aligned} \tag{5.20}$$

Inequalities (5.18) and (5.20) allow establishing the uniform asymptotic stability of the trivial solution of the neutral differential equation (5.14).

Remark 5.2.16 Since the negativity of the Lyapunov functional candidate does not use any information about the delay size, the asymptotic stability property holds for any positive delay.

5.2.2.1 Illustrative Example: Scalar Case

The following result provides delay-independent stability conditions for a scalar neutral-type time-delay system of the form:

$$\dot{x}(t) - d\dot{x}(t-\tau) = a_0 x(t) + a_1 x(t-\tau), \tag{5.21}$$

with a_0, a_1 and $d \in \mathbb{R}$.

Proposition 5.2.17 *The scalar neutral system* (5.21) *is delay-independent asymptotically stable if:*

(i) $a_0 < 0$,
(ii) $|d| < 1$,
(iii) $|a_1| < |a_0|$.

Proof Statements (i), (ii) and (iii) of Proposition 5.2.17 follow directly from Theorem 5.2.15 by replacing the matrices D, A_0, and A_1 by the scalars d, a_0 and a_1.

Remark 5.2.18 The stability conditions of Proposition 5.2.17 coincide with the ones presented in [125] which are derived through a frequency-domain approach.

5.3 Delay Effects on Stability

This section presents some important results on the so-called *delay margin* of neutral-type time-delay systems; the notion of delay margin is referred to the critical delay values at which a system becomes unstable [119].

Due to the study case under consideration (torsional-axial drilling model of the form (5.7)), the results presented below are developed considering a neutral-type system with two commensurate delays; however, they can be easily extended to a system with q (commensurate) delays. It is worth mentioning that the stability analysis proposed in the next paragraphs follows closely the results developed in [104].

Consider a linear time-invariant neutral-type time-delay system described by the following differential-difference equation:

$$y^{(n)}(t) + \sum_{k=1}^{2} d_k\, y^{(n)}(t - k\tau) + \sum_{i=0}^{n-1}\sum_{k=0}^{2} a_{ki}\, y^{(i)}(t - k\tau) = 0, \quad \tau \geqslant 0, \tag{5.22}$$

where the coefficients d_k, and a_{ki}, $k = 0,\ 1,\ 2$, $i = 0,\ 1,\ \cdots,\ n-1$ are known.

The state-space representation is given by:

$$\begin{aligned} &\dot{x}(t) - D_1\,\dot{x}(t-\tau) - D_2\,\dot{x}(t-2\tau) \\ &\quad = A_0\,x(t) + A_1\,x(t-\tau) + A_2\,x(t-2\tau), \quad \tau \geqslant 0, \end{aligned} \tag{5.23}$$

where D_k and A_k, $k = 1,\ 2$ are given matrices.

The characteristic function of the system (5.22) is given in [125]:

$$a\left(\lambda,\ \mathrm{e}^{-\tau\lambda}\right) = \sum_{k=0}^{2} a_k(\lambda)\mathrm{e}^{-k\tau\lambda}, \qquad \tau \geqslant 0, \tag{5.24}$$

with

$$a_0(\lambda) = \lambda^n + \sum_{i=0}^{n-1} a_{0i}\lambda^i, \quad a_1(\lambda) = d_k\,\lambda^n + \sum_{i=0}^{n-1} a_{ki}\lambda^i.$$

System (5.22) is stable for a given $\tau \geqslant 0$ if and only if for some $\varepsilon < 0$,

$$a\left(\lambda,\, e^{-\tau\lambda}\right) \neq 0, \quad \forall\lambda \in \bar{\mathbb{C}}_{\varepsilon+}, \tag{5.25}$$

where $\mathbb{C}_{\varepsilon+} = \{\lambda : \operatorname{Re}(\lambda) > \varepsilon\}$.

The system is said to be delay-independent stable if condition (5.25) is satisfied for all $\tau \geqslant 0$.

Similarly, the characteristic quasipolynomial of the system (5.23) is

$$p\left(\lambda,\, e^{-\tau\lambda}\right) = \det\left(\lambda I - D_1\,\lambda\,\mathrm{e}^{-\tau\lambda} - D_2\,\lambda\,\mathrm{e}^{-2\tau\lambda} - \sum_{k=0}^{2} A_k \mathrm{e}^{-k\tau\lambda}\right), \quad \tau \geqslant 0,$$

and the system is stable for a given $\tau \geqslant 0$ if and only if for some $\varepsilon < 0$,

$$p\left(\lambda,\, e^{-\tau\lambda}\right) \neq 0, \quad \forall\lambda \in \bar{\mathbb{C}}_{\varepsilon+}, \tag{5.26}$$

and it is delay-independent stable if (5.26) holds for all $\tau \geqslant 0$.

As discussed before, the stability of a neutral delay system requires that its neutral part be stable; for the systems (5.22) and (5.23), this implies the stability of the difference equations

$$y(t) + d_1 y(t-\tau) + d_2 y(t-2\tau) = 0, \tag{5.27}$$

and

$$x(t) - D_1\, x(t-\tau) - D_2\, x(t-2\tau) = 0. \tag{5.28}$$

It is well known that if these equations admit stable solutions for some $\tau \geqslant 0$, then, they will be stable for all $\tau \geqslant 0$, i.e., they are delay-independent stable [125].

The following facts provide necessary and sufficient conditions for the stability of the difference equations (5.27) and (5.28).

Fact 1 *Equation* (5.27) *is stable for all* $\tau \geqslant 0$ *if and only if*

$$\rho(N_d) < 1, \tag{5.29}$$

where

$$N_d := \begin{bmatrix} -d_1 & -d_2 \\ 1 & 0 \end{bmatrix}.$$

Equivalently, Eq. (5.27) *is stable for all* $\tau \geqslant 0$ *if and only if the polynomial*

$$d(z) = z^2 + d_1 z + d_2.$$

is Schur stable; i.e., $d(z)$ *has all its zeros in* $\mathbb{D}$.
Fact 2 *Equation* (5.28) *is stable for all* $\tau \geqslant 0$ *if and only if*

$$\rho(N_s) < 1, \tag{5.30}$$

where

$$N_s := \begin{bmatrix} D_1 & D_2 \\ I & 0 \end{bmatrix}.$$

Hereinafter, we assume that conditions (5.29) and (5.30) hold; under these assumptions, (5.25) and (5.26) must hold only for $\mathbb{C}_+$, (i.e., $\mathbb{C}_{\varepsilon+}$ can be replaced by $\mathbb{C}_+$). We also consider that systems (5.22) and (5.23) are stable for $\tau = 0$.

We aim at finding the delay margin for systems (5.22) and (5.23), defined as

$$\tau_d := \inf\left\{\tau : a\left(\lambda, e^{-\tau\lambda}\right) = 0 \text{ for some } \lambda \in \bar{\mathbb{C}}_+\right\}$$
$$\tau_s := \inf\left\{\tau : p\left(\lambda, e^{-\tau\lambda}\right) = 0 \text{ for some } \lambda \in \bar{\mathbb{C}}_+\right\}.$$

Remark 5.3.1 Further discussions on the continuity properties of the spectrum of linear neutral-type time-delay systems with respect to the delay parameters can be found in [196].

5.3.1 Differential-Difference Equations

Consider the characteristic function (5.24) of a differential-difference equation of the form (5.22). If condition (5.29) holds, the zeros of $a\left(\lambda, e^{-\tau\lambda}\right)$ continuously vary with $\tau \geqslant 0$, then the delay margin τ_d is reduced to

$$\tau_d = \inf\left\{\tau : a\left(j\omega, e^{-j\tau\omega}\right) = 0 \text{ for some } \omega \in \mathbb{R}_+\right\}. \tag{5.31}$$

Notice that we only consider $\omega \in \mathbb{R}_+$ since the complex zeros of $a\left(\lambda, e^{-\tau\lambda}\right)$ are conjugate symmetric. The values of $\omega \in \mathbb{R}_+$ satisfying $a\left(j\omega, e^{-\tau\omega}\right) = 0$ are called crossing frequencies of $a(\lambda, e^{-\tau\lambda})$. From (5.31), we can see that the delay margin τ_d can be obtained after determining the crossing frequencies, or alternatively, when the zeros (λ, z) of the bivariate polynomial $a(\lambda, z)$ are such that $\lambda \in \partial\mathbb{C}_+$ and $z \in \partial\mathbb{D}$. The following result provides conditions for determining the delay margin τ_d.

Theorem 5.3.2 *Assume that system* (5.22) *is stable for* $\tau = 0$*, and that condition* (5.29) *is satisfied. Define*

$$T_n := \begin{bmatrix} 1 & 0 \\ d_1 & 1 \end{bmatrix}, \quad T_i := \begin{bmatrix} a_{0i} & 0 \\ a_{1i} & a_{0i} \end{bmatrix}, i = 0, 1, \ldots, n-1,$$
$$H_n := \begin{bmatrix} d_2 & d_1 \\ 0 & d_2 \end{bmatrix}, \quad H_i := \begin{bmatrix} a_{2i} & a_{1i} \\ 0 & a_{2i} \end{bmatrix}, i = 0, 1, \ldots, n-1,$$
$$P_i := \begin{bmatrix} (j)^i T_i & (j)^i H_i \\ (-j)^i H_i^T & (-j)^i T_i^T \end{bmatrix}, \; i = 0,\; 1, \ldots,\; n.$$

Then P_n is invertible. Define P as

$$P := \begin{bmatrix} 0 & I & \cdots & 0 \\ \vdots & \vdots & \ddots & \vdots \\ 0 & 0 & \cdots & I \\ -P_n^{-1}P_0 & -P_n^{-1}P_1 & \cdots & -P_n^{-1}P_{n-1} \end{bmatrix}.$$

Then, $\tau_d = \infty$ if $\sigma(P) \cap \mathbb{R}_+ = \emptyset$, or $\sigma(P) \cap \mathbb{R}_+ = \{0\}$. Otherwise, let $\sigma(P) \cap \mathbb{R}_+ = \{\omega_k : \omega_k \neq 0,\; k = 1, \ldots, l,\; l \leqslant 4n\}$, and define

$$F(\lambda) := \begin{bmatrix} 0 & 1 \\ -a_0(\lambda) & -a_1(\lambda) \end{bmatrix},$$
$$G(\lambda) := \; diag\,(1 \;\; 1 \;\; a_2(\lambda)).$$

If $\sigma(F(j\omega_k),\; G(j\omega_k)) \cap \partial\mathbb{D} = \emptyset$ for all $k = 1, \ldots, l$, then $\tau_d = \infty$; otherwise,

$$\tau_d = \min_k \min_i \frac{\alpha_k^{(i)}}{\omega_k},$$

with $\sigma(F(j\omega_k),\; G(j\omega_k)) \cap \partial\mathbb{D} = \{e^{-j\alpha_k^{(i)}} : \alpha_k^{(i)} \in [0,\; 2\pi],\; i = 1,\; 2\}$.

For a fixed $\lambda \in \mathbb{C}$, let $a(\lambda, z)$ be a bivariate polynomial. The Schur-Cohn matrix $\Delta(\lambda)$ associated to the polynomial $a(\lambda, z)$ is defined as follows [22],

$$\Delta(\lambda) := \begin{bmatrix} \Delta_1(\lambda) & \Delta_2(\lambda) \\ \Delta_2^H(\lambda) & \Delta_1^H(\lambda) \end{bmatrix},$$

where

$$\Delta_1(\lambda) := \begin{bmatrix} a_0(\lambda) & 0 \\ a_1(\lambda) & a_0(\lambda) \end{bmatrix}, \quad \Delta_2(\lambda) := \begin{bmatrix} a_2(\lambda) & a_1(\lambda) \\ 0 & a_2(\lambda) \end{bmatrix}.$$

According to the *Orlando formula* [22], the determinant of $\Delta(\lambda)$ is given by

$$\det(\Delta(\lambda)) = |a_2(\lambda)|^4 \prod_{i,j=1}^{2} (1 - z_i \bar{z}_j), \tag{5.32}$$

where z_i, $i = 1, 2$ are the zeros of the polynomial $a(\lambda, z)$, for a given λ. Notice that for any $\lambda \in \mathbb{C}$, $\det(\Delta(\lambda)) = 0$ is satisfied if $a(\lambda, z)$ has all its zeros in $\mathbb{D}^c$. Then, consequently, we can determine all $\omega \in \mathbb{R}_+$ such that $a(j\omega, z) = 0$ for some $z \in \partial\mathbb{D}$ by finding the solutions of $\det(\Delta(j\omega)) = 0$.

Proof We first prove that P_n is invertible. To this end, consider the polynomial $d(z) = z^2 + d_1 z + d_2$. The Schur-Cohn matrix corresponding to this polynomial coincides with

$$\tilde{P}_n = \begin{bmatrix} T_n & H_n \\ H_n^T & T_n^T \end{bmatrix}.$$

Hence, in view of (5.32),

$$\det(\tilde{P}_n) = (-1)^n \prod_{i,j=1}^{n} (1 - z_i \bar{z}_j),$$

where z_i are the zeros of $d(z)$. Since if condition (5.29) is satisfied $d(z)$ is stable, if follows that $\det(\tilde{P}_n) \neq 0$. Note that

$$P_n = \begin{bmatrix} (j)^n T_n & (j)^n H_n \\ (-j)^n H_n^T & (-j)^n T_n^T \end{bmatrix} = \begin{bmatrix} (j)^n I & 0 \\ 0 & (-j)^n I \end{bmatrix} \tilde{P}_n,$$

we conclude that $\det(P_n) \neq 0$. For the remainder of the proof, observe that

$$\det(\omega I - P) = \det(P_n^{-1}) \det\left[\Delta(j\omega)\right],$$

which follows readily from the *Schur determinant formula* [132]. The rest of the proof then seeks to determine the crossing frequencies and the corresponding roots in $\partial\mathbb{D}$.

5.3.2 State-Space Models

Consider now the state-space representation (5.23) of a neutral-type system. We aim at determining the crossing frequencies of $p(\lambda, e^{-\tau\lambda})$, i.e., the critical frequencies $\omega \in \mathbb{R}_+$ at which the characteristic quasipolynomial $p(j\omega, e^{-j\tau\omega})$ satisfies $p(j\omega, e^{-j\tau\omega}) = 0$.

Denote

$$\mathscr{A}(z) := \left(I - \sum_{k=1}^{2} D_k z^k\right)^{-1} \sum_{k=0}^{2} A_k z^k.$$

We seek to find all $z_k \in \partial\mathbb{D}$ such that $\sigma(\mathscr{A}(z_k)) \cap \partial\mathbb{C}_+ \neq \emptyset$; in this case, there will exist $j\omega_k \in \partial\mathbb{C}_+$ and $z_k \in \partial\mathbb{D}$ such that $j\omega_k \in \sigma(\mathscr{A}(z_k))$, or

$$\det(j\omega_k I - \mathscr{A}(z_k)) = 0. \tag{5.33}$$

By finding all $\omega_k \in \mathbb{R}_+$ and $z_k \in \partial\mathbb{D}$, we may compute the delay margin τ_s.

Notice that the condition (5.33) is equivalent to

$$\det\left[\mathscr{A}(z_k) \oplus \mathscr{A}^H(z_k)\right] = 0. \tag{5.34}$$

Let

$$A(z) = \sum_{k=0}^{2} A_k z^k, \qquad D(z) = I - \sum_{k=1}^{2} D_k z^k.$$

According to the Kronecker product properties, it follows that for any $z \in \partial\mathbb{D}$,

$$\begin{aligned}
\det\left[\mathscr{A}(z) \oplus \mathscr{A}^H(z)\right] &= \det\left[\mathscr{A}(z) \otimes I + I \otimes \mathscr{A}^H(z)\right] \\
&= \det\left[\left(D^{-1}(z)A(z)\right) \otimes I + I \otimes \left(A^H(z)D^{-H}(z)\right)\right] \\
&= \det\left[(D(z) \otimes I)^{-1}(A(z) \otimes I) + \left(I \otimes A^H(z)\right)\left(I \otimes D^H(z)\right)^{-1}\right] \\
&= \det\left[(D(z) \otimes I)^{-1}\right] \det\left[A(z) \otimes D^H(z) + D(z) \otimes A^H(z)\right] \det\left[\left(I \otimes D^H(z)\right)^{-1}\right].
\end{aligned}$$

The results on the delay margin are stated in the following theorem:

Theorem 5.3.3 *Assume that system* (5.23) *is stable for* $\tau = 0$, *and that the condition* (5.30) *is satisfied.*

Let

$$H_k = \sum_{i=\max\{0,k-2\}}^{\min\{k,2\}} \left[A_{k-i} \otimes D_{2-i}^T + D_{k-i} \otimes A_{2-i}^T\right], \quad k = 0, 1, \ldots, 4,$$

$$Q_k = \begin{cases} I \otimes A_{2-k}^T - H_k & k = 0, 1, \\ A_0 \oplus A_0^T - H_2 & k = 2, \\ A_{k-2} \otimes I - H_k & k = 3, 4 \end{cases}$$

with $D_0 = 0$.

Define also

$$U := \begin{bmatrix} I & & & \\ & \ddots & & \\ & & I & \\ & & & Q_4 \end{bmatrix}, \quad V := \begin{bmatrix} 0 & I & \cdots & 0 \\ \vdots & \vdots & \ddots & \vdots \\ 0 & 0 & \cdots & I \\ -Q_0 & -Q_1 & \cdots & -Q_3 \end{bmatrix}.$$

Then, $\tau_s = \infty$ if $\sigma(V, U) \cap \partial\mathbb{D} = \emptyset$. If, however, $\sigma(V, U) \cap \partial\mathbb{D} \neq \emptyset$ and $\sigma(\mathscr{A}(z_k)) = \{0\}$ for all $z_k \in \sigma(V, U) \cap \partial\mathbb{D}$, then $\tau_s = \infty$ as well. Otherwise, let $\sigma(V, U) \cap \partial\mathbb{D} = \{e^{j\alpha_k} : \alpha_k \in [0, 2\pi],\ k = 1, \ldots, m,\ m \leqslant 2n^4\}$. If $\sigma(\mathscr{A}(e^{-j\alpha_k})) \cap \partial\mathbb{C}_+ = \emptyset$ for all $k = 1, \ldots, m$, then $\tau_s = \infty$; otherwise

$$\tau_s = \min_k \min_i \frac{\alpha_k}{\omega_k^{(i)}},$$

with $\omega_k^{(i)} \in \mathbb{R}_+$, $\omega_k^{(i)} \neq 0$ and $j\omega_k^{(i)} \in \sigma(\mathscr{A}(e^{-j\alpha_k})) \cap \partial\mathbb{C}_+$ for $i = 1, \ldots, l$, $l \leqslant m$.

Proof Under the assumption (5.30), $\det[D(z)] \neq 0$ for any $z \in \partial\mathbb{D}$. Hence, in view of the above derivation,

$$\det\left[\mathscr{A}(z) \oplus \mathscr{A}^H(z)\right] \neq 0 \tag{5.35}$$

for any $z \in \partial\mathbb{D}$ if and only if

$$\det\left[A(z) \otimes D^H(z) + D(z) \otimes A^H(z)\right] \neq 0.$$

It is easy to show that for any $z \in \partial\mathbb{D}$,

$$A(z) \otimes D^H(z) + D(z) \otimes A^H(z) = z^{-2} \sum_{k=0}^{4} Q_k z^k.$$

As a consequence, condition (5.35) holds if and only if

$$\det\left(\sum_{k=0}^{4} Q_k z^k\right) \neq 0. \tag{5.36}$$

However, according to the Schur formula ([119, p. 50]),

$$\det\left(\sum_{k=1}^{4} Q_k z^k\right) = \det(zU - V).$$

The rest of the proof follows.

5.3.3 Scalar Example Revisited

In this section we provide an example to illustrate the delay margin results.

Example 5.1 Consider the first-order neutral system

$$\dot{x}(t) - d\dot{x}(t - \tau) = a_0 x(t) + a_1 x(t - \tau). \tag{5.37}$$

To ensure the stability of the difference operator, we assume $|d| < 1$.

For the retarded case ($d = 0$), the system is delay-independent stable if and only if $a_0 + a_1 < 0$ and $a_0 \geqslant |a_1|$ and its delay margin is given by ([119, p. 40]):

$$\tau_r = \frac{\cos^{-1}(\frac{|a_0|}{-a_1})}{\sqrt{a_1^2 - a_0^2}}.$$

For the neutral case ($d \neq 0$, $|d| < 1$) the delay margin can be obtained by using Theorem 5.3.2.

Matrix P is defined as

$$P = -\frac{j}{1-d^2}\begin{bmatrix} a_0 - da_1 & a_1 - da_0 \\ -a_1 + da_0 & -a_0 + da_1 \end{bmatrix},$$

which has a single positive eigenvalue

$$\omega^* = \sqrt{\frac{a_1^2 - a_0^2}{1-d^2}}. \tag{5.38}$$

The matrix pair $(F(\lambda),\ G(\lambda))$ is given as $F(\lambda) = -\lambda + a_0$, $G(\lambda) = -(d\lambda + a_1)$, whose generalized eigenvalue for $\lambda = j\omega^*$ is given by

$$\lambda^* = -\frac{-j\omega^* + a_0}{jd\omega^* + a_1}.$$

It is easy to show that $\lambda^* \in \partial\mathbb{D}$. Let $\lambda^* = e^{-j\alpha^*}$, note that

$$\alpha^* = \pi - \angle\left(\frac{-j\omega^* + a_0}{jd\omega^* + a_1}\right). \tag{5.39}$$

The delay margin can then be determined as $\tau_d = \alpha^*/\omega^*$.

Figures 5.1, 5.2 and 5.3 show how the delay margin τ_d varies with the system parameters for some given values of d, a_0, and a_1.

Observe that for $|a_1| < |a_0|$, $\tau_d = 0$, i.e., there is no delay value at which the system becomes unstable (the system is delay-independent asymptotically stable, see statement (iii) of Proposition 5.2.17).

5.3.4 Drilling Model

In this section we use the result derived in Example 5.1 about the delay margin of a scalar neutral system to analyze the asymptotic stability of the drilling system.

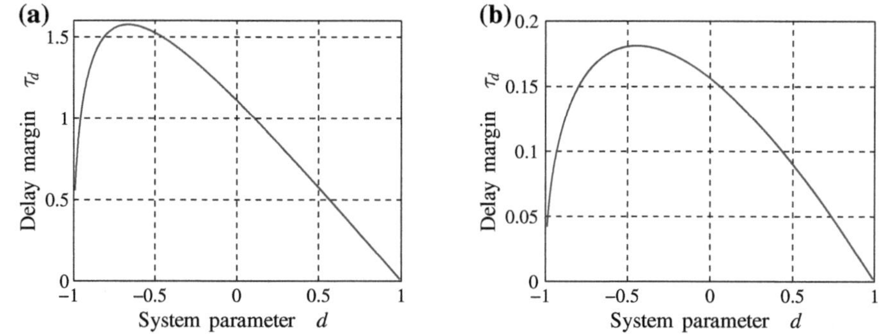

Fig. 5.1 Delay margin τ_d versus parameter d of a neutral-type time-delay system of the form (5.37) for: **a** $a_0 = \sqrt{2}/2, a_1 = -1$; **b** $a_0 = 0.1, a_1 = -10$

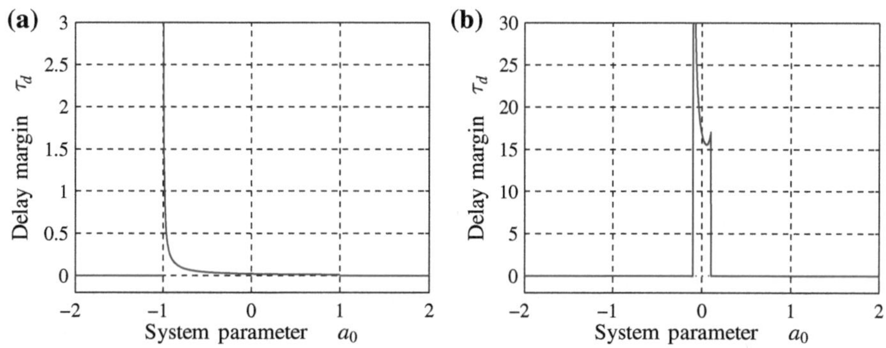

Fig. 5.2 Delay margin τ_d versus parameter a_0 of a neutral-type time-delay system of the form (5.37) for: **a** $d = 0.99, a_1 = -1$; **b** $d = -0.7, a_1 = -0.1$

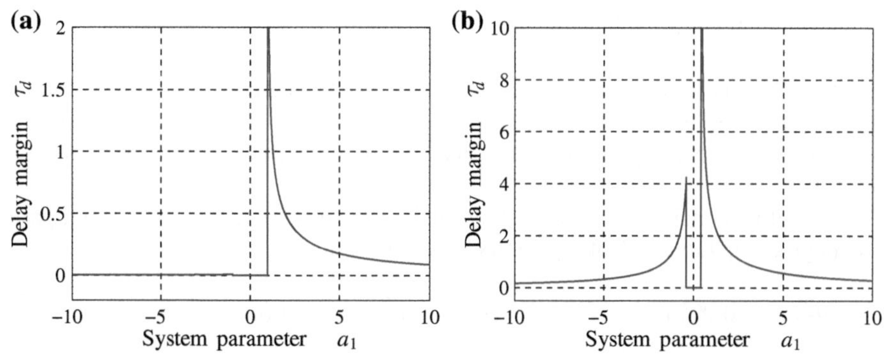

Fig. 5.3 Delay margin τ_d versus parameter a_1 of a neutral-type time-delay system of the form (5.37) for: **a** $d = 0.99, a_0 = 0.99$; **b** $d = -0.7, a_0 = 0.4$

Consider the matrix representation of the drilling model (5.6) given in (5.7). By taking the torque on bit function as $T(\dot{\Phi}_b(t)) = k_t \dot{\Phi}_b(t)$, the system constants are defined as:

$$d_{11}^1 = \Upsilon \qquad d_{22}^2 = \tilde{\Upsilon} \qquad a_{11}^0 = -\Psi - \frac{k_t}{I_B} \qquad a_{12}^0 = -\frac{k_t}{M_B}$$
$$a_{22}^0 = -\tilde{\Psi} \qquad a_{11}^1 = -\Upsilon\Psi - \frac{\Upsilon k_t}{I_B} \qquad a_{21}^2 = \frac{\tilde{\Upsilon} k_t}{M_B} \qquad a_{22}^2 = -\tilde{\Upsilon}\tilde{\Psi}.$$

Proposition 5.3.4 *The drilling model described by a NFDE of the form* (5.7)*, under a frictional torque on bit model given by* $T(\dot{\Phi}_b(t)) = k_t \dot{\Phi}_b(t)$*, is asymptotically stable.*

Proof As stated in Proposition 5.2.4, the characteristic quasipolynomial associated with the drilling system is given by:

$$p\left(\lambda, \mathrm{e}^{-\tau_1\lambda}, \mathrm{e}^{-\tau_2\lambda}\right) = p_1\left(\lambda, \mathrm{e}^{-\tau_1\lambda}\right) p_2\left(\lambda, \mathrm{e}^{-\tau_2\lambda}\right),$$

where $\tau_1 = 2\tau$, $\tau_2 = 2\tilde{\tau}$, p_1 and p_2 are given by:

$$p_1\left(\lambda, \mathrm{e}^{-\tau_1\lambda}\right) = \left(1 - d_{11}^1 \mathrm{e}^{-\tau_1\lambda}\right)\lambda - a_{11}^0 + a_{11}^1 \mathrm{e}^{-\tau_1\lambda} \tag{5.40}$$

and

$$p_2\left(\lambda, \mathrm{e}^{-\tau_2\lambda}\right) = \left(1 - d_{22}^2 \mathrm{e}^{-\tau_2\lambda}\right)\lambda - a_{22}^0 + a_{22}^2 \mathrm{e}^{-\tau_2\lambda}. \tag{5.41}$$

Notice that the characteristic quasipolynomial of the drilling system can be seen as the product of the quasipolynomials of two scalar "subsystems." Then, the asymptotic stability of the drilling system can be analyzed by computing the delay margin of each "subsystem."

First, let us consider the quasipolynomial $p_1\left(\lambda, \mathrm{e}^{-\tau_1\lambda}\right)$. Considering $k_t = 0.1$ and the system parameters given in Table C.1 of Appendix C, we have that

$$d_{11}^1 = 0.7393, \quad a_{11}^0 = -3.3689, \qquad a_{11}^1 = -2.4907, \quad \tau_1 = 0.7445,$$

and the delay margin is such that $\tau_d = \alpha^*/\omega^* = 0$ (where ω^* and α^* are defined in (5.38) and (5.39), with $a_0 = a_{11}^0$, $a_1 = a_{11}^1$, and $d = d_{11}^1$), which means that the "subsystem" corresponding to the quasipolynomial $p_1\left(\lambda, \mathrm{e}^{-\tau_1\lambda}\right)$ is stable (see Fig. 5.4).

Now, let us consider the quasipolynomial $p_2\left(\lambda, \mathrm{e}^{-\tau_2\lambda}\right)$. Using the system parameters given in Table C.1 of Appendix C, we have that

$$d_{22}^2 = -0.9971, \qquad a_{22}^0 = -3.500, \qquad a_{22}^2 = 3.4900, \qquad \tau_2 = 0.4688,$$

and the delay margin is $\tau_d = \alpha^*/\omega^* = 0$ (ω^* and α^* are defined in (5.38) and (5.39), with $a_0 = a_{22}^0$, $a_1 = a_{22}^2$, and $d = d_{22}^2$), then, the "subsystem" corresponding to $p_2\left(\lambda, \mathrm{e}^{-\tau_2\lambda}\right)$ is asymptotically stable (see Fig. 5.5).

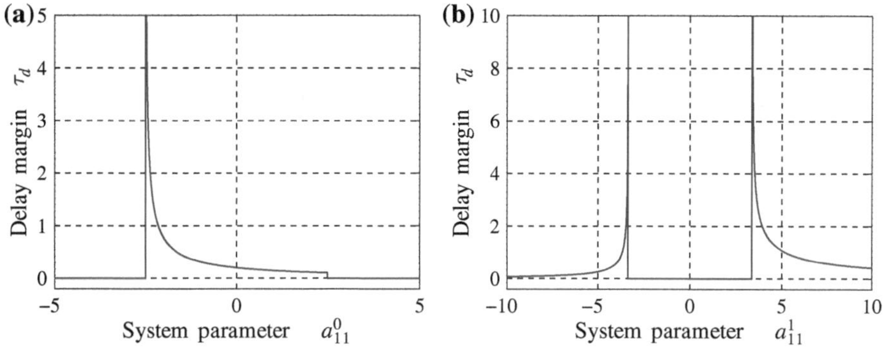

Fig. 5.4 Delay margin of the scalar "subsystem" of the drilling model whose characteristic quasi-polynomial $p_1\left(\lambda, e^{-\tau_1\lambda}\right)$ is given in (5.40). **a** τ_d versus a_{11}^0 for $d_{11}^1 = 0.7393$ and $a_{11}^1 = -2.4907$. **b** τ_d versus a_{11}^1 for $d_{11}^1 = 0.7393$ and $a_{11}^0 = -3.3689$

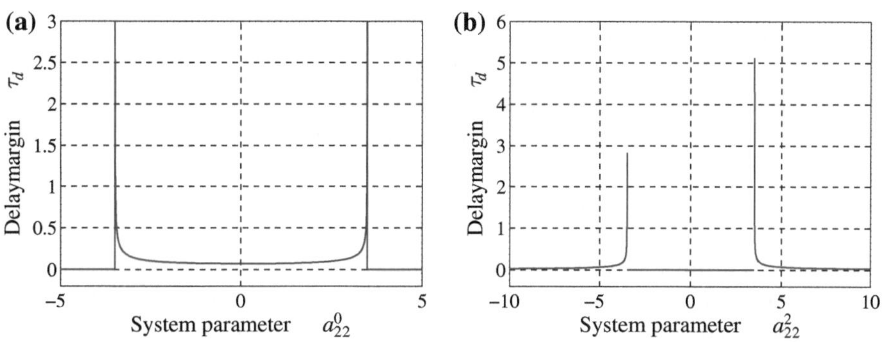

Fig. 5.5 Delay margin of the scalar "subsystem" of the drilling model whose characteristic quasi-polynomial $p_2\left(\lambda, e^{-\tau_2\lambda}\right)$ is given in (5.41). **a** τ_d versus a_{22}^0 for $d_{22}^2 = -0.9971$ and $a_{22}^2 = 3.4900$. **b** τ_d versus a_{22}^2 for $d_{22}^2 = -0.9971$ and $a_{22}^0 = -3.500$

Remark 5.3.5 The proof of Proposition 5.3.4 can be alternatively developed by using Proposition 5.2.17. Notice that since conditions (i), (ii), and (iii) of Proposition 5.2.17 are satisfied for both "subsystems" associated with $p_1\left(\lambda, e^{-\tau_1\lambda}\right)$ and $p_2\left(\lambda, e^{-\tau_2\lambda}\right)$, the drilling system described by (5.7), under the ToB model $T(\dot{\Phi}_b(t)) = k_t\dot{\Phi}_b(t)$, is delay-independent asymptotically stable.

We conclude that the drilling system is formally stable (Proposition 5.2.7) and furthermore, it is asymptotically stable for $T(\dot{\Phi}_b(t)) = k_t\dot{\Phi}_b(t)$ (Proposition 5.3.4). However, it is usually assumed that the system unstability leading to the occurrence of drilling vibrations arises from the bit-rock interaction which should be modeled by an appropriate nonlinear torque on bit function (see Chap. 3).

The next chapter provides a stability analysis of the drilling system under a nonlinear ToB model to approximate the phenomena leading to undesirable vibrations.

5.4 Notes and References

Extensive research effort has been devoted to the study of FDE; the stability analysis of time-delay systems has been addressed in several research papers and survey contributions. See for instance [159] which gives an overview of the different techniques to analyze the delay effects on the system stability. The survey paper [149] investigates the stability and robust stability of uncertain linear time-delay systems. In [242], particular problems of time-delay systems such as the constructive use of delayed inputs, the digital implementation of distributed delays, the control via the delay, and the handling of information related to the delay value are discussed. A review on existing algebraic tools for the analysis of FDE leading to constructive control algorithms is presented in [185].

Some of the main results on the stability of linear neutral-type time-delay systems have been presented in this chapter; two analysis frameworks were considered: frequency-domain approach [144] and time-domain techniques [150]. Supplementary theoretical tools can be found in the books of Jack K. Hale; introductory concepts are given in [126], and further developments in [124].

The delay margin analysis of a linear NFFD described by a differential-difference equation or in its state-space form was also studied in this chapter. Some techniques to compute the critical delay value at which the system loses its stability were introduced. Since the proposed techniques require only the computation of eigenvalues and generalized eigenvalues, the delay margin can be computed efficiently and with high numerical precision. As an application example, we have investigated the delay margin of the drilling system coupled to a linear ToB model; it has been shown that, under this circumstance, the system response exhibits stable properties; however, as will be discussed, the bit-rock frictional interface, modeled by a nonlinear function depending on the bit velocity, leads to the system asymptotic instability.

The basic notions of the general theory for NFDE reviewed in this chapter are useful in analyzing the drilling system dynamics. The results presented in the following chapter uses some frequency-domain concepts to develop a bifurcation analysis which allows characterizing the qualitative response of the system. The stability concepts in the time-domain framework will be applied in Chap. 7 to investigate the ultimate boundedness of the system trajectories, i.e., the practical stability of the system.

Chapter 6
Bifurcation Analysis of the Drilling System

Bifurcation theory is concerned with the study of changes in the qualitative structure of the solutions of a given family of differential equations. In dynamical systems, a bifurcation occurs when a small smooth alteration of the parameter values (the bifurcation parameters) causes a sudden qualitative or topological change in its behavior [35]. Bifurcations occur in both continuous systems (described by ODEs, DDEs or PDEs), and discrete systems (described by maps). The term bifurcation was introduced in 1885 by Henri Poincaré [231].

Bifurcations can be classified into two main categories: *global bifurcations*, which occur when larger invariant sets of the system *collide* with each other, or with the equilibrium point of the system; and *local bifurcations*, which can be analyzed through variations of the local stability properties of the equilibrium point. Some examples of bifurcations of local type are: Pitchfork bifurcation, Andronov-Hopf bifurcation, and Bogdanov-Takens bifurcation. Appendix A provides a short summary of each one of them.

In order to characterize the qualitative dynamic response of the rotary drilling system, this chapter addresses its local bifurcation analysis. Based on the *center manifold theorem* [52] and *normal forms theory* [121], a set of neutral-type time-delay equations that models the coupled axial-torsional drilling vibrations is reduced via spectral projections to a finite-dimensional system described by an ODE which simplifies the analysis task.

6.1 Local Bifurcation Analysis

Consider the following system of equations describing the coupled axial-torsional drilling vibrations:

$$\begin{cases} \frac{\partial^2 U}{\partial s^2}(s,t) &= c^2 \frac{\partial^2 U}{\partial t^2}(s,t), \quad c = \sqrt{\frac{\rho_a}{E}} \\ E\Gamma \frac{\partial U}{\partial s}(0,t) &= \alpha \frac{\partial U}{\partial t}(0,t) - u_H(t) \\ E\Gamma \frac{\partial U}{\partial s}(L,t) &= -M_B \frac{\partial^2 U}{\partial t^2}(L,t) + F\left(\frac{\partial U}{\partial t}(L,t)\right) \end{cases}$$

M.B. Saldivar Márquez et al., *Analysis and Control of Oilwell Drilling Vibrations*, Advances in Industrial Control, DOI 10.1007/978-3-319-15747-4_6

and

$$\begin{cases} \frac{\partial^2 \Phi}{\partial s^2}(s,t) = \tilde{c}^2 \frac{\partial^2 \Phi}{\partial t^2}(s,t), \quad \tilde{c} = \sqrt{\frac{\rho_a}{G}} \\ GJ\frac{\partial \Phi}{\partial s}(0,t) = \beta \frac{\partial \Phi}{\partial t}(0,t) - u_T(t) \\ GJ\frac{\partial \Phi}{\partial s}(L,t) = -I_B \frac{\partial^2 \Phi}{\partial t^2}(L,t) - \tilde{F}\big(\frac{\partial U}{\partial t}(L,t)\big) \end{cases}$$

where $U(s,t)$ and $\Phi(s,t)$ are the rotary angle and the longitudinal position, respectively. The system is controlled through the upward hook force u_H and the rotary table motor torque u_T. The nonlinear aspect of the model is considered by taking functions F and $\tilde{F}$ of the form:

$$z \mapsto p\bar{k}\,z/(\bar{k}^2 z^2 + \zeta),$$

where the parameters $\bar{k}$, ζ ($0 < \zeta \ll 1$ and $0 < \bar{k} < 1$) are positive integers responsible of the sharpness of the friction force function and p is acting on its amplitude.

As explained in Sect. 2.3, the above PDE model can be transformed into the following pair of coupled neutral-type time-delay equations:

$$\begin{aligned} \ddot{U}_b(t) - \tilde{\Upsilon}\ddot{U}_b(t-2\tilde{\tau}) = &-\tilde{\Psi}\dot{U}_b(t) - \tilde{\Upsilon}\tilde{\Psi}\dot{U}_b(t-2\tilde{\tau}) \\ &+ \frac{1}{M_B}F\big(\dot{U}_b(t)\big) - \frac{1}{M_B}\tilde{\Upsilon}F\big(\dot{U}_b(t-2\tilde{\tau})\big) + \tilde{\Pi}u_H(t-\tilde{\tau}), \end{aligned}$$

$$\begin{aligned} \ddot{\Phi}_b(t) - \Upsilon\ddot{\Phi}_b(t-2\tau) = &-\Psi\dot{\Phi}_b(t) - \Upsilon\Psi\dot{\Phi}_b(t-2\tau) \\ &+ \frac{1}{I_B}\tilde{F}\big(\dot{U}_b(t)\big) - \frac{1}{I_B}\Upsilon\tilde{F}\big(\dot{U}_b(t-2\tau)\big) + \Pi u_T(t-\tau), \end{aligned}$$

where

$$\begin{aligned} \tilde{\Pi} &= \frac{2\tilde{\Psi}}{\alpha + cE\Gamma}, & \tilde{\Upsilon} &= \frac{\alpha - cE\Gamma}{\alpha + cE\Gamma}, & \tilde{\Psi} &= \frac{cE\Gamma}{M_B}, & \tilde{\tau} &= cL, \\ \Pi &= \frac{2\Psi}{\beta + \tilde{c}GJ}, & \Upsilon &= \frac{\beta - \tilde{c}GJ}{\beta + \tilde{c}GJ}, & \Psi &= \frac{\tilde{c}GJ}{I_B}, & \tau &= \tilde{c}L, \end{aligned}$$

and $\dot{U}_b(t)$, $\dot{\Phi}_b(t)$ are the the axial and angular velocities at the bottom extremity, respectively.

A normalization of the above model yields the following dimensionless system:

$$\begin{cases} \ddot{U}_b(t) - \tilde{\Upsilon}_n \ddot{U}_b(t-2) = -\tilde{\Psi}_n \dot{U}_b(t) - \tilde{\Upsilon}_n \tilde{\Psi}_n \dot{U}_b(t-2) + \dfrac{1}{M_B} F\big(\dot{U}_b(t)\big) \\ \qquad - \dfrac{1}{M_B} \tilde{\Upsilon}_n F\big(\dot{U}_b(t-2)\big) + \tilde{\Pi}_n u_H(t-1), \\ \ddot{\Phi}_b(t) - \Upsilon_n \ddot{\Phi}_b(t-2\tau) = -\Psi_n \dot{\Phi}_b(t) - \Upsilon_n \Psi_n \dot{\Phi}_b(t-2\tau) + \dfrac{1}{I_B} \tilde{F}\big(\dot{U}_b(t)\big) \\ \qquad - \dfrac{1}{I_B} \Upsilon_n \tilde{F}\big(\dot{U}_b(t-2\tau)\big) + \Pi_n u_T(t-\tau), \end{cases} \tag{6.1}$$

where τ is the ratio of the speeds $\tau = \tilde{c}/c$ and

$$\tilde{\Pi}_n = \frac{2\tilde{\Psi}_n}{\alpha+1}, \quad \tilde{\Upsilon}_n = \frac{\alpha-1}{\alpha+1}, \quad \tilde{\Psi}_n = \frac{1}{M_B},$$
$$\Pi_n = \frac{2\tilde{c}GJ}{I_B\,(cE\Gamma\beta + \tilde{c}GJ)}, \quad \Upsilon_n = \frac{cE\Gamma\beta - \tilde{c}GJ}{cE\Gamma\beta + \tilde{c}GJ}, \quad \Psi_n = \frac{\tilde{c}GJ}{cE\Gamma I_B}.$$

First, we analyze the uncontrolled system ($u_H = u_T = 0$). Denote by x_1 and x_2 the axial and angular velocities at the bottom extremity ($\dot{U}_b$ and $\dot{\Phi}_b$, respectively). The state-space matrix representation of the system is written as follows:

$$\begin{cases} \dot{x}(t) = D_1\,\dot{x}(t-2) + D_2\,\dot{x}(t-2\tilde{c}/c) + A_0\,x(t) + A_1\,x(t-2) \\ \qquad +A_2\,x(t-2\tilde{c}/c) + \mathscr{F}(x(t), x(t-2), x(t-2\tilde{c}/c)) \end{cases} \tag{6.2}$$

where $x = (x_1, x_2)^T$, $\mathscr{F} : \mathbb{R}^n \times \mathbb{R}^n \times \mathbb{R}^n \to \mathbb{R}^n$ is a mapping representing the nonlinear part of system (6.1), and the matrices $D_1,\ D_2,\ A_0,\ A_1,\ A_2$ are given by:

$$D_1 = \begin{bmatrix} \tilde{\Upsilon}_n & 0 \\ 0 & 0 \end{bmatrix}, \quad D_2 = \begin{bmatrix} 0 & 0 \\ 0 & \Upsilon_n \end{bmatrix},$$

$$A_0 = \begin{bmatrix} -\frac{\tilde{\Psi}_n(p\bar{k}+\zeta)}{\zeta} & 0 \\ \frac{p\bar{k}}{J\zeta} & -\Psi_n \end{bmatrix}, \quad A_1 = \begin{bmatrix} -\frac{\tilde{\Upsilon}_n\tilde{\Psi}_n(p\bar{k}+\zeta)}{\zeta} & 0 \\ 0 & 0 \end{bmatrix}, \quad A_2 = \begin{bmatrix} 0 & 0 \\ -\frac{\Upsilon_n p\bar{k}}{\zeta I_B} & -\Upsilon_n\Psi_n \end{bmatrix}.$$

The characteristic equation is a powerful tool for analyzing stability of the steady-state solution of functional differential equation. In what follows we discuss the stability conditions of the drilling system based in the bifurcation parameters p (parameter related to the amplitude of the friction forces F and $\tilde{F}$) and α (viscous friction coefficient at the top extremity) [39]. The numerical values of the model parameters are given in Table C.1 of Appendix C. Appendix A provides a brief discussion on local bifurcations theory.

Here the friction force amplitude p as well as the mass at the BHA M_B are left free and are considered as the bifurcation parameters. When $p = p_c$ then zero is an eigenvalue with algebraic and geometric multiplicity 1. Moreover, zero is the only eigenvalue with zero real part and the remaining eigenvalues have negative real parts. Furthermore, there exists a *Pitchfork bifurcation*, which comes from the $\mathbb{Z}_2$ symmetry structure of the system. When in addition, the mass at the BHA M_B reaches some critical value $M_B = M^*$, then zero is an eigenvalue of algebraic multiplicity 2 and of geometric multiplicity 1. Moreover, under such conditions zero is the only eigenvalue with zero real part and the remaining eigenvalues have negative real parts. Furthermore, it is worth mentioning that the largest algebraic multiplicity of such an eigenvalue at the origin is equal to two in this case [38]. Next, the zero eigenvalue is non-semisimple and the singularity is of *Bogdanov-Takens* type. Finally, although there are no characteristic roots with positive real parts, the system (6.2) is formally stable (see Definition 5.2.6), but not asymptotically stable.

As discussed in Chap. 5 (Proposition 5.2.4), the characteristic equation of system (6.2) is given by

$$
\begin{aligned}
det(\Delta) &= F_1(\lambda)\, F_2(\lambda, p) \\
&= \left(\lambda(1 - \mathrm{e}^{-\lambda\tau_2}\Upsilon_n) - \Upsilon_n + \mathrm{e}^{-\lambda\tau_2}\Upsilon_n\Psi_n\right) \\
&\quad \times \left(\lambda(1 - \mathrm{e}^{-2\lambda}\tilde{\Upsilon}_n) + \frac{\tilde{\Upsilon}_n\tilde{\Psi}_n\,(p\bar{k} + \zeta)}{\zeta}\mathrm{e}^{-2\lambda} + \frac{\tilde{\Psi}_n\,(p\bar{k} + \zeta)}{\zeta}\right).
\end{aligned}
$$

For the sake of simplicity, let us consider separately the factors $F_1(\lambda)$ and $F_2(\lambda, p)$. The first factor is given by:

$$
F_1(\lambda) = \left(\lambda(1 - \mathrm{e}^{-\lambda\tau_2}\Upsilon_n) - \Upsilon_n + \mathrm{e}^{-\lambda\tau_2}\Upsilon_n\Psi_n\right).
$$

Notice that $F_1(\lambda)$ is a scalar first-order quasipolynomial of neutral type. It is easy to prove that the associated continuous-time difference equation is asymptotically stable. Indeed, the scalar quasipolynomial satisfy the conditions of the Pontryagin theorem leading to prove that all spectral values have negative real part as emphasized in Proposition 5.2.8 (for further insights on the proof see [305]). Thus, one concludes that there are no imaginary crossing roots for the factor F_1. Consider now the second factor $F_2(\lambda, p)$ defined as:

$$
F_2(\lambda, p) = \left(\lambda(1 - \mathrm{e}^{-2\lambda}\tilde{\Upsilon}_n) + \frac{\tilde{\Upsilon}_n\tilde{\Psi}_n\,(p\bar{k} + \zeta)}{\zeta}\mathrm{e}^{-2\lambda} + \frac{\tilde{\Psi}_n\,(p\bar{k} + \zeta)}{\zeta}\right).
$$

Similarly, it is easy to show that apart from $\omega_0 = 0$, there are no spectral values with zero real part. Moreover, $p = p_c = 6.66749$ is the only possible value of p leading to a spectral value in zero. Considering the parameterization $\mu = \alpha - 30\,p$ and using the model parameters given in Table C.1 of Appendix C, F_2 can be written as:

$$F_2(\lambda, M_B, p) = \left(-M_B + 0.99\, M_B \mathrm{e}^{-2\lambda}\right)\lambda + (-29.7\,p - 0.99)\,\mathrm{e}^{-2\lambda} + 30\,p - 1$$

A simple substitution of $\lambda = 0$ into F_2 shows that $p = p_c$ leads to a first spectral value on the imaginary axis $\lambda_1 = 0$; a Pitchfork bifurcation comes from the $\mathbb{Z}_2$ symmetry structure of the system. If additionally $(p = p_c)$ $M_B = M^*$, then the first derivative of F_2 vanish at $\lambda_2 = 0$. Since the null space $\mathcal{N}(\lambda_0\, Id - A)$ have only one eigenvector:

$$v_0 = \begin{bmatrix} 1 \\ 4.853 \times 10^{11} \end{bmatrix},$$

there exists then, a double root of non-semisimple type at zero. Moreover, in such a particular case, and for the same reasons as for the first factor, the remaining roots of F_2 have negative real parts.

In conclusion, the system is formally stable but not asymptotically stable (although there are no characteristic roots with positive real parts) and the singularity is of *Bogdanov-Takens* type [165].

Remark 6.1 The multiplicity of the root at the origin can not exceed two for the parameter values given in Table C.1. However, it is worthy of note that, for general second-order systems of neutral type the maximal multiplicity of the characteristic root at the origin is less than the degree of the generic quasipolynomial.

6.2 Model Reduction

Generally speaking, Functional Differential Equations (FDE) share some properties with Ordinary Differential Equations (ODE). This section presents a normal form theory-based technique allowing to approximate a neutral delay differential equation (NDDE) by an ODE. The transformation method, based in the *Center Manifold Theorem* [52], simplifies the system structure, preserving the qualitative dynamic of the system in a neighborhood of the equilibrium state.

Consider the general form of a discrete time-delay autonomous first-order nonlinear system of neutral-type:

$$\frac{d}{dt}[x(t) + \sum_{k=1}^{n} A_k\, x(t-\tau_k)] = \sum_{k=0}^{n} B_k\, x(t-\tau_k) + \mathcal{F}(x(t), \ldots, x(t-\tau_n)), \tag{6.3}$$

where A_i, B_j are $n \times n$ real valued matrices (there is at least one matrix $A_k \neq 0$ for some $k \in \{1, 2, \ldots, n\}$). The time-delays are such that $\tau_0 = 0$, $\tau_i < \tau_j$ for $i < j$ and $\tau_n = r$.

System (6.3) can be written in compact form as follows:

$$\frac{d}{dt}\mathcal{D}x_t = \mathcal{L}x_t + \mathcal{F}(x_t), \tag{6.4}$$

where $x_t \in C = C([-r, 0], \mathbb{R}^n)$, $x_t(\theta) = x(t+\theta)$. The bounded linear operators $\mathscr{D}$ and $\mathscr{L}$ are such that $\mathscr{D}\phi = \phi(0) + \sum_{k=1}^{n} A_k \phi(-\tau_k)$ and $\mathscr{L}\phi = \sum_{k=0}^{n} B_k \phi(-\tau_k)$, $\mathscr{F}$ is a sufficiently smooth function mapping C into $\mathbb{R}^n$ with $\mathscr{F}(0) = D\mathscr{F}(0) = 0$. The linear operators $\mathscr{D}$ and $\mathscr{L}$ can be written in the integral form as $\mathscr{L}\phi = \int_{-r}^{0} d\eta(\theta)\phi(\theta)$ and $\mathscr{D}\phi = \phi(0) + \int_{-r}^{0} d\mu(\theta)\phi(\theta)$, where μ and η are two real valued $n \times n$ matrices.

The linear part of system (6.4) is given by:

$$\frac{d}{dt}\mathscr{D}x_t = \mathscr{L}x_t. \tag{6.5}$$

Let $\mathscr{T}(t)(\phi) = x_t(.\,,\phi)$, with $x_t(.\,,\phi)(\theta) = x(t+\theta,\phi)$ for $\theta \in [-r, 0]$, the solution operator associated with the linear system. $\mathscr{T}(t)$ is a strongly continuous semigroup with the infinitesimal generator given by $\mathscr{A} = \frac{d\phi}{d\theta}$ with the domain

$$\mathscr{D}om(\mathscr{A}) = \left\{\phi \in C : \frac{d\phi}{d\theta} \in C, \mathscr{D}\frac{d\phi}{d\theta} = \mathscr{L}\phi\right\}.$$

The spectrum of $\mathscr{A}$, $\sigma(\mathscr{A}) = \sigma_p(\mathscr{A})$, consists of complex values $\lambda \in \mathbb{C}$ satisfying the characteristic equation: $p(\lambda) = \det\Delta(\lambda) = 0$. See [196] for further details.

Denote by $\mathscr{M}_\lambda$, the eigenspace associated with $\lambda \in \sigma(\mathscr{A})$. We define $C^* = C([-r, 0], \mathbb{R}^{n*})$, where $\mathbb{R}^{n*}$ is the space of n-dimensional row vectors. Consider the bilinear form on $C^* \times C$, proposed in [126]:

$$\begin{aligned}(\psi, \phi) = \phi(0)\,\psi(0) - \int_{-r}^{0} d\left[\int_{0}^{\theta} \psi(\tau - \theta)d\mu(\tau)\right] \\ + \int_{-r}^{0}\int_{0}^{\theta} \psi(\tau - \theta)d\eta(\theta)\phi(\tau)d\tau.\end{aligned} \tag{6.6}$$

Let $\mathscr{A}^T$ be the transposed operator of $\mathscr{A}$, i.e., $(\psi, \mathscr{A}\phi) = (\mathscr{A}^T\psi, \phi)$. The following result leads to an appropriate decomposition of the space C.

Theorem 6.2.1 ([124]) *Let Λ be a nonempty finite set containing the eigenvalues of $\mathscr{A}$ and let $P = span\{\mathscr{M}_\lambda(\mathscr{A}),\ \lambda \in \Lambda\}$ and $P^T = span\{\mathscr{M}_\lambda(\mathscr{A}^T),\ \lambda \in \Lambda\}$. Then P is invariant under $\mathscr{T}(t)$, $t \geq 0$ and there exists a space $\mathscr{Q}$, also invariant under $\mathscr{T}(t)$ such that $C = P \bigoplus Q$. Furthermore, if $\Phi = (\phi_1, \ldots, \phi_m)$ forms a basis of P, and $\Psi = col(\psi_1, \ldots, \psi_m)$ is a basis of P^T in C^*, such that $(\Phi, \Psi) = Id$, then*

$$\begin{aligned} Q &= \{\phi \in C \setminus (\Psi, \phi) = 0\}, \\ P &= \{\phi \in C \setminus \exists b \in \mathbb{R}^m : \phi = \Phi b\}. \end{aligned} \tag{6.7}$$

Moreover, $\mathscr{T}(t)\Phi = \Phi\, e^{Bt}$, where B is a $m \times m$ matrix such that $\sigma(B) = \Lambda$.

Denote by BC, the extension of the space C containing continuous functions on $[-r, 0)$ with a possible jump discontinuity at 0. A given function $\xi \in BC$ can be written as: $\xi = \varphi + X_0\,\alpha$, where $\varphi \in C$, $\alpha \in \mathbb{R}^n$ and $X_0(\theta) = 0$ for $-r \leq \theta < 0$, $X_0(0) = Id_{n\times n}$. The Hale-Verduyn Lunel bilinear form (6.6) can be extended to the space $C^* \times BC$ by $(\psi, X_0) = \psi(0)$, and the infinitesimal generator $\mathscr{A}$ is extended to an operator $\tilde{\mathscr{A}}$ (defined in C^1) into the space BC as follows:

$$\tilde{\mathscr{A}}\phi = \mathscr{A}\phi + X_0[\mathscr{L}\phi - \mathscr{D}\phi']. \tag{6.8}$$

On the basis of the above considerations, Eq. (6.4) can be written as an abstract ODE [88], as follows:

$$\dot{x}_t = \tilde{\mathscr{A}}x_t + X_0\mathscr{F}(x_t). \tag{6.9}$$

The projection $\Pi : BC \to P$, satisfying $\Pi(\varphi + X_0\alpha) = \Phi[(\Psi, \varphi) + \Psi(0)\alpha]$, yields $x_t = \Phi y(t) + z_t$, where $y(t) \in \mathbb{R}^m$. Equation (6.4) can be represented as:

$$\begin{cases} \dot{y} &= By + \Psi(0)F(\Phi y + z) \\ \dot{z} &= \tilde{\mathscr{A}}_Q + (I - \Pi)X_0\mathscr{F}(\Phi y + z). \end{cases} \tag{6.10}$$

After writing z as a function of y, we focus only on the first equation.

The center manifold of a dynamical system is composed by orbits whose behavior around the equilibrium point is not managed by the attraction of the stable manifold (given by the eigenspace of eigenvectors corresponding to eigenvalues with negative real) nor by the repulsion of the unstable manifold. The analysis of the center manifold requires the equilibrium point to be hyperbolic. Center manifolds play an important role in bifurcation theory because interesting system behavior takes place on it.

6.3 Center Manifold and Normal Forms Theory

The center manifold is a powerful tool to analyze the dynamic behavior of a given system in a neighborhood of a nonhyperbolic equilibrium point x^*.

Definition 6.3.1 Consider a C^1 map $h : \mathbb{R} \mapsto Q$. The graph of h is said to be a local manifold associated with system (6.4), if $h(0) = Dh(0) = 0$.

Remark 6.3.2 There exists a neighborhood V of $0 \in \mathbb{R}^n$ such that for each $\xi \in V$, $\delta = \delta(\xi) > 0$. The solution x of system (6.4) with initial data $\Phi\xi + h(\xi)$ exists on the interval $]-\delta - r, \delta[$; it is given by $x_t = \Phi y(t) + h(y(t))$ for $t \in [0, \delta[$, where $y(t)$ is the unique solution of the ODE:

$$\begin{aligned} \dot{y} &= By + \Psi(0)F(\Phi y + h(y)), \\ y(0) &= \xi. \end{aligned} \tag{6.11}$$

Only a few works have investigated the center manifold that arise when considering different matrices B; see for instance [6], where the characterization of the function h in Eq. (6.11) is detailed. On this basis, it is possible to decompose the eigenspace into the subspace containing all imaginary eigenvalues (having real part equal to zero) and the one with the remaining spectral values (that are assumed to have negative real parts).

In what follows, we adopt the formulations introduced in [48] for the study of the center manifold, originally developed for delay differential equations.

It is worth of note that $y \in \mathbb{R}^2$ for the most common singularities; for instance, the (one parameter) Andronov-Hopf bifurcation and the (two-parameter) Bogdanov-Takens bifurcation. See Appendix A.

It is well known that, *normal forms theory* is useful in analyzing local dynamics in the neighborhood of singular points. Among other problems, local bifurcation and stability analysis take advantage of it.

Let $x = (x_1, \ldots, x_n) \in \mathbb{R}^n$, and let $f(x_1, \ldots, x_n)$ be a polynomial vector with components in $\mathbb{R}[x_1, \ldots, x_n]$. Consider the general n-dimensional system of ODE:

$$\dot{x} = Lx + f(x) = Lx + f_2(x) + f_3(x) + \cdots, \tag{6.12}$$

where L is the Jacobian matrix associated with system (6.12), Lx represents the linear part of the system, and $f_k(x)$ denotes the kth homogeneous polynomial vector. We assume that the system admits an equilibrium at the origin "o". The basic idea of the normal form theory is to find a near-identity transformation:

$$x = y + h(y) = y + h_2(y) + h_3(y) + \cdots + h_k(y) + \cdots, \tag{6.13}$$

so that, the resulting system,

$$\dot{y} = Ly + g(x) = Ly + g_2(y) + g_3(y) + \cdots + g_k(y) + \cdots, \tag{6.14}$$

be as simple as possible. In that sense, the terms that are not essential in the local dynamic behavior are removed from the analytical expression of the vector field. Let us denote by $h_k(y)$ and $g_k(y)$, the kth homogeneous polynomial vectors of y. According to *Takens normal form theory* [121], we define the following operators:

$$L_k : H_k \to H_k, \quad U_k \in H_k \mapsto L_k(U_k) = [U_k, u_1] \in H_k, \tag{6.15}$$

where $u_1 = Ly$ is the linear part of the vector field, and H_k denotes a linear vector space containing the kth homogeneous polynomial vector of $y = (y_1, \ldots, y_n)$. The operator [., .], called *Lie Bracket*, is defined as:

$$[U_k, u_1] = LU_k - D(U_k)u_1,$$

where D denotes the *Fréchet derivative*.

The next step is to determine the spaces R_k and K_k; the range of H_k; and the complementary space of R_k, so that $H_k = R_k + K_k$. Now, bases for K_k and R_k can be chosen. The normal form theorem establishes a transformation of the analytic expression of the vector field, see for instance [121], where a detailed analysis of the quadratic and cubic cases is given.

Consequently, a homogeneous polynomial vector $f_k \in H_k$ can be divided into two parts; one of them can be spanned in K_k, and the remaining one in R_k. Normal form theory suggests that the part belonging to R_k can be disregarded and the one of K_k can be retained in normal form. Through this method, algebraic equations are obtained from (6.12)–(6.14).

Next, we aim to apply the above ideas to analyze the qualitative behavior of the drilling system.

6.3.1 Drilling System Analysis

In [87], a decomposition method to compute the normal form of a singular delay system linearly dependent on one parameter (of the class studied in [127]) is proposed. We aim to extend the proposed techniques to the case of NFDE systems. To this end, system (6.2) is rewritten as

$$\begin{aligned} \frac{d}{dt}\mathscr{D}\, x_t :&= \mathscr{L}_0\, x_t + \tilde{\mathscr{F}}_{p,\, M_B}(x_t) \\ &= \mathscr{L}_0\, x_t + (\mathscr{L} - \mathscr{L}_0)\, x_t + \mathscr{F}_{\mu, p_\epsilon}(x_t), \end{aligned} \tag{6.16}$$

where $\tilde{\mathscr{F}}$ is regarded as a perturbation, $\mathscr{L}_0 = \mathscr{L}|_{\{p=p_c,\, M_B=M^*\}}$, and

$$\mathscr{F}_p = \begin{bmatrix} -0.006750\, px_1^3(t) + 0.006682 px_1^3(t-2) \\ -1.875\, px_1^3(t) + 1.874998\, px_1^3(t-3.176) \end{bmatrix}.$$

Clearly,

$$\frac{d}{dt}\mathscr{D}\, x_t = \mathscr{L}_0\, x_t, \tag{6.17}$$

corresponds to the perturbation-free system.

Following [48], we compute first the evolution equation associated to the center manifold of system (6.17).

Considering the drilling system parameters given in Table C.1, the matrix Φ, depending on θ, is defined as:

$$\Phi(\theta) = \begin{bmatrix} 1-\theta & 1 \\ 4.853 \times 10^{11} - 4.853 \times 10^{11}\theta & 4.853 \times 10^{11} \end{bmatrix},$$

where $\theta \in [-3.176, 0]$. The adjoint linear equation associated to system (6.2) is given by:

$$\dot{u}(t) = D_1\dot{u}(t+2) + D_2\dot{u}\left(t + 2\tilde{c}/c\right) - A_0 u(t) - A_1 u(t+2) - A_2 u\left(t + 2\tilde{c}/c\right). \tag{6.18}$$

Consider the following basis for the generalized eigenspace corresponding to the double eigenvalue $\lambda_0 = 0$ evaluated at $\theta = 0$,

$$\Psi(\theta) = \begin{bmatrix} -86050 & 0 \\ -8.810 \times 10^{12} + 0.4924583\xi & 0 \end{bmatrix}, \quad \xi \in [0, 3.176].$$

The associated bilinear form is given by:

$$\begin{aligned}(\psi, \varphi) = {} & \psi(0)(\varphi(0) - D_1\varphi(-2) - D_2\varphi(-3.176)) \\ & + \int_{-2}^{0} \psi(\xi + 2)A_1\varphi(\xi)d\xi \\ & + \int_{-3.176}^{0} \psi(\xi + 3.176)A_2\varphi(\xi)d\xi \\ & - \int_{-2}^{0} \psi'(\xi + 2)D_1\varphi(\xi)d\xi \\ & - \int_{-3.176}^{0} \psi'(\xi + 3.176)D_2\varphi(\xi)d\xi.\end{aligned}$$

Notice that, according to the bilinear form, we have that $(\Psi, \Phi) = Id$. Hence, the space C can be decomposed as $C = P \bigoplus Q$, where $P = \{\varphi = \Phi z;\ z \in \mathbb{R}^2\}$ and $Q = \{\varphi \in C;\ (\Psi, \varphi) = 0\}$. It is important to emphasize that the subspaces P and Q are invariant under the semigroup $T(t)$. Matrix B (referred in 6.11), satisfying $\mathscr{A}\Phi = \Phi B$, is given by

$$B = \begin{bmatrix} 0 & 0 \\ -1 & 0 \end{bmatrix}. \tag{6.19}$$

Consider the following decomposition $x_t = \Phi y(t) + z(t)$, where $z(t) \in Q$, $y(t) \in \mathbb{R}^2$, $z(t) = h(y(t))$ and h is some analytic function such that $h : P \to Q$. The explicit solution corresponding to the center manifold is determined by:

$$\dot{y}(t) = By(t) + \Psi(0)\mathscr{F}[\Phi(\theta)y(t) + h(\theta, y(t))], \tag{6.20}$$

$$\begin{aligned}&\tfrac{\partial h}{dy}\{By + \Psi(0)\mathscr{F}[\Phi(\theta)y + h]\} + \Phi(\theta)\Psi(0)\mathscr{F}[\Phi(\theta)y + h] \\ &= \begin{cases} \frac{\partial h}{d\theta}, & -3.176 \leqslant \theta \leqslant 0, \\ \mathscr{L}(h(\theta, y)) + \mathscr{F}[\Phi(\theta)y + h(\theta, y)], & \theta = 0, \end{cases}\end{aligned} \tag{6.21}$$

where $h = h(\theta, y)$, and $\tilde{\mathscr{F}}$ is defined in (6.16). Further details are given in [48].

It is easy to show that, the evolution of the solutions of system (6.17) on the center manifold can be determined by solving (6.21) (subject to $p = p_c$, $M_B = M^*$) for

$h(\theta, y)$, and (6.20) for $y(t)$ (considering the truncation order). Notice that $\mathscr{F}$ is an odd function, which implies that it is the computation of h is not required to obtain a cubic truncation.

Based on these assessments, and considering $p = p_c$, $M_B = M^*$, the system of neutral-type time-delay equations (6.1), describing the coupled axial-torsional drilling dynamics, is reduced to the following third-order ODE:

$$\dot{y}(t) = \begin{bmatrix} 0 & 0 \\ -1 & 0 \end{bmatrix} \begin{bmatrix} y_1 \\ y_2 \end{bmatrix} + \begin{bmatrix} \begin{pmatrix} 0.65\, y_1{}^3 + 1.025\, y_1{}^2 y_2 \\ -0.140\, y_1 y_2{}^2 + 0.102\, y_2{}^3 \end{pmatrix} \\ \begin{pmatrix} 2.54\, y_1{}^3 - 1.554\, y_1{}^2 y_2 \\ +0.267\, y_1 y_2{}^2 - 0.0004\, y_2{}^3 \end{pmatrix} \end{bmatrix}. \tag{6.22}$$

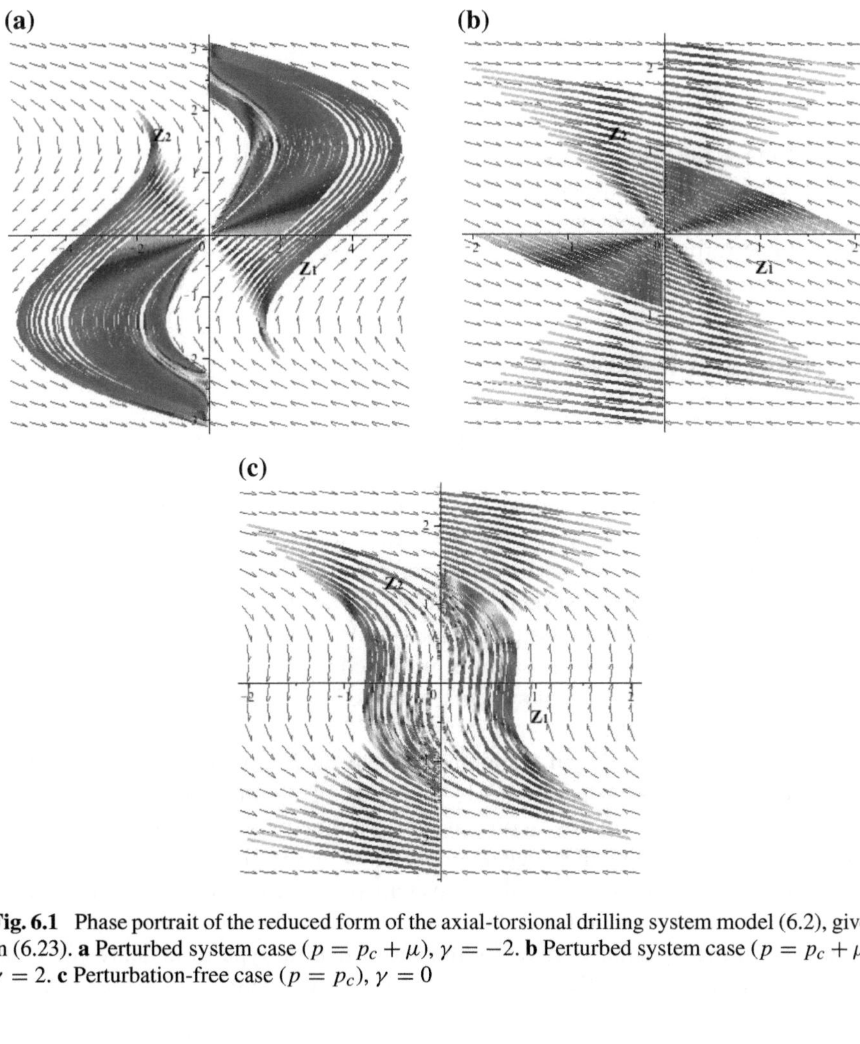

Fig. 6.1 Phase portrait of the reduced form of the axial-torsional drilling system model (6.2), given in (6.23). **a** Perturbed system case ($p = p_c + \mu$), $\gamma = -2$. **b** Perturbed system case ($p = p_c + \mu$), $\gamma = 2$. **c** Perturbation-free case ($p = p_c$), $\gamma = 0$

In order to analyze the parameter bifurcations, we compute the evolution equation of the trajectories of the perturbed system (6.16) on the center manifold. Now, p and M_B are given, respectively, by $p = p_c + \mu_1$ and $M_B = M^* + \mu_2$ where $\mu_{1,2}$ are small parameters (recall that the approximation associated to the perturbation-free system (6.17) was developed with $p = p_c$ and $M_B = M^*$). The introduction of a time scaling as well as a small scaling parameter (blow-up parameter) allows zooming the neighborhood of the singularity, see for instance [271].

The cubic normal form reduction of system (6.2):

$$\begin{cases} \dot{z}_1 = \gamma z_1 + \delta z_2 - z_1 {z_2}^2 \\ \dot{z}_2 = -z_1 \end{cases} \tag{6.23}$$

is obtained by considering the change of coordinates,

$$\mu_1 = 1326.69991\, \gamma r^2, \; p = p_c + \mu_1, \; \mu_2 = 17,87\, \delta r,$$
$$M_B = M^* + \mu_2, \; y_1 = r^2 z_1, \; y_2 = r z_2,$$

and the time scaling defined by $t_{old} = r t_{new}$.

Figure 6.1 shows phase space portraits of the reduced model (6.23) on the center manifold, in different scenarios: perturbation-free system ($p = p_c$) and system subject to perturbations ($p = p_c + \mu_1$), for $\gamma = -2$ and $\gamma = 2$.

6.4 Notes and References

Bifurcation theory provides a framework for understanding the behavior of dynamical systems, playing a key role in the study of several real-world problems. For instance, in the context of biological systems, the ability of making dramatic changes in the system output is essential for proper body functioning; bifurcations are therefore ubiquitous in biological networks [9]. A different application example is given in [127], where physiological systems, modeled by delay differential equations with double-zero eigenvalue singularity, are analyzed in the light of Bogdanov-Takens bifurcations. Another application field is presented in [154], where bifurcations arising in mechanical and physical systems, modeled by nonlinear partial differential equations, are investigated.

A wide range of tools and techniques of the qualitative theory of differential equations and bifurcation principles to the study of nonlinear oscillations can be found in 120. Detailed studies on one and two-parameter bifurcations in continuous and discrete time dynamical systems are presented in [165].

Qualitative analysis of the drilling system presented in this chapter is based on a reduced model. The infinite-dimensional representation of the coupled axial-torsional drilling dynamics, given by the set of neutral-type time-delay equations (6.1), is reduced to an ODE allowing a simplified qualitative system analysis.

The model transformation is developed by using normal forms theory concepts. On that basis, the stability of the steady-state of system (6.2) is investigated.

An alternative stability analysis of the drilling system developed in the frequency-domain framework is presented in [207]; it uses the lumped parameter model describing the coupled axial-torsional drilling dynamics given in (2.6), coupled to the bit-rock interface model discussed in Sect. 4.3.2. The steady-state solution of the model is considered and its stability to small perturbations is analyzed. Some stability charts are derived from such analysis, thus deducing the stable operating regime in the WoB-rotary speed parameter plane. It is concluded that large speeds are eventually stable for all weights on bit, but such large speeds may not be practically feasible. As we shall see in subsequent chapters, an increase of the rotary velocity constitutes a well-known empirical practice to avoid drilling oscillations.

A different analysis approach to derive operational guidelines for avoiding the stick-slip phenomenon is presented in [211]; it uses a n-dimensional lumped parameter model similar to the one given in (2.1). The model is assumed to be coupled to a frictional torque on bit approximated by a combination of the switch model studied in Sect. 3.2.3 (Dry friction plus Karnopp's model) and the model given in (3.7)–(3.9). The identification of key-drilling parameters ranges for which nondesired torsional oscillations are present is carried out by analyzing Hopf bifurcations in the vicinity of the system equilibrium point when rotary velocities are greater than zero. Changes in drillstring behavior are studied through variations in three parameters: The weight on the bit, the rotary speed at the top-rotary drillstring system, and the torque given by the surface motor. The intersection of the region of parameters where no Hopf bifurcations is present with the region where no stuck bit is possible provides a good estimation of the system parameters which provide safe drilling operations. See also [236], where an optimum range of operating parameters with different WOB and revolutions per minute combinations were provided to ensure the highest possible ROP.

The bifurcation analysis developed in [211], constitutes the basis of the sliding mode control design proposed in [213]. The underlying idea of avoiding the stick-slip phenomenon is to drive the rotary velocities of drillstring components to specified values. A discontinuous lumped parameter torsional model of four degrees of freedom, coupled to the torque on bit model given in (3.14) with the friction coefficient given in (3.15) is considered.

Bifurcation analysis methods to characterize the system response are carried out through frequency-domain techniques. Alternative strategies to investigate the stability and the dynamic response of the drilling system are developed through the temporal approach. The following chapter addresses a dissipativity analysis of the system without control actions. This analysis, developed within the time-domain framework, allows concluding on the system's practical stability.

Chapter 7
Ultimate Boundedness Analysis

Physically-motivated systems are generally subject to nonlinearities and uncertainties, for these systems, classical stability definitions (e.g., asymptotic or exponential stability in the sense of Lyapunov) can be too restrictive. Namely, the state of a system may be mathematically unstable in the sense of Lyapunov, but the response oscillates close enough to the equilibrium, to be considered as acceptable. In many stabilization problems, the aim is to bring states close to certain sets rather than to a particular state. In this situation, the notion of asymptotic or exponential stability is not suitable. More appropriate performance specifications, from an engineering point of view, are given by the definition of ultimate boundedness with a fixed bound [148], also referred to as practical stability. Ultimate boundedness not only provides information on the stability of the system, but also characterizes its transient behavior with estimates of the bounds on the system trajectories [7].

In this chapter, the ultimate boundedness analysis of a class of perturbed distributed parameter systems is investigated. Through Lyapunov techniques, we derive LMI-type conditions to establish ultimate bounds on the system response.

Based on the proposed methodologies, the drilling system described by the wave equation with nonlinear boundary conditions is analyzed. A proposal of Lyapunov functional allows establishing the conditions under which the nongrowth of energy in the system is guaranteed.

7.1 Preliminary Results

Consider a system of the form:

$$\frac{\partial^2\theta}{\partial t^2}(x,t) = a\frac{\partial^2\theta}{\partial x^2}(x,t) + d\frac{\partial\theta}{\partial t}(x,t),\ t > t_0,\ \ 0 < x < 1, \tag{7.1}$$

M.B. Saldivar Márquez et al., *Analysis and Control of Oilwell Drilling Vibrations*,
Advances in Industrial Control, DOI 10.1007/978-3-319-15747-4_7

with the boundary conditions:

$$\begin{aligned} &\theta(0,t) = 0, \\ &\frac{\partial \theta}{\partial x}(1,t) = -k\frac{\partial \theta}{\partial t}(1,t) + r\omega(t),\ t > 0, \end{aligned} \tag{7.2}$$

and initial conditions:

$$\begin{aligned} \theta(x,0) &= \zeta(x), \\ \frac{\partial \theta}{\partial x}(x,0) &= \dot{\zeta}(x) \in L_2(0,1), \\ \frac{\partial \theta}{\partial t}(x,0) &= \zeta_1(x) \in L_2(0,1). \end{aligned} \tag{7.3}$$

Assume that k satisfies $0 < k_0 \leqslant k \leqslant k_1$. The unknown function $\omega(t)$ represents external disturbances and modeling errors, we consider that $\omega(t)$ satisfies:

$$\omega^2(t) \leqslant \bar{\varepsilon}.$$

When the disturbance term $\omega(t)$ is not identically zero, we cannot prove exponential stability of the solution. However, it is possible to prove ultimate boundedness on the solutions [148]. The following lemma constitutes the basis of the ultimate boundedness analysis.

Lemma 7.1.1 ([102]) *Let $V : [0,\infty) \to \mathbb{R}_+$ be an absolutely continuous function. If there exists $\sigma > 0$, $\bar{\beta} > 0$ such that the derivative of V satisfies almost everywhere the inequality*

$$\frac{d}{dt}V(t) + \sigma V(t) - \bar{\beta}\omega^2(t) \leqslant 0,$$

then it follows that for all $\omega(t)$ satisfying $\omega^2(t) \leqslant \bar{\varepsilon}$, the inequality

$$V(t) \leqslant e^{-\sigma(t-t_0)}V(t_0) + (1 - e^{-\sigma(t-t_0)})\frac{\bar{\beta}\bar{\varepsilon}}{\sigma},$$

holds.

Proof Multiplying by $e^{\sigma(\theta-t)}$ the inequality $\frac{d}{dt}V + \sigma V \leqslant \bar{\beta}\omega^2$ and integrating further from t_0 to "t", we have

$$\int_{t_0}^{t} \frac{d}{ds}(e^{\sigma(s-t)}V(s))ds \leqslant \bar{\beta}\int_{t_0}^{t} e^{\sigma(s-t)}\omega^2(s)ds,$$

thus,

$$V(t) - e^{-\sigma(t-t_0)}V(t_0) \leqslant \frac{\bar{\beta}}{\sigma}(1 - e^{-\sigma(t-t_0)})\bar{\varepsilon},$$

and the result follows.

For later use, we recall the following result.

Lemma 7.1.2 ([301]) *Let $z \in W^{1,2}([a,b],\mathbb{R})$ be a scalar function satisfying $z(a) = 0$. Then*

$$max_{\delta\in[a,b]}z^2(\delta) \leqslant (b-a)\int_a^b (z'(\delta))^2 d\delta. \tag{7.4}$$

7.2 Ultimate Boundedness Conditions

In order to fulfill the requirements of Lemma 7.1.1, we consider the Lyapunov functional

$$V\left(\frac{\partial\theta}{\partial x}(\cdot,t), \frac{\partial\theta}{\partial t}(\cdot,t)\right) = pa\int_0^1 \left(\frac{\partial\theta}{\partial x}(x,t)\right)^2 dx + p\int_0^1 \left(\frac{\partial\theta}{\partial t}(x,t)\right)^2 dx$$
$$+2\chi\int_0^1 x\frac{\partial\theta}{\partial x}(x,t)\frac{\partial\theta}{\partial t}(x,t)dx$$

proposed in [216], with constants $p > 0$ and small enough $\chi > 0$. In [100], the following LMI

$$\begin{bmatrix} ap & \chi \\ \chi & p \end{bmatrix} > 0 \tag{7.5}$$

was introduced to guarantee that the Lyapunov functional V satisfies $V > 0$ for

$$\int_0^1 \left(\frac{\partial\theta}{\partial x}(x,t)\right)^2 dx + \int_0^1 \left(\frac{\partial\theta}{\partial t}(x,t)\right)^2 dx > 0.$$

Theorem 7.2.1 *Given $\sigma > 0$, if there exist $p > 0$ and $\chi > 0$ such that (7.5) and the following LMIs*

$$\begin{bmatrix} -2ak_ip + \chi & 0 & 0 & -a\chi k_ir + apr & ak_i\chi \\ * & \psi_2 & (\sigma+d)\chi & 0 & 0 \\ * & * & \psi_3 & 0 & 0 \\ * & * & * & -\bar{\beta} + \chi ar^2 & 0 \\ * & * & * & * & -a\chi \end{bmatrix} < 0,\ i = 0, 1, \tag{7.6}$$

where

$$\psi_2 = -a\chi + \sigma ap, \\ \psi_3 = -\chi + 2pd + \sigma p, \tag{7.7}$$

are feasible, then the solutions of the boundary-value problem (7.1), (7.2) *with initial conditions* (7.3) *satisfy the inequality*

$$\begin{aligned} max_{x\in[0,1]}\theta^2(x,t) &\leqslant \int_0^1 \left(\frac{\partial\theta}{\partial x}(x,t)\right)^2 dx + \int_0^1 \left(\frac{\partial\theta}{\partial t}(x,t)\right)^2 dx \\ &\leqslant \frac{\alpha_2}{\alpha_1} e^{-\sigma(t-t_0)} \int_0^1 \left(\zeta_1^2(x)+\dot{\zeta}^2(x)\right) dx + \frac{\bar{\beta}\bar{\varepsilon}}{\alpha_1\sigma}, \end{aligned} \tag{7.8}$$

where

$$\alpha_1 = \lambda_{\min}\begin{bmatrix} ap & 0 \\ 0 & p \end{bmatrix}, \quad \alpha_2 = \lambda_{\max}\begin{bmatrix} ap & \chi \\ \chi & p \end{bmatrix}. \tag{7.9}$$

Proof As the LMI

$$\begin{bmatrix} ap & \chi x \\ \chi x & p \end{bmatrix} > 0$$

is affine in $x \in [0, 1]$, it follows from Schur complements and Rayleigh's Theorem that

$$\begin{aligned} \alpha_1 \left[\int_0^1 \left(\frac{\partial\theta}{\partial x}(x,t)\right)^2 dx + \int_0^1 \left(\frac{\partial\theta}{\partial t}(x,t)\right)^2 dx\right] &\leqslant V\left(\frac{\partial\theta}{\partial x}(\cdot,t), \frac{\partial\theta}{\partial t}(\cdot,t)\right) \\ &\leqslant \alpha_2 \left[\int_0^1 \left(\frac{\partial\theta}{\partial x}(x,t)\right)^2 dx + \int_0^1 \left(\frac{\partial\theta}{\partial t}(x,t)\right)^2 dx\right], \end{aligned} \tag{7.10}$$

with α_1 and α_2 satisfying (7.9).

Next, we find $\frac{d}{dt}V$. Following [100], we derive

$$\begin{aligned} \frac{d}{dt}\left(2\int_0^1 x\frac{\partial\theta}{\partial t}(x,t)\frac{\partial\theta}{\partial x}(x,t)dx\right) &= 2\int_0^1 x\frac{\partial^2\theta}{\partial t^2}(x,t)\frac{\partial\theta}{\partial x}(x,t)dx \\ &\quad +2\int_0^1 x\frac{\partial\theta}{\partial t}(x,t)\frac{\partial^2\theta}{\partial x\partial t}(x,t)dx \\ &= 2a\int_0^1 x\frac{\partial^2\theta}{\partial x^2}(x,t)\frac{\partial\theta}{\partial x}(x,t)dx \\ &\quad +2\int_0^1 x\frac{\partial\theta}{\partial t}(x,t)\frac{\partial^2\theta}{\partial x\partial t}(x,t)dx \\ &\quad +2d\int_0^1 x\frac{\partial\theta}{\partial t}(x,t)\frac{\partial\theta}{\partial x}(x,t)dx. \end{aligned}$$

Integration by parts gives

$$2\int_0^1 x\frac{\partial\theta}{\partial t}(x,t)\frac{\partial^2\theta}{\partial x\partial t}(x,t)dx = -2\int_0^1 x\frac{\partial^2\theta}{\partial x\partial t}(x,t)\frac{\partial\theta}{\partial t}(x,t)dx$$
$$-2\int_0^1\left(\frac{\partial\theta}{\partial t}(x,t)\right)^2 dx$$
$$+2\left(\frac{\partial\theta}{\partial t}(1,t)\right)^2,$$

and

$$2\int_0^1 x\frac{\partial\theta}{\partial t}(x,t)\frac{\partial^2\theta}{\partial x\partial t}dx = -\int_0^1\left(\frac{\partial\theta}{\partial t}(x,t)\right)^2 dx + \left(\frac{\partial\theta}{\partial t}(1,t)\right)^2.$$

Similarly,

$$2\int_0^1 x\frac{\partial^2\theta}{\partial x^2}(x,t)\frac{\partial\theta}{\partial x}(x,t)dx = -\int_0^1\left(\frac{\partial\theta}{\partial x}(x,t)\right)^2 dx + \left(\frac{\partial\theta}{\partial x}(1,t)\right)^2.$$

Substitution of the boundary condition yields

$$2\frac{d}{dt}\left(\int_0^1 x\frac{\partial\theta}{\partial t}(x,t)\frac{\partial\theta}{\partial x}(x,t)dx\right) = -\int_0^1\left(\frac{\partial\theta}{\partial t}(x,t)\right)^2 - a\int_0^1\left(\frac{\partial\theta}{\partial x}(x,t)\right)^2 dx$$
$$+\left(\frac{\partial\theta}{\partial t}(1,t)\right)^2 + a\left(-k\frac{\partial\theta}{\partial t}(1,t)+r\omega(t)\right)^2$$
$$+2d\int_0^1 x\frac{\partial\theta}{\partial t}(x,t)\frac{\partial\theta}{\partial x}(x,t)dx.$$

Thus, differentiating V along (7.1), we obtain:

$$\frac{d}{dt}V = 2pa\int_0^1 \frac{\partial\theta}{\partial x}(x,t)\frac{\partial^2\theta}{\partial t\partial x}(x,t)dx + 2p\int_0^1\frac{\partial\theta}{\partial t}(x,t)\frac{\partial^2\theta}{\partial t^2}(x,t)dx$$
$$+2\chi\frac{d}{dt}\left(\int_0^1 x\frac{\partial\theta}{\partial t}(x,t)\frac{\partial\theta}{\partial x}(x,t)dx\right)$$
$$= 2p\int_0^1\left(a\frac{\partial\theta}{\partial x}(x,t)\frac{\partial^2\theta}{\partial t\partial x}(x,t) + a\frac{\partial\theta}{\partial t}(x,t)\frac{\partial^2\theta}{\partial x^2}(x,t)\right)dx$$
$$+2pd\int_0^1\frac{\partial\theta}{\partial t}(x,t)\frac{\partial\theta}{\partial t}(x,t)dx + 2\chi\frac{d}{dt}\left(\int_0^1 x\frac{\partial\theta}{\partial t}(x,t)\frac{\partial\theta}{\partial x}(x,t)dx\right).$$

Then, integrating by parts and substituting the boundary condition (7.2), we obtain

$$\int_0^1 \frac{\partial\theta}{\partial t}(x,t)\frac{\partial^2\theta}{\partial x^2}(x,t)dx = \left.\frac{\partial\theta}{\partial t}(x,t)\frac{\partial\theta}{\partial x}(x,t)\right|_0^1 - \int_0^1 \frac{\partial^2\theta}{\partial t\partial x}(x,t)\frac{\partial\theta}{\partial x}(x,t)dx$$
$$= \frac{\partial\theta}{\partial t}(1,t)\left(-k\frac{\partial\theta}{\partial t}(1,t)+r\omega(t)\right)$$
$$-\int_0^1 \frac{\partial^2\theta}{\partial t\partial x}(x,t)\frac{\partial\theta}{\partial x}(x,t)dx.$$

Therefore

$$\frac{d}{dt}V = -2apk\left(\frac{\partial\theta}{\partial t}(1,t)\right)^2 + 2ap\frac{\partial\theta}{\partial t}(1,t)r\omega(t) + 2pd\int_0^1 \frac{\partial\theta}{\partial t}(x,t)\frac{\partial\theta}{\partial t}(x,t)dx$$
$$-\chi\int_0^1\left(\frac{\partial\theta}{\partial t}(x,t)\right)^2 dx - a\chi\int_0^1\left(\frac{\partial\theta}{\partial x}(x,t)\right)^2 dx + \chi\left(\frac{\partial\theta}{\partial t}(1,t)\right)^2$$
$$+a\chi\left(\frac{\partial\theta}{\partial x}(1,t)\right)^2 + 2d\chi\int_0^1 x\frac{\partial\theta}{\partial t}(x,t)\frac{\partial\theta}{\partial x}(x,t)dx.$$

It follows that

$$\frac{d}{dt}V + \sigma V - \bar{\beta}\omega^2 = -2apk\left(\frac{\partial\theta}{\partial t}(1,t)\right)^2 + 2apr\frac{\partial\theta}{\partial t}(1,t)\omega(t)$$
$$+ 2pd\int_0^1 \frac{\partial\theta}{\partial t}(x,t)\frac{\partial\theta}{\partial t}(x,t)dx - \chi\int_0^1\left(\frac{\partial\theta}{\partial t}(x,t)\right)^2 dx$$
$$- a\chi\int_0^1\left(\frac{\partial\theta}{\partial x}(x,t)\right)^2 dx + \chi\left(\frac{\partial\theta}{\partial t}(1,t)\right)^2$$
$$+ a\chi\left(-k\frac{\partial\theta}{\partial t}(1,t)+r\omega(t)\right)^2 + 2d\chi\int_0^1 x\frac{\partial\theta}{\partial t}(x,t)\frac{\partial\theta}{\partial x}(x,t)dx$$
$$- \bar{\beta}\omega^2(t) + \sigma pa\int_0^1\left(\frac{\partial\theta}{\partial x}(x,t)\right)^2 dx$$
$$+ \sigma p\int_0^1\left(\frac{\partial\theta}{\partial t}(x,t)\right)^2 dx + 2\sigma\chi\int_0^1 x\frac{\partial\theta}{\partial x}(x,t)\frac{\partial\theta}{\partial t}(x,t)dx.$$

By setting $\vartheta^T(x,t) = \left[\frac{\partial\theta}{\partial t}(1,t)\ \frac{\partial\theta}{\partial x}(x,t)\ \frac{\partial\theta}{\partial t}(x,t)\ \omega(t)\right]$, we conclude that

$$\frac{d}{dt}V + \sigma V - \bar{\beta}\omega^2 = \int_0^1 \vartheta^T(x,t)\Psi\vartheta(x,t)dx < 0,$$

if

$$\Psi = \begin{bmatrix} -2akp + (1+ak^2)\chi & 0 & 0 & -a\chi kr + apr \\ * & \psi_2 & (\sigma+d)\chi\sigma & 0 \\ * & * & \psi_3 & 0 \\ * & * & * & -\bar{\beta} + \chi ar^2 \end{bmatrix} < 0. \tag{7.11}$$

Applying Schur complements to $ak^2\chi$ in (7.11) and using the affinity of the resulting LMI in $x \in [0, 1]$ and $k \in [k_0, k_1]$, it is easy to see that (7.11) holds if (7.6) is feasible.

Then, if (7.6) is feasible, it follows from (7.10) and Lemma 7.1.1 that

$$\begin{aligned} \alpha_1\left[\int_0^1 \left(\frac{\partial\theta}{\partial x}(x,t)\right)^2 dx + \int_0^1 \left(\frac{\partial\theta}{\partial t}(x,t)\right)^2 dx\right] &\leqslant V\left(\frac{\partial\theta}{\partial x}(\cdot,t), \frac{\partial\theta}{\partial t}(\cdot,t)\right) \\ &\leqslant V\left(\frac{\partial\theta}{\partial x}(\cdot,t_0), \frac{\partial\theta}{\partial t}(\cdot,t_0)\right)\mathrm{e}^{-\sigma(t-t_0)} \\ &\quad + \frac{\bar{\beta}}{\sigma}(1 - e^{-\sigma(t-t_0)})\bar{\varepsilon}^2 \\ &\leqslant \alpha_2 \mathrm{e}^{-\sigma(t-t_0)} \int_0^1 [\zeta_1^2(x) + \dot{\zeta}^2(x)]dx \\ &\quad + \frac{\bar{\beta}}{\sigma}(1 - e^{-\sigma(t-t_0)})\bar{\varepsilon}^2. \end{aligned}$$

In addition, it follows from (7.4) that

$$\begin{aligned} max_{x\in[0,1]}\theta^2(x,t) \leqslant \int_0^1 \left(\frac{\partial\theta}{\partial x}(x,t)\right)^2 dx &\leqslant \int_0^1 \left(\frac{\partial\theta}{\partial x}(x,t)\right)^2 dx \\ &+ \int_0^1 \left(\frac{\partial\theta}{\partial x}(x,t)\right)^2 dx. \end{aligned}$$

Remark 7.2.2 Inequality (7.8) means that the distributed system (7.1) with boundary conditions (7.2) is input-to-state stable. The conditions for exponential stability of the disturbance free system given in Theorem 7.2.1 coincide with the ones from [100].

7.2.1 Illustrative Numerical Example

Consider the system (7.1) with boundary conditions defined in (7.2) for

$$a = 1.0043, \quad k = r = 0.0033, \quad d = 0.$$

Table 7.1 Numerical results determining ultimate bounds on the response of the distributed drilling system obtained through Theorem 7.2.1

Case	1	2	3	4	5
σ	0.16	0.12	0.08	0.02	0.0002
$\bar{\beta}$	3.2521	1.0707	1.2145	1.5221	1.7951
α_1	5.0009	1.0934	1.2657	1.6273	1.9328
α_2	5.9854	1.3019	1.5074	1.9383	2.3023

The LMI-type conditions of Theorem 7.2.1 lead to the pairs $(\sigma, \bar{\beta})$ given in Table 7.1.

From the above table, we can obtain the following ultimate bound for $\sigma = 0.08$:

$$\int_0^1 \left[\left(\frac{\partial\theta}{\partial x}(x,t)\right)^2 + \left(\frac{\partial\theta}{\partial t}(x,t)\right)^2\right] dx \leqslant 1.1909\mathrm{e}^{-0.08t} \int_0^1 \left[\zeta_1^2(x) + \dot{\zeta}^2(x)\right] dx$$
$$+11.9944\bar{\varepsilon}^2.$$

By choosing $d = -1.0526$ (instead of $d = 0$), the LMI-type conditions of Theorem 7.2.1, give rise to:

$$\int_0^1 \left[\left(\frac{\partial\theta}{\partial x}(x,t)\right)^2 + \left(\frac{\partial\theta}{\partial t}(x,t)\right)^2\right] dx \leqslant 1.1854\mathrm{e}^{-0.08t} \int_0^1 [\zeta_1^2(x) + \dot{\zeta}^2(x)] dx$$
$$+18.8654\bar{\varepsilon}^2.$$

7.3 Nongrowth of Energy in the Drilling System

Based on the above ideas, we can establish a condition under which the nongrowth of energy of the drilling system is ensured.

Consider the wave equation (2.12) describing the torsional drilling vibrations coupled to the boundary conditions (2.16).

The torque on bit can be modeled by the following nonlinear function inspired by the decaying friction model (3.7)–(3.9):

$$T\left(\frac{\partial\Phi}{\partial t}(1,t)\right) = c_b \frac{\partial\Phi}{\partial t}(1,t) + W_{ob} R_b \mu_b \left(\frac{\partial\Phi}{\partial t}(1,t)\right) \mathrm{sgn}\left(\frac{\partial\Phi}{\partial t}(1,t)\right),$$

where the bit dry friction coefficient is given by:

$$\mu_b\left(\frac{\partial\Phi}{\partial t}(1,t)\right) = \mu_{cb} + (\mu_{sb} - \mu_{cb})\mathrm{e}^{-\gamma_b \left|\frac{\partial\Phi}{\partial t}(1,t)\right|}.$$

Notation is defined as for model (3.14) with the friction coefficient given in (3.15).

The torsional drilling dynamics are then described the following normalized model:

$$\frac{\partial^2 \Phi}{\partial t^2}(s,t) = a\frac{\partial^2 \Phi}{\partial s^2}(s,t) + d\frac{\partial \Phi}{\partial t}(s,t), \; t > t_0, \; 0 < s < 1, \tag{7.12}$$

where

$$a = \frac{GJ}{IL^2}, \quad d = \frac{\gamma}{I},$$

coupled to the mixed boundary conditions:

$$\frac{\partial \Phi}{\partial s}(0,t) = g\left(\frac{\partial \Phi}{\partial t}(0,t) - \Omega(t)\right), \quad s \in (0,1), \quad t > 0, \tag{7.13}$$

$$\frac{\partial \Phi}{\partial s}(1,t) = -k\frac{\partial \Phi}{\partial t}(1,t) - q\mu_b\left(\frac{\partial \Phi}{\partial t}(1,t)\right)\mathrm{sgn}\left(\frac{\partial \Phi}{\partial t}(1,t)\right) - h\frac{\partial^2 \Phi}{\partial t^2}(1,t),$$

where $\Omega(t)$ is the angular velocity coming from the rotary table and

$$g = \frac{\beta L}{GJ}, \quad k = \frac{c_b L}{GJ}, \quad q = \frac{W_{ob} R_b L}{GJ}, \quad h = \frac{I_B L}{GJ}.$$

Consider the following energy function:

$$V(t) = a\int_0^1 \left(\frac{\partial \Phi}{\partial s}(s,t)\right)^2 ds + \int_0^1 \left(\frac{\partial \Phi}{\partial t}(s,t)\right)^2 ds + ah\left(\frac{\partial \Phi}{\partial t}(1,t)\right)^2. \tag{7.14}$$

By differentiating (7.14), we obtain:

$$\begin{aligned}\frac{d}{dt}V(t) = {}& 2a\int_0^1 \frac{\partial \Phi}{\partial s}(s,t)\frac{\partial^2 \Phi}{\partial t \partial s}(s,t)ds + 2\int_0^1 \frac{\partial \Phi}{\partial t}(s,t)\frac{\partial^2 \Phi}{\partial t^2}(s,t)ds \\ & + 2ah\frac{\partial \Phi}{\partial t}(1,t)\frac{\partial^2 \Phi}{\partial t^2}(1,t),\end{aligned}$$

substituting equation (7.12) yields:

$$\begin{aligned}\frac{d}{dt}V(t) = {}& 2a\int_0^1 \frac{\partial \Phi}{\partial s}(s,t)\frac{\partial^2 \Phi}{\partial t \partial s}(s,t)ds + 2a\int_0^1 \frac{\partial \Phi}{\partial t}(s,t)\frac{\partial^2 \Phi}{\partial s^2}(s,t)ds \\ & - 2d\int_0^1 \frac{\partial \Phi}{\partial t}(s,t)\frac{\partial \Phi}{\partial t}(s,t)ds + 2ah\frac{\partial \Phi}{\partial t}(1,t)\frac{\partial^2 \Phi}{\partial t^2}(1,t).\end{aligned}$$

Integrating by parts and substituting the boundary conditions gives

$$\begin{aligned}\int_0^1 \frac{\partial \Phi}{\partial t}(s,t)\frac{\partial^2 \Phi}{\partial s^2}(s,t)ds &= \frac{\partial \Phi}{\partial t}(s,t)\frac{\partial \Phi}{\partial s}(s,t)\Big|_0^1 - \int_0^1 \frac{\partial \Phi}{\partial s}(s,t)\frac{\partial^2 \Phi}{\partial t \partial s}(s,t)ds\\ &= \frac{\partial \Phi}{\partial t}(1,t)\left(-k\frac{\partial \Phi}{\partial t}(1,t) - h\frac{\partial^2 \Phi}{\partial t^2}(1,t)\right)\\ &\quad \frac{\partial \Phi}{\partial t}(1,t)\left(-q\mu_b\left(\frac{\partial \Phi}{\partial t}(1,t)\right)\mathrm{sgn}\left(\frac{\partial \Phi}{\partial t}(1,t)\right)\right)\\ &\quad -g\frac{\partial \Phi}{\partial t}(0,t)\left(\frac{\partial \Phi}{\partial t}(0,t) - \Omega(t)\right)\\ &\quad -\int_0^1 \frac{\partial \Phi}{\partial s}(s,t)\frac{\partial^2 \Phi}{\partial t \partial s}(s,t)ds.\end{aligned}$$

Hence,

$$\begin{aligned}\frac{d}{dt}V(t) &= 2a\int_0^1 \frac{\partial \Phi}{\partial s}(s,t)\frac{\partial^2 \Phi}{\partial t \partial s}(s,t)ds\\ &\quad +2a\frac{\partial \Phi}{\partial t}(1,t)\left(-k\frac{\partial \Phi}{\partial t}(1,t) - h\frac{\partial^2 \Phi}{\partial t^2}(1,t)\right)\\ &\quad +2a\frac{\partial \Phi}{\partial t}(1,t)\left(-q\mu_b\left(\frac{\partial \Phi}{\partial t}(1,t)\right)\mathrm{sgn}\left(\frac{\partial \Phi}{\partial t}(1,t)\right)\right)\\ &\quad -2ag\frac{\partial \Phi}{\partial t}(0,t)\left(\frac{\partial \Phi}{\partial t}(0,t) - \Omega(t)\right)\\ &\quad -2a\int_0^1 \frac{\partial \Phi}{\partial s}(s,t)\frac{\partial^2 \Phi}{\partial t \partial s}(s,t)ds - 2d\int_0^1 \frac{\partial \Phi}{\partial t}(s,t)\frac{\partial \Phi}{\partial t}(s,t)ds\\ &\quad +2ah\frac{\partial \Phi}{\partial t}(1,t)\frac{\partial^2 \Phi}{\partial t^2}(1,t),\end{aligned}$$

since

$$\frac{\partial \Phi}{\partial t}(1,t)\mu_b\left(\frac{\partial \Phi}{\partial t}(1,t)\right)\mathrm{sgn}\left(\frac{\partial \Phi}{\partial t}(1,t)\right) = \mu_b\left(\frac{\partial \Phi}{\partial t}(1,t)\right)\left|\frac{\partial \Phi}{\partial t}(1,t)\right|,$$

we have

$$\begin{aligned}\frac{d}{dt}V(t) &= -2aq\mu_b\left(\frac{\partial \Phi}{\partial t}(1,t)\right)\left|\frac{\partial \Phi}{\partial t}(1,t)\right| - 2d\int_0^1\left(\frac{\partial \Phi}{\partial t}(s,t)\right)^2 ds\\ &\quad -2ag\frac{\partial \Phi}{\partial t}(0,t)\left(\frac{\partial \Phi}{\partial t}(0,t) - \Omega(t)\right) - 2ak\left(\frac{\partial \Phi}{\partial t}(1,t)\right)^2.\end{aligned} \tag{7.15}$$

In order to ensure the dissipativity of the system, the angular velocity $\Omega(t)$ should allow the negativity of (7.15).

Choosing $\Omega(t)$ as follows:

$$\Omega(t) = (1 - c_1)\frac{\partial \Phi}{\partial t}(0, t) + 2c_1\frac{\partial \Phi}{\partial t}(1, t) - c_1\frac{\left(\frac{\partial \Phi}{\partial t}(1, t)\right)^2}{\frac{\partial \Phi}{\partial t}(0, t)}, \tag{7.16}$$

where $c_1 > 0$ is a free design parameter, we obtain that

$$\begin{aligned}\frac{d}{dt}V(t) = &-2aq\left(\frac{\partial \Phi}{\partial t}(1, t)\right)\left|\frac{\partial \Phi}{\partial t}(1, t)\right| - 2d\int_0^1\left(\frac{\partial \Phi}{\partial t}(s, t)\right)^2 ds\\ &-2agc_1\left(\frac{\partial \Phi}{\partial t}(1, t) - \frac{\partial \Phi}{\partial t}(0, t)\right)^2 - 2ak\left(\frac{\partial \Phi}{\partial t}(1, t)\right)^2.\end{aligned}$$

Taking into account that $\mu_b\left(\frac{\partial \Phi}{\partial t}(1, t)\right) > 0$ and a, q, k, d and g are positive constants we find that $\frac{d}{dt}V(t) \leq 0$. The energy dissipation of the drilling system is established:

Proposition 7.3.1 *For all solutions of* (7.12) *under the boundary conditions* (7.13), *the energy given in* (7.14) *does not grow if the angular velocity $\Omega(t)$ can be chosen such that* (7.16) *is satisfied.*

7.4 Notes and References

The concept of ultimate boundedness (or practical stability) was introduced in 1961 by La Salle and Lefschetz [168]. In 1973, Grujic retakes this concept to analyze nonlinear nonautonomous systems [118]. In [167], a solid theoretical background on practical stability of nonlinear systems is provided.

Although the ultimate boundedness definition provides a less strict concept of stability, there are a small number of contributions in this direction. In the framework of delay free systems, the following contributions can be mentioned. In [198] a practical stability result for dynamic systems depending on a small parameter is displayed, as well as a practical stability analysis of fast time-varying systems studied in averaging theory, and of highly oscillatory systems. An extension of Lyapunov's second method to investigate the practical stability of the solution process of stochastic hybrid parabolic partial differential equations of Itô-type under Markovian structural perturbations is presented in [7]. In the framework of time-delay systems the following contributions can be mentioned: sufficient conditions for the practical stability of a class of time-delay systems are provided in [297] for the retarded case, and in [296] for neutral-type systems. The practical stability of the solutions of nonlinear impulsive functional differential equations based on the method of vector

Lyapunov functions and on differential inequalities for piecewise continuous functions is studied in [277]. The design of an observer for a class of nonlinear systems with unknown, time-varying, bounded delays, on both state and input and sufficient conditions to prove the practical stability of the observer is proposed in [113].

The ultimate boundedness property of the response of the drilling system described by the wave equation model is studied in [251]. An approach based on a difference equation derived through the d'Alembert transformation can be found in [103]. Although, the nongrowth of energy was established even in presence of torsional vibrations, this property is not sufficient to guarantee an optimal drilling operation; then, control action must be taken into account.

Subsequent chapters are focused on the design of control strategies to eliminate undesirable drilling vibrations. The first issue that will be reviewed is the empirical-based control; the following chapter provides a summary of the most popular practical strategies used in the field to overcome the stick-slip, bit-bounce, and whirling phenomena. Some stabilizing feedback control strategies of our own authorship will be presented from Chap. 9 onwards.

It is worthy of mention that the notions presented in this chapter will be applied to the design feedback controllers which guarantee the practical stability of the closed loop drilling system, leading to a satisfactory perforation process (see Chap. 12).

Part III
Control

Chapter 8
Field Observations and Empirical Drilling Control

This chapter discusses the most important phenomena observed in oil drilling platforms. Several destabilizing dynamics occur when the bit comes in contact with the drilling surface giving rise to the occurrence of drillstring vibrations. Three main types of drillstring vibrations are identified:

- longitudinal (whirling),
- axial (bit-bounce),
- torsional (stick-slip),

the latter ones being the most frequently observed.

A detailed description of each vibration mode is presented in this chapter. Failures induced by drillstring vibrations lead to premature wear of the system components, breakage of drilling bits, unscrewing of pipe connections, wastage of energy, reduction of the rate of penetration, ..., which clearly increases the operating costs. Detection guidelines and detrimental consequences of the stick-slip, bit-bounce, and whirling are herein described.

Helpful guidelines to reduce undesirable drillstring behaviors have been determined through practical experience in the field of borehole drilling. Most of the empirical strategies that qualitatively capture the drillers' expertise consist in the alignment of different drilling parameters, such as the rotary speed provided by the rotary table, the weight on bit, and the drilling fluid.

The most popular empirical methods used in practice to avoid the stick-slip vibrations are evaluated through numerical simulations of the torsional drilling model described by a neutral-type time-delay equation. An indirect validation of the proposed model is achieved through the simulation results in close agreement with field observations regarding the stick-slip and bit-bounce.

Besides the alignment of drilling parameters, certain devices that absorb the vibration energy are usually implemented; a brief review of the most popular tools used in oil platforms to dampen drillstring vibrations is presented in this chapter. Furthermore, the different methods used for acquiring, monitoring, and transmitting data from the downhole to the surface, and the innovative automated systems that improve the perforation process are also addressed.

M.B. Saldivar Márquez et al., *Analysis and Control of Oilwell Drilling Vibrations*, Advances in Industrial Control, DOI 10.1007/978-3-319-15747-4_8

8.1 Vibration-Induced Failures

Drilling vibration is a leading cause of the drillstring component failures. Many factors that contribute to drillstring vibrations: mass imbalance, misalignment and kinks or bends, nutation or wobbling of the rotor within the stator in mud motors, cutting action of the drill bit, and friction factor between the column and borehole wall [79] and between the bit and the drilling surface. Drillstring oscillations are detrimental to the drilling process; they may cause [107]:

- Premature wear and damage of the drilling equipment resulting in fatigue-induced failures.
- Mechanical overload.
- Decrease of the rate of penetration resulting in an increase of drill operating costs [66].
- Interferences on the measurements performed during the drilling process and damage to measurement equipment [169].
- Substantial waste of energy.
- BHA instability and reduction of the directional control [79].
- Stress Corrosion Cracking (SCC) resulting in fissure of screw connections, rupture of drilling pipes, weakening of the drill bit and damage of downhole tools.

Although drill bits and drillstrings can exhibit complex motions, such as backward or forward whirl, these motions can be simplified into three basic vibration modes (as illustrated in Fig. 8.1) [18]:

- Torsional (stick-slip): motion causing twist-off and torque,
- Axial (longitudinal): motion along drillstring axis,
- Lateral (transverse): side-to-side motion.

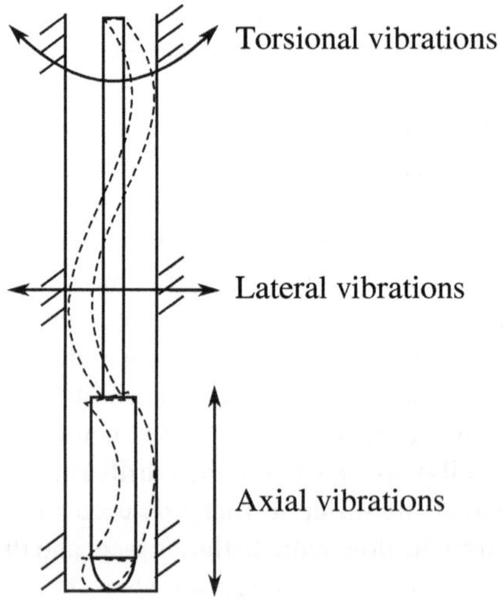

Fig. 8.1 Drilling vibrations

The destructive nature of drilling oscillations depends on the direction in which the vibration takes place. Torsional vibrations strongly affect threaded pipe connections and lead to failures in the process of adding/removing pipe sections. Axial vibrations have detrimental effects on the BHA components (downhole tools, drill collars and bit). Lateral vibrations lead to the propagation of fatigue-induced fissures in screwed joints. In all cases, string vibrations reduce the rate of penetration.

A brief discussion on certain particularities of drillstring vibrations is presented below.

8.1.1 Stick-slip

The stick-slip phenomenon appears due to downhole conditions, such as significant drag, tight hole, and formation characteristics. It occurs when the bit is stalled in the formation while the rotary table continues to rotate. When the trapped torsional energy (similar to a wound-up spring) reaches a level that the bit can no longer resist, the bit suddenly comes loose, rotating and whipping at very high speeds. This stick-slip behavior can generate a torsional wave that travels up the drillstring to the rotary top system. Because of the high inertia of the rotary table, the latter acts like a fixed end to the drillstring and reflects the torsional wave back down the drillstring to the bit. The bit may stall again, and the torsional wave cycle repeats [209].

The typical environment for stick-slip is in high-angle wells with aggressive PDC bits and high WoB, when the downhole frictional torque exceeds the rotary torque [18].

The stick-slip is the most frequently occurring and harmful vibration mode. As the severity of stick-slip is intensified, the stick period increases; when the bit is released, the rotary acceleration of the drillstring is largely increased. This unstable behavior triggers axial drillstring vibrations.

Stick-slip vibrations have a typical frequency less than 2 Hz.

Detection Guidelines

Some indicators that could suggest the presence of torsional vibrations are: increase up to 20 % of the torque at the upper extremity, significant variations of the drillstring angular velocity, clogging of the rotary table and the top drive, unmatched top and bottom angular velocities, unscrew of pipe joints, increase of the torque of screwed joints, among others.

Detrimental Consequences

Common problems associated with the occurrence of stick-slip oscillations are: rate of penetration reduction, BHA components breakage, impact-induced bit damage, screwed joints rupture, PDC bit damage, connection overtorque, drillstring twist-off, etc.

Empirical Control

There are several empirical strategies to avoid the stick-slip; among them we can mention the following ones:

- *Software solution* [18]. Modifying the drilling parameters to mitigate stick-slip has been shown to be a quite effective method. In [224], the influence of drilling parameters such as the angular velocity at the surface and the WoB on stick-slip vibrations is investigated; it was shown that reducing the WoB eases the stick-slip but reduces the ROP. A well-known technique to avoid torsional vibrations is described as follows: when stick-slip occurs, the angular velocity should be increased or the WoB must be decreased by about 15 %. If stick-slip persists, the rotary table must be stopped and then restarted under a higher speed and/or lower WoB.
  A software package that provides real-time display of the magnitude of torsional vibration by utilizing Fourier spectral analysis on the data from drilling sensors was used during drilling operations on two wells off the East Coast of Trinidad; the obtained results are reported in [237].
- *Hardware solution* [18]. Other solutions to eliminate torsional vibration are: reducing bit torque through a less aggressive bit design,[1,2] reducing BHA torque through better wellbore quality and reducing the number of stabilizers and gauge sizes, reducing drillstring torque through better hole cleaning and mud lubricant.

Some other practical solutions to eliminate the stick-slip are: increasing the drilling mud lubricity, increasing the mud flow to reduce friction,[3] using downhole drilling motors, controlling the depth of cut and minimizing the variable torque [67], improving the cutting device design [135], and implementing a torsional impact hammer[4] [303].

8.1.2 Bit-bounce

Longitudinal drillstring vibration, resulting in a repetitive loss of contact between the bit and the drilling surface (bit-bounce phenomenon), is usually induced by torsional drilling oscillations. The bit-bounce phenomenon is more likely to develop when drilling hard surfaces with roller cone bits or when the rock formation has changing characteristics. Significant WoB fluctuations also contribute to the occurrence of bit-bouncing.

Axial vibrations have typical frequencies between 1 and 10 Hz.

---

[1] In [228], some general optimization guidelines to improve PDC bit design are suggested.

[2] A study presented in [230] showed that the torsional vibration is reduced as much as 50 % using hybrid bits instead of conventional ones.

[3] In [146], the effects of drilling mud flow on drilling vibrations are investigated.

[4] See [69], where a report on the drilling performance when using a torsional impact hammer is presented.

Detection Guidelines

Axial vibrations are detected if one or more of the following characteristics are observed: vibration of the pipe sections at the upper extremity, irregular movements of the drillstring along its axis, significant fluctuation of the WoB.

Detrimental Consequences

The impact loads from the axial vibration may cause reduction of the rate of penetration and damage to the bit cutters, bearings, drillstring, and downhole tools. Furthermore, they provoke failures to the surface hoisting equipment.

Empirical Control

Some of the most popular techniques to eliminate axial vibrations can be classified as follows:

- *Software Solution.* The *resonance effect* is also a source of axial instabilities. This effect occurs when the rotary speed is close to one of the natural frequencies of the BHA, this velocity is often called *critical*. In order to avoid the resonance, a proper angular velocity must be chosen. In [57], a method to find the natural frequency of axial vibration is developed through an algorithm designed using the programming language *Delphi*. The strategy is shown to be effective in reducing the drill fatigue by optimizing the drilling parameters. It is worth mentioning that the modification of the drilling parameters, such as reducing the WoB and increasing the angular velocity also helps avoiding the bit-bounce.
- *Hardware solution.* Some popular techniques to avoid axial vibrations are the use of unaggressive drill bits[5] and the implementation of shock sub devices.[6] According to [18], during a drilling operation, the modification of the hardware (such as the use of a less aggressive bits and/or a shock sub) results in a more effective elimination of axial vibrations in comparison with the technique of modifying the drilling parameters.

8.1.3 Whirling

Whirling phenomenon is a lateral instability characterized by impacts between the drillstring and the borehole. Lateral vibration gives rise to high shock loads and high cyclic bending stresses as the BHA impacts the wellbore wall.

Three different types of whirling can occur: backward whirl, forward whirl, and chaotic whirl. Backward whirl is the most severe one; it induces high-frequency large-magnitude bending moment fluctuations resulting in high rates of components

[5] The drill bit is less aggressive when the cutters are in retracted position compared to when the cutters are in extended position.

[6] A shock sub is a drillstring component that is designed to absorb and dampen the variable axial dynamic loads produced by the drill bit during normal drilling operations by interrupting the harmonic cycle created by the bit.

and connections fatigue. Forward whirl is the result of imbalance in a pipe section causing centrifugally induced bowing of the drillstring which leads to one-sided wear of components. In the worst-case scenario, the drilling rod alternately moves backwards and forwards generating chaotic whirling.

Lateral vibrations have typical frequencies between 10 and 50 Hz.

Detection Guidelines

Lateral vibration is hardly detectable at the surface since it may be suddenly attenuated. However, an amplification of harmonics in the frequency spectrum and an increase of the torque on bit and of the torque at the upper end may suggest the presence of whirling.

Detrimental Consequences

Whirling phenomenon reduces the rate of penetration, creates an eccentric borehole, damages several system components (bit, stabilizers, pipe joints, MWD equipment, actuators) [18], causes twist-off and washout, and provokes the premature wear of the bit.

Empirical Control

In [18], the following software and hardware-based techniques to avoid the whirling phenomenon are investigated:

- *Software Solution* [18]. The resonance is one of the main causes of transversal irregular movement of the drillstring, which leads to self-excited high-magnitude vibration. In [18] a finite element-based method to predict the critical speeds is proposed. It is important to point out that resonance is not the only cause of lateral drilling vibration. Thus, avoiding critical speeds should be considered only one portion of a vibration mitigation process.
- *Hardware solution* [18]. BHA eccentricity and buckling are common sources of lateral vibrations. Downhole tools, such as bicentered bits and roller reamers, have to be carefully chosen to minimize the eccentricity forces.

Some additional strategies to eliminate lateral vibrations are: modifying the drilling parameters (e.g., WoB reduction, angular velocity increase), increasing the drilling mud lubricity, and inserting stabilizers to increase BHA rigidity[7] or implementing instead roller reamers[8] [279].

An alternative method to reduce lateral vibrations consists in optimizing the BHA configuration. In [308], the bit-BHA interactions are shown to be minimized through the determination of the optimum zone which is obtained by predetermining the critical WOB and rotations per minute triggering drilling vibrations; the critical

[7] See [192], where the effects of stabilizer placement is investigated.

[8] In [276], it was shown that the use of roller reamers reduces the occurrence of stick-slip vibrations, furthermore, replacing stabilizers with roller reamers is an effective approach to decoupling the stick-slip and the whirling phenomena. The roller reamer plays a role in introducing a low-friction bearing between the drillstring and borehole [312].

values represent the boundaries of different drilling conditions, such as the maximum torque limited by rig and minimum ROP specified by operators. A similar approach is presented in [21].

8.1.4 Vibration Detection Methods

There are three methods for detecting drilling vibrations [18]:

Downhole vibration sensors. Sensors located at the bottom end are the most effective method to identify drilling vibrations since they can detect axial, torsional, and even lateral vibrations, which are hardly detected at the surface because, generally, they do not propagate along the drillstring.

Surface sensors. Surface sensors provide limited information on drilling vibration, in fact, they provide data associated only with the bit-bounce and the stick-slip phenomena. They are used only when downhole vibration sensors are not available or when the MWD data quality is poor.

Post-run inspection. Drilling vibrations can be detected through post-run inspection. As explained before, there are several detection guidelines that suggest that an specific drilling vibration has occurred and proper actions should be considered to prevent system failures.

8.2 Drilling Vibration Simulations

Through mathematical models, fluctuations of the bit velocity which characterize the stick-slip phenomenon can be reproduced. In order to get a graphical representation of the drilling behavior, the neutral-type time-delay model (2.32) is simulated. The torque on bit model used for the system simulations is the one inspired by the Karnopp friction law (3.7)–(3.9):

$$T\left(\dot{\Phi}_b(t)\right) = c_b\dot{\Phi}_b(t) + W_{ob}R_b\left(\mu_{cb} + (\mu_{sb} - \mu_{cb})\mathrm{e}^{-\gamma_b|\dot{\Phi}_b(t)|}\right)\mathrm{sgn}\left(\dot{\Phi}_b(t)\right), \tag{8.1}$$

where R_b denotes the bit radius, W_{ob} is the weight on bit, μ_{cb} and μ_{sb} are the coefficients of Coulomb and static friction.

The model parameters used in simulations are given in Table C.1 of Appendix C.

Figure 8.2 highlights the fact that model (2.32) subject to the frictional torque model (8.1) accurately reproduces the occurrence of stick-slip vibrations under certain operating conditions.

The stick-slip phenomenon occurs when a section of the rotating drillstring is momentarily caught by friction against the borehole, then released. The bit might eventually get stuck and then, after accumulating energy in terms of torsion, be suddenly released, the collar rotation speeds up dramatically and large centrifugal

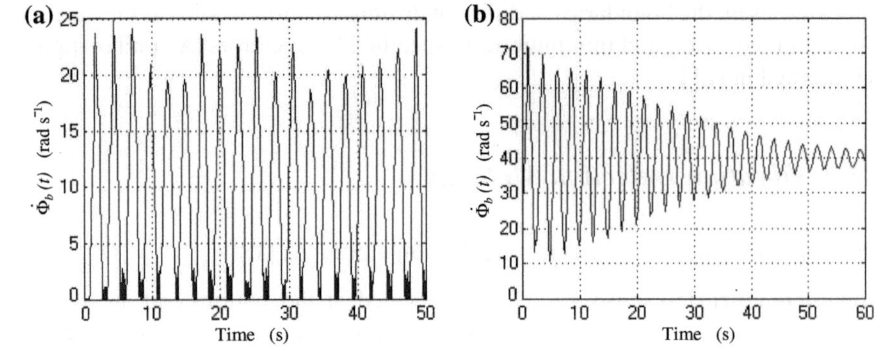

Fig. 8.2 Simulation of the torsional model (2.32) subject to the frictional torque (8.1). **a** Angular velocity at the bottom extremity $\dot{\Phi}_b(t)$ for $\Omega_0 = 10\,\mathrm{rad\,s^{-1}}$ (stick-slip). **b** Angular velocity at the bottom extremity $\dot{\Phi}_b(t)$ for $\Omega_0 = 40\,\mathrm{rad\,s^{-1}}$

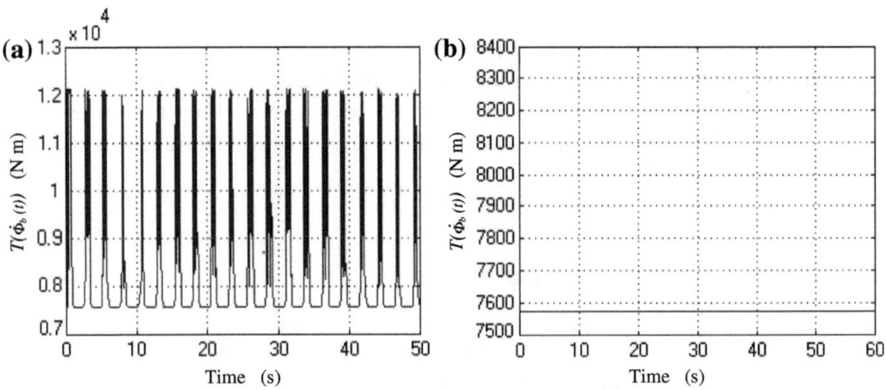

Fig. 8.3 Simulation of the torsional model (2.32) subject to the frictional torque (8.1). **a** Torque on bit $T(\dot{\Phi}_b(t))$ for $\Omega_0 = 10\,\mathrm{rad\,s^{-1}}$ (torque fluctuations driven by stick-slip). **b** Torque on bit $T(\dot{\Phi}_b(t))$ for $\Omega_0 = 40\,\mathrm{rad\,s^{-1}}$

accelerations occur. Figure 8.2a shows the stick phases, during which the rotation stops completely, and the slip ones, during which the angular velocity of the tool increases up to more than twice the nominal angular velocity. Figure 8.3a shows that, as reported in real wells, stick-slip vibrations cause important torque fluctuations [162].

It is assumed that the rotary table and the bit are rotating at the desired speed when the bit is off bottom. When the bit starts to interact with the formation, the system will inevitably be disturbed with the possibility of axial and torsional vibrations occurring simultaneously.

Figure 8.2b shows that an increase of the angular velocity Ω_0 provided by the rotary table leads to the reduction of the stick-slip. The qualitative agreement between this result and the field data observed in [128] is remarkable.

Due to the momentarily stopping of the bit, the torque reaches very high values, as shown in Fig. 8.3. When the bit starts slipping, the energy stored in the drillstring is released, causing very large torsional vibrations. During this time, the bit remains in contact with the formation; some of the energy stored during stick phase is transferred into axial vibrations.

As mentioned before, axial vibrations are characterized by an intermittent loss of contact between the bit and the rock, so that the drill area is severely hit. This bit-bouncing behavior is provoked by the high bit speed resulting from the stick-slip torsional motion.

A suitable mathematical description should consider the intimate coupling of longitudinal and rotational drilling dynamics. A simulation of the model (2.2) subject to the frictional torque (8.1) depending on the torsional bit velocity confirms that fluctuations of the bit rotary velocity excite axial vibrations. Figures 8.4 and 8.5 show irregular variations of the longitudinal bit position characterizing the occurrence of bit-bounce.

Loss of contact between the bit and the drilling surface arises from the axial resonance of the string. While the bit is off bottom, the torque on bit is null; consequently, the critical frequencies of lateral and axial vibrations change [56]. Without control actions, these large amplitude vibrations remain with continuous energy exchange between various modes of vibration. It is obvious that this behavior will eventually result in failures due to very large and often cyclic stresses present at different sections of the drillstring.

From experimental results and simulation of models, it is clear that most of the downhole vibrations are driven by bit motion. Therefore, for an effective control of vibrations, the measurement and feedback of certain variables at the bit level are required.

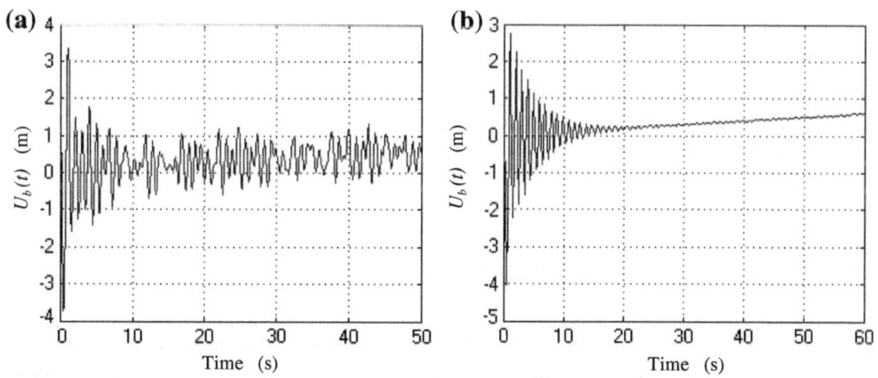

Fig. 8.4 Simulation of the axial model (2.2) subject to the frictional torque (8.1). **a** Axial displacement of the drillstring $U_b(t)$ for $\Omega_0 = 10\,\mathrm{rad\,s^{-1}}$ (bit-bounce). **b** Axial displacement of the drillstring $U_b(t)$ for $\Omega_0 = 40\,\mathrm{rad\,s^{-1}}$

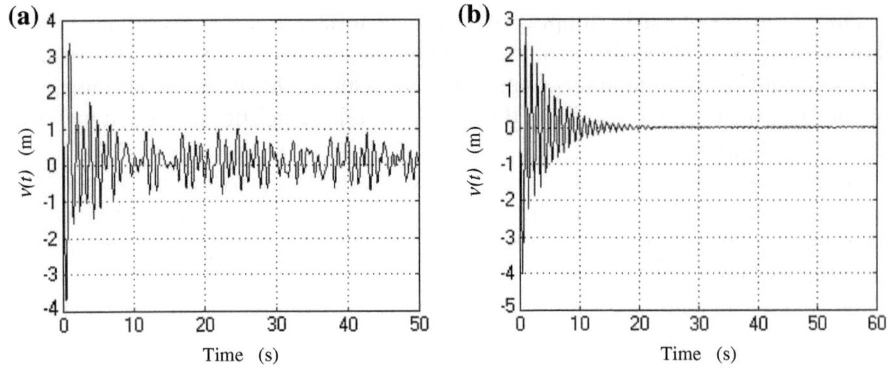

Fig. 8.5 Simulation of the axial model (2.2) subject to the frictional torque (8.1). **a** Variable $\nu = U_b - \rho_0 t$ for $\Omega_0 = 10\,\mathrm{rad\,s^{-1}}$ (bit-bounce). **b** Variable $\nu = U_b - \rho_0 t$ for $\Omega_0 = 40\,\mathrm{rad\,s^{-1}}$

8.3 Practical Strategies to Reduce Drilling Vibrations

Drilling vibrations are usually avoided or mitigated through empirical methods based on the expertise of drilling operators, most of them involve a proper tuning of certain surface-controlled drilling parameters, such as the weight on bit, the flow of drilling fluid through the drill pipe, the drillstring rotational speed and the density and viscosity of the drilling fluid.

In this section, the main strategies based on drillers' experience to reduce torsional vibrations are evaluated through simulations of the neutral-type time-delay model given in (2.32) subject to the frictional torque (8.1). Furthermore, a brief review of the physical devices that absorb or preclude the reflection of the vibration waves, used in conjunction with the empirical control methods to reduce drilling vibrations, is presented.

Simulations presented hereinafter are developed using the variable step *MATLAB-SIMULINK* solver ode45 (*Dormand Prince Method*). They provide an indirect validation of the axial and torsional models (2.2) and (2.32), coupled through the frictional torque (8.1).

8.3.1 Decreasing the Weight on Bit

According to drillers' experience, for a fixed angular velocity at the surface, a reduction of the weight on bit makes the stick-slip phenomenon disappear; on the contrary, the bit may stop if the WoB is excessive [261]. Figure 8.6 shows the system response when the weight on bit is linearly decreased from 97347 N to 31649 N and the angular velocity is $\Omega_0 = 10\,\mathrm{rad\,s^{-1}}$. An important reduction of stick-slip oscillations is observed.

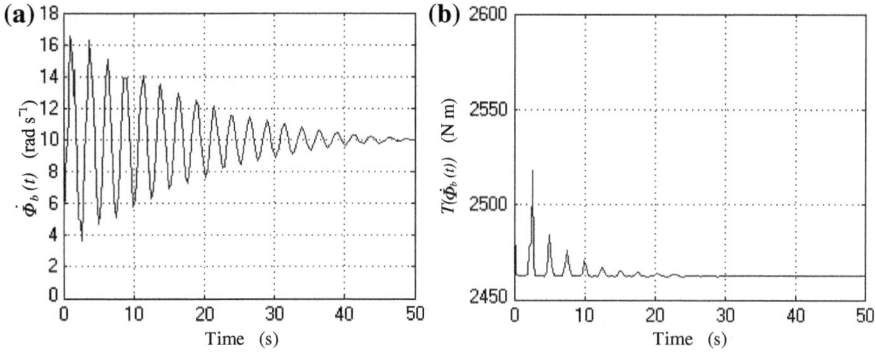

Fig. 8.6 Simulation of the torsional model (2.32) subject to the frictional torque (8.1). Reduction of stick-slip phenomenon by decreasing the weight on bit from 97347 N to 31649 N. **a** Angular velocity at the bottom extremity $\dot{\Phi}_b(t)$. **b** Torque on bit $T(\dot{\Phi}_b(t))$

To ensure an optimal perforation rate, it is necessary to maintain certain amount of weight on the cutting device. A shortcoming of the proposed empirical method is that, the reduction of torsional oscillations may require a substantial decrement of the WoB, which implies an inefficient drilling rate or even the stopping of the perforation process.

8.3.2 Increasing the Angular Velocity at the Upper Part

Experimental results indicate that the stick-slip phenomenon is avoided by increasing the angular velocity provided by the rotary table. The simulation results of Fig. 8.7 show that the stick-slip vibrations are reduced by means of a linear increase of the

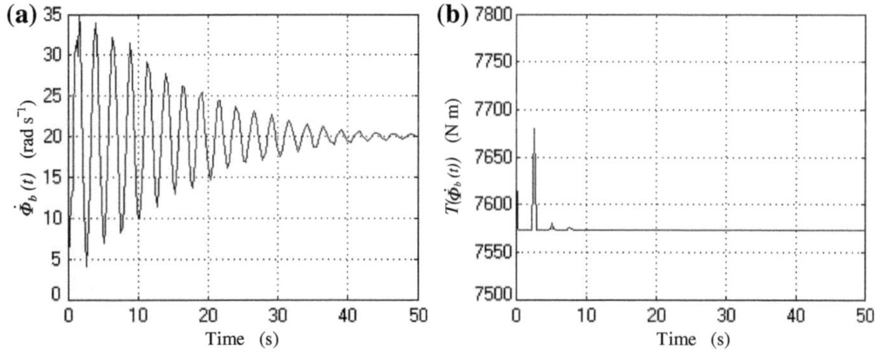

Fig. 8.7 Simulation of the torsional model (2.32) subject to the frictional torque (8.1). Reduction of stick-slip phenomenon by increasing Ω_0 from 10 rad s^{-1} to 20 rad s^{-1}. **a** Angular velocity at the bottom extremity $\dot{\Phi}_b(t)$. **b** Torque on bit $T(\dot{\Phi}_b(t))$

velocity at the top extremity from $\Omega_0 = 10\,\text{rad}\,\text{s}^{-1}$ to $\Omega_0 = 20\,\text{rad}\,\text{s}^{-1}$, while the weight on bit remains constant ($W_{ob} = 97347\,\text{N}$).

One of the disadvantages of the strategy is that a substantial increment of the angular speed induces lateral problems, such as irregular rotation, which causes repeated collisions between the rod and the borehole walls producing an eccentric hole. The shocks may damage the drilling system components and deteriorate the borehole wall affecting the overall drilling direction.

8.3.3 Introducing a Variation Law of the Weight on Bit

From field data experience and from simulations of the drilling model, it is concluded that the manipulation of the weight on bit can be a solution for stick-slip oscillations even for low angular velocities. Increasing velocities at the rotary top driving system may lead to lateral vibrations, which is why the weight on bit manipulation can be an alternative solution to attenuate stick-slip oscillations.

The weight on bit variation law, proposed in [209], is given by:

$$W_{ob}(\dot{\Phi}_b(t)) = K_w \left|\dot{\Phi}_b(t)\right| + W_{ob_0}, \tag{8.2}$$

with $W_{ob_0} > 0$ and $W_{ob} > W_{ob_0}$. The variation law (8.2) implies a weight on bit reduction for a decreasing $\dot{\Phi}_b(t)$. Furthermore, considering that low values of WoB would make drilling stop, the constant value W_{ob_0} guarantees a desirable rate of penetration.

The simulation results of Fig. 8.8 show the successful stick-slip elimination by means of the introduction of the variation law (8.2). We have considered a drilling

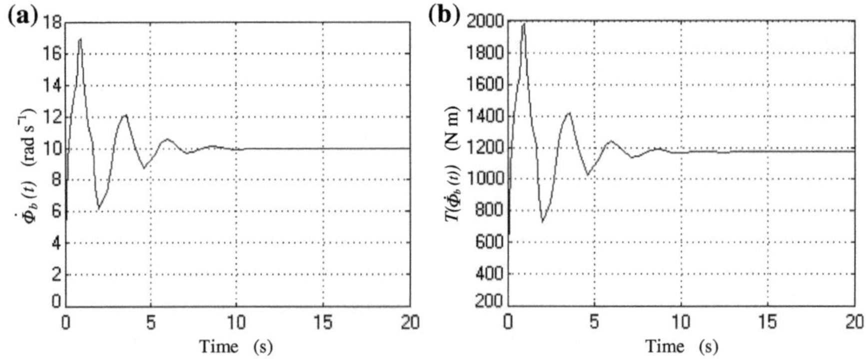

Fig. 8.8 Simulation of the torsional model (2.32) subject to the frictional torque (8.1). Reduction of stick-slip oscillations by means of the variation law (8.2). **a** Angular velocity at the bottom extremity $\dot{\Phi}_b(t)$. **b** Torque on bit $T(\dot{\Phi}_b(t))$

parameters combination for which torsional oscillations occur, i.e., $\Omega_0 = 10\,\text{rad s}^{-1}$ and $W_{ob_0} = 97347\,\text{N}$.

The weight on bit variation law constitutes an efficient strategy to counteract torsional oscillations; in fact, several existing control strategies are based on the manipulation of this variable. However, due to large bit speed variations driven by stick-slip, the implementation of the variation law (8.2) may demand abrupt WoB fluctuations. Since the WoB is provided by the drill collars, the proposed strategy would require a sudden addition and removal of drill collar sections, which may result infeasible. Even in the case that torsional vibrations are not severe, and the variation law can be implemented, there exists the possibility of inducing longitudinal vibrations.

8.3.4 Absorbing the Vibration Energy

An effective mitigation of drilling vibrations is achieved by allowing the absorption of the energy arising from drilling vibrations. This can be done through different methods, for example, by increasing the damping at the down end through the modification of the drilling fluid characteristics or with the inclusion of vibration absorbers at the BHA, such as the shock sub and the active vibration damper, to attenuate torsional vibrations generated at the bit and prevent them from traveling up and back down the drillstring [209]. Another technique to absorb the vibration energy consists in implementing a Soft Torque Rotary System (STRS) which dampens the propagating waves at the top extremity and precludes their reflection. These strategies are discussed below.

Implementing a shock sub device. The shock sub is an engineered device to increase bit penetration and to prolong bit life as well as reduce drill collar fatigue, vibration, and impact loads. This tool can be run at the bit or in the drill collar string as required by the location personnel. In areas where severe drilling causes the bit to vibrate and the string to jump, this tool will eliminate the problem by smoothing out these movements and forcing the bit to stay on the bottom, thereby increasing the penetration rate and reducing trip time caused by bit wear and failure during rough drilling. All internal parts of the tool are lubricated and in the event of fluid loss, the tool still functions as the mechanical spring mechanism will function from lubrication by drilling fluid. Very little maintenance is required as there are just a few moving parts. Assembly and disassembly of the tool can be completed in a short time. The shock sub is most beneficial when drilling in hard rock, broken formations, and intermittent hard and soft streaks. Reducing the impact loads helps to increase ROP, improve borehole quality, and extend the life of the cutting structure, bearings, connections, and surface equipment.

There are mainly three types of shock sub devices:

- *Mechanical hydraulic*. The mechanical hydraulic shock sub is a two-function shock sub that absorbs the pulsate and shock from drill tools by compressing

spring and oil to store the energy. It is used for reducing vibrations caused by hard formation drilling and keeping the drill bit firmly on the bottom, so that it helps to reduce drill string connection fatigue and prolong drill string life.
- *Floating*. The floating shock sub is utilized in both slanted and vertical drilling operations to reduce vibration and wear of the drill string and drill rig components. It has two important functions. First of all, it provides a means of making up and breaking out threaded connections without damaging threads. The second is that it prolongs bearing and gear life of the rotary top drive.
- *Bumper*. The drilling bumper sub is an important component for deepwater drilling operations, where drillstring oscillations is a major problem. This tool provides six feet of reliable telescopic movement, without placing any limitations on drillstring torque capacity, tensile strength, or hydraulic capability.

Several industrial companies offer this device, for instance, *Schlumberger*, *Stabil Drill*, *JA Oilfield Manufacturing*, *Toro Downhole Tools*, *Dynomax Drilling Tools*, just to mention a few.

Several analyses of the effect of a shock sub on drilling vibrations have been carried out. See, for example, [160], where a mathematical investigation is made into the longitudinal vibrations of a drillstring, with and without a shock sub. In [114], a dynamic finite element model (FEM) and an analytical elastodynamic model, both including a shock sub device, have been developed to study the complex nonlinear coupled axial–lateral dynamics of a drillstring. The design parameters of a shock sub absorber to be placed above the bit for the purpose of minimizing vibrations are analyzed in [83].

Implementing an Active Vibration Damper sub (**AVD**). AVD sub is a stand-alone downhole tool introduced by *APS Technology* that autonomously adapts to changing downhole BHA motion in real time to minimize axial and torsional drill-string vibration. The AVD improves the ROP and doubles the bit life due to reduced vibration. Other downhole drill string components, like MWD/LWD tools, also benefit from lower vibration.

Structurally, the AVD is similar to a shock sub, with the addition of a damper section that has programmable stiffness. The damper chamber is filled with a magne-torheological fluid that has electronically controlled viscosity. An integrated motion sensor measures displacement several times per second and changes the damping factor over a 7-to-1 range based on observed drilling conditions. By keeping tool string damping in the right range for current drilling conditions, the AVD significantly reduces vibration, maintaining the bit in better contact with the formation and increasing ROP.

The AVD may be run as a self-contained drilling tool with no calibration or other rig maintenance required. In this mode the AVD records vibration data for later download.

In [60], the AVD was tested by drilling blocks of hard concrete using a tricone bit. It was concluded that the AVD is likely to provide significant time and cost savings, particularly in deep wells arising not only from the increased instantaneous ROP, but also from fewer trips for bit or equipment changes, and lower costs for replacing damaged MWD tools, motors, or other expensive components.

Modifying the drilling fluid characteristics. As explained before, the drilling fluid or drilling mud has the function of transporting cutting materials from bit to surface through the annulus between the drill pipe and the borehole wall. The drilling fluid aids the cutting process by jetting action, it also cools and lubricates the bit. It is well known that an increase of damping of drilling mud reduces stick-slip oscillations. See, for instance [313], where self-excited stick-slip oscillations under the influence of damping of drilling mud and active damping system are analyzed.

An increase of the damping at the rock-bit interface through the manipulation of the drilling mud properties can be approximated by modifying the value of the viscous damping coefficient c_b of the frictional torque model (8.1). Figures 8.9 and 8.10 show an important attenuation of torsional drilling oscillations for high values of c_b.

The management of drilling fluid characteristics is a recommended strategy to be applied in conjunction with a more sophisticated methodology to avoid drillstring vibrations. In practice, the single application of this empirical method is insufficient to suppress the stick-slip.

Implementing the Soft Torque Rotary System (**STRS**). The STRS is a rotary drive control system that helps reducing drilling torsional vibrations. It allows the rotary table speed to respond to dynamic torque oscillations in such a way that the rotary table absorbs or dampens the vibrations and avoids their reflection. The system was developed by the *Shell* company in the early 1990s. The very first system was developed for DC drilling drives using analog torque feedback; however, with the introduction of AC top drives in the drilling industry, a redesign of the STRS was required. An overview of the design developments is presented in [249].

The STRS is a PI-like speed controller which involves a pair of gains κ_i (drive stiffness in Nm/rad), κ_p (drive damping in Nms rad^{-1}) that must be properly tuned. Depending on the drillstring and BHA configuration, the parameters κ_i and κ_p are

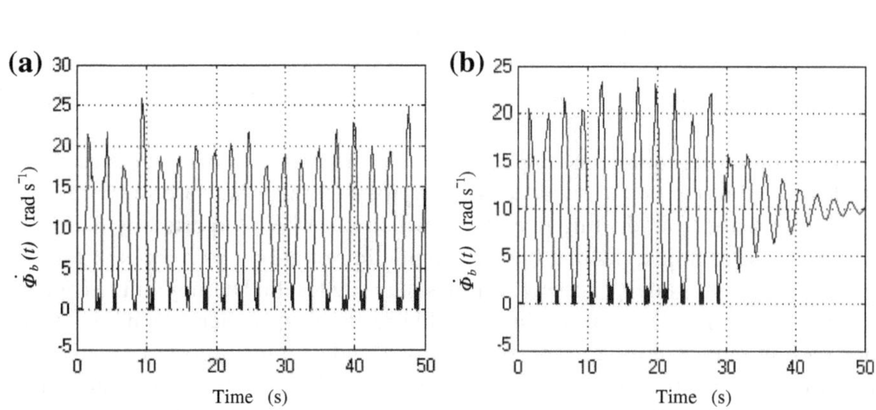

Fig. 8.9 Simulation of the torsional model (2.32) subject to the frictional torque (8.1). Reduction of stick-slip oscillations by increasing the damping at the down end. **a** Angular velocity at the bottom extremity $\dot{\Phi}_b(t)$ for $c_b = 0.8$. **b** Angular velocity at the bottom extremity $\dot{\Phi}_b(t)$ for $c_b = 20$

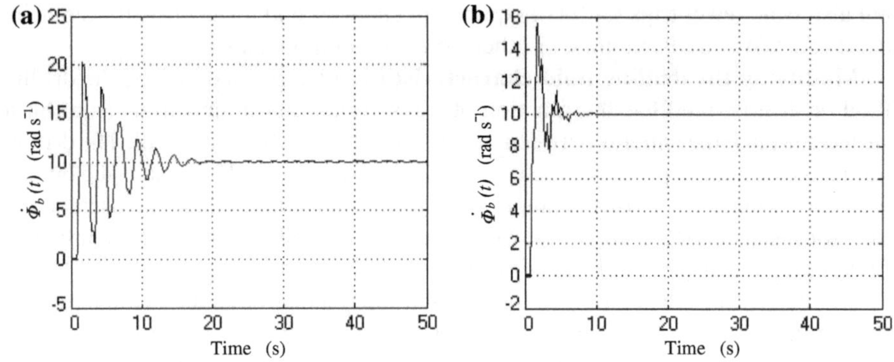

Fig. 8.10 Simulation of the torsional model (2.32) subject to the frictional torque (8.1). Reduction of stick-slip oscillations by increasing the damping at the down end. **a** Angular velocity at the bottom extremity $\dot{\Phi}_b(t)$ for $c_b = 50$. **b** Angular velocity at the bottom extremity $\dot{\Phi}_b(t)$ for $c_b = 150$

calculated. This control method is able to reduce the stick-slip vibrations when the κ_i and κ_p take the ideal values (which must be far from the critical velocity to avoid the resonance effect) or are close to them.

In Chap. 9, this methodology is studied and its performance is evaluated through simulations of the proposed drilling model.

8.4 Data Acquisition and Monitoring Systems

An appropriate drilling operation requires certain devices for supervising, analyzing, displaying, recording, and retrieving information.

The main parameters to be acquired and monitored are: the drilling and mud flow rates, the hook load, the hole depth, the pump pressure, the bottom hole torque, the rotary speed, the mud tank level, the pump strokes, the weight on bit, the hoisting speed, and other important mud properties such as the density, temperature, and salinity [18].

The monitoring equipment helps to detect most of the problems associated with the drillstring vibrations such as well kicks and pipe sticking. There are several problem indicators reported in drilling rate charts which must be interpreted by the drilling operator. For example, drilling breaks provide information on changes of lithology and formation pressures; excessive torques may suggest bit bearing failures or a high concentration of drilled cuttings in the wellbore annulus; an increase in hook load or decrease in mud returning to the surface could indicate that a lost circulation zone has been encountered; a sudden increase in pit level indicates that formation fluids are entering the wellbore and, hence, that blowout is eminent [18].

An adequate drilling performance requires a suitable maintenance of the WoB, the rotary speed and the properties and flow rates of the drilling mud. In short, the monitoring system is vitally important for the entire drilling process; the existing advanced technology has allowed the monitoring, recording, analysis, storage, and retrieval of drilling data to become a routine part of the drilling operation process.

Acquiring and monitoring methods have evolved over the years. Next, we present a brief chronological review of the main techniques of sensing and supervising drilling data.

The first drillstring vibration analyses, conducted in the 1960s by Bailey and Finnie of *Shell Development Company* [20, 89], were carried out using a surface measurement package placed beneath the rotary table to measure torque, axial force, rotation, and axial displacement while drilling [171].

In 1968, researches from *Esso Research Company* reported the use of a downhole recording tool to measure downhole forces and motions [70]. The tool could measure axial, torsional, and bending forces and moments; axial, lateral, and angular accelerations; internal and external pressures.

In 1986, the French oil company *Elf Aquitaine* started the DYNAFOR research project with the purpose of improving drilling performance through a better knowledge of dynamic phenomena. A surface measurement device called the *Dynamètre* was developed enabling measurement of tension, torque, and accelerations in three directions at the top of the drillstring.

Measurements While Drilling (MWD) systems, developed in the 1980s, enable operators to save time by acquiring formation evaluation and drilling optimization data during drilling operations, which also helps keep wells on track to minimize failures. The studies developed by Aarrestad and Kyllingstad of *Rogaland Research Institute* about the coupling between bit torque and axial load at the bit [1] were conducted through a hardwire MWD tool.

In 1989, Cook et al. presented the first real-time downhole Root Mean Square (RMS) measurements of forces, accelerations, and fluid pressures [63]. The quantities as RMS values were transmitted to surface using mud-pulse telemetry.

In the late 1980s the *Institute Français du Pétrole* designed the TRAFOR system,[9] a research tool to improve knowledge about drilling and how to model it. The TRAFOR system consists of a downhole measurement sub, called the *Télévigile*, and a surface measurement device known as the *Survigile*. The signals of the Télévigile and Survigile are gathered by a computer and synchronized. The Télévigile is connected to the surface equipment through an electric wire. Downhole WOB, TOB, accelerations in three directions and bending moments in two directions are measured. Torque, tension, and rotary speed are measured at surface. The great merit of the TRAFOR system is the ability to measure both downhole and surface data at real time.

In 1990, Besaisow et al. presented the *Advanced Drillstring Analysis and Measurement System (ADAMS)* which only measures surface data [30]. The surface

[9] In 1994, Pavone and Desplans gave a description of the TRAFOR system and some experimental results [226]. The most remarkable result of this contribution is the relation found between the TOB and the bit rotary speed. The characteristic is clearly showing a stick phase and lower torques for higher speeds (a negative slope). The authors proposed a Proportional–Integral–Differential (PID) control and the use of an anti stick-slip tool to prevent stick-slip.

measurements had a large spectral content, and the authors provided a variety of explanations for the peaks that were observed.

Dubinsky, Henneuse and Kirkman, in 1992, gave a historical overview of Russian, European, and American research in surface monitoring of downhole vibrations [78]. The Russians aimed their attention in the 1970s on optimizing the performance of turbines. A method was developed to detect stick-slip vibrations while using a turbine by inspecting the auto-spectral density of the surface axial accelerations. In the United States the goal was to reduce drillstring failures. The authors suggested that downhole dysfunctions are always observable at surface.

Emerging technology has allowed the development of innovative systems for measuring and supervising drilling data. A technological solution that combines the ability to measure and monitor the downhole drilling environment and to actively control and damping drilling-induced vibrations is the so-called Drilling Vibration Monitoring & Control System (DVMCS). The DVMCS increases the ROP by keeping the bit in contact with the cutting surface, prologues the bit and MWD/LWD sensors life by eliminating shock and vibration damage and reduces the number of trips needed to complete a well. A report on the design, modeling, and laboratory testing of the DVMCS can be found in [59].

Data acquisition devices (sensors) can be wired or wireless. The most frequently encountered are the wired ones; however, they are costly, not best suited for harsh environments and are difficult to deploy and maintain. An increasing research effort devoted to the design of wireless sensor equipment, that introduces significant benefits in cost, ease of deployment, flexibility and convenience, has been recently developed. See, for instance [173], where the design of an oil drilling wireless data acquisition system consisting of a low power consumption single chip and a wireless data transmission chip is presented. Among the benefits that this system provides, we can mention its structural simplicity, reliability, high transmission rate, low cost, and low power consumption characteristics. In [5], a survey on wireless sensor network applications is provided; furthermore, it presents a discussion on the performance results from certain case studies and on the future prospects and research challenges.

A recent field study in which downhole vibrations were measured by use of a newly available bit monitoring device is provided in [170]; the focus of the study was to understand the primary source of the PDC bits damage, which as expected is the stick-slip phenomenon.

8.5 Data Transmission Methods

Transmission systems provide bottom measurement information to the drill operator. Clearly, monitoring of the system variables for control purposes requires the transmission of measurements. It is worth mentioning that the transmission models must take into consideration the presence of time delays; some discussions on this topic can be found in [23], where a lumped parameter model describing the torsional drilling dynamics coupled to a delayed measurement model is studied.

There are several methods to transfer bottom information to the drilling operator, the most popular ones are:

Mud-pulse telemetry (**MPT**). It is a binary coding system that transmits the information via the mud that goes through the drilling system; the data is represented by pressure pulses [140]. A stepper-motor-based device (pulse actuator) and a main valve constrains the flow and generates pressure-pulse sequences [283]. Pressure variations, captured by a piezoelectric mechanism, are analyzed by a microcontroller. Clearly, due to the irregular nature of the mud flow, the low frequency vibrations originated by mud pumps and pulsation dampeners, the data is disturbed and attenuated by noise. It is worthy of mention that MPT technology offers cost-effective data transmission, although its rate of transmission is low (1 or 2 bps) [77]. Mud-pulse velocity decreases with mud density disturbances, gas content, and mud compressibility. The method becomes less effective with increasing well depth. Pulse waves travel through the borehole at 1,200 m/s [180], hence the measure arrives with some delay that increases up to $\tau_{max} \approx 6.6$ s. A novel high-speed telemetry system with measurements along the string is presented in [292]; the broadband network provides downhole information including bit whirl, stick-slip, and axial and lateral vibrations in real time.

Acoustic waves. Acoustic transmission is an effective method to emit pulses to the surface. The propagation velocity of acoustic waves is at least three times superior to transmission velocity offered by the mud-pulse telemetry [58]; the transmission rate is about 6 bps. Acoustic waves are produced by torsional contractions generated by magnetorestrictive rings set inside the drillstring [76]. The maximum delay of data transmission is $\tau_{max} \approx 2.2$ s. It is also noteworthy that there exists an attenuation of around 4 dB/300 m [75]. However, this problem is solved with the inclusion of repeaters at any joint at each 10–15 m of the pipe section. Implementation of repeaters does not imply considerable additional delays since the repeater's amplification occurs almost instantaneously. The telemetry system transfer signals directly to the surface through the channel. Generally, there is an embedded sensor which measure the angular bit velocity $\dot{\Phi}_b$. A noise measurement $S(t)$ is added to the data and then coded all together to be transmitted through the acoustic channel G. At the top extremity, a receiver reads the encoded signal with the noise $N(t)$ [272]. In addition, a digital algorithm is used to decode this data and make it accessible to drill operators. Figure 8.11 shows a schematic model of this data transmission method [23].

Wired drill pipe (WDP) technology. Wired drill pipe technology is an innovative method capable to transmit high-resolution data at speeds of up to 57,600 bps,

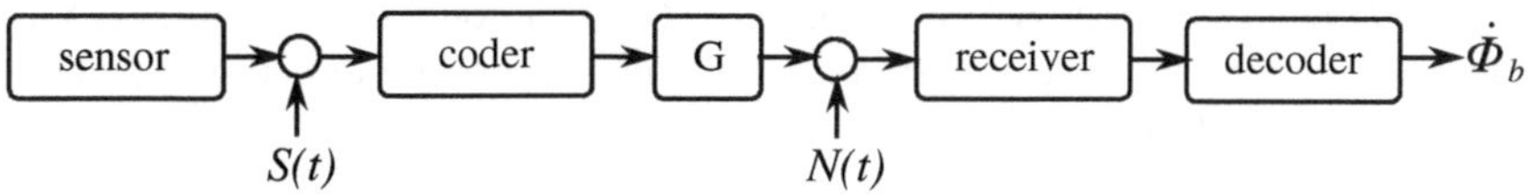

Fig. 8.11 Testbed schematics with sonar pulses

much faster than conventional telemetry systems [306]. It provides real-time visibility of the drilling process, borehole and formation data, which significantly improves real-time control strategies.

A WDP system consists of four basic elements: an inductive coil that electrically connects neighboring pipes together; a high-strength data cable that transmits data through each tubular; an electronic network that control the flow of data through the system; and the tools used at surface to control and monitor system performance.

WDP technology offers robust, reliable operation and is virtually transparent to standard rig procedures [115]. It has been deployed worldwide in both offshore and land wells, providing numerous benefits, for instance [191]:

- Directional drillers are able to accurately land wells in complex reservoir sections, even in the presence of recurrently changing formation characteristics.
- High-resolution density images are used to identify borehole geometry characteristics, leading to real-time adjustments and reducing nonproductive time in some historically complex formations in offshore wells.
- Downhole tool communication time is radically reduced, allowing a decrement on the order of 10 % in total drilling time.
- Drilling efficiency is improved by using WDP to enable implementation of automated control systems. An improvement of up to 80 % in the rate of penetration may be achieved.

Over the last decade, more than 110 wells have been drilled with wired drill pipes. WDP technology has been applied in well depths of up to 6,400 m, at temperatures of up to 150 °C, and in a variety of fluid environments.

8.6 Automated Drilling Systems

Since the beginning of the drilling industry for hydrocarbons extraction, processes have been manually operated by drilling specialists; however, nowadays, with the help of autonomous computer-controlled systems, the drilling automation is possible.

Automation has been widely used in several industries: aeronautics, automobile manufacturing, utility and power generation, and general processing. It is well known that industrial processes are greatly improved when human intervention is minimized.

Drilling automation seeks to improve the process, achieving optimized rates of penetration, consistent hole quality and overall drilling performance, which allow decreasing the perforation time and increasing the workers safety. *Apache* (*APA*), *National Oilwell Varco* (*NOV*), and *Statoil* (*STO*) are among the companies working on technology that will take humans out of the most repetitive, dangerous, and time-consuming parts of oil field work.

Some of the most popular automated systems of the drilling industry are:

DrillTronics. DrillTronics is a real-time system for monitoring, diagnostics, and control of the drilling process. DrillTronics provides several automated functions which allow reducing nonproductive time and optimizing drilling operations. DrillTronics controls actively the draw-work, top drive, and mud pumps to account for the dynamic behavior of the well during drilling operations. Some of its main features are [244]:

- It comprises a software modeling with algorithms that reflect the wellbore behavior and its interaction with the drilling equipment.
- The models are driven in real time by drilling data that are logged at high acquisition rate. The system is continuously calibrated through analyzing the data with advanced filtering techniques.
- Real-time diagnosis of the drilling process is obtained from comparing measured data with model predictions.
- An integrated drilling simulator is developed by linking the modules together, and combining it with an ROP model.

eDrilling. eDrilling system allows real-time drilling simulation, 3D visualization and control from a remote drilling expert center. The concept uses all available real-time drilling data in combination with real-time modeling to monitor and optimize the drilling process. Some elements of the eDrilling system are [244]:

- An advanced and fast Integrated simulator which is capable to model the different drilling subprocesses dynamically, and also the interaction between these subprocesses in real time.
- Automatic quality check and corrections of drilling data; making them suitable for processing by computer models.
- Real-time supervision methodology for the drilling process using time based drilling data.
- Methodology for diagnosis of the drilling state and conditions (obtained from comparing model predictions with measured data).
- A Virtual Wellbore, with advanced visualization of the downhole process.

Robotic Drilling System (RDS). Recently, the *Norway's Robotic Drilling Systems company* (formerly *Seabed Rig*) developed an innovative autonomous robotic drilling rig for unmanned drilling operations. The RDS sets new standards with increased safety and cost-effective planning and drilling and can be implemented on existing, as well as new drilling structures, both offshore and on land. The unmanned system utilizes autonomous robotic working operations that can be remotely controlled from an interactive 3D interface. In [275], some of the drilling process efficiency improvements when using robotics are highlighted; a discussion on how robotized drilling systems improve safety and cost savings is provided.

GetSMART. The *Institut Français du Pétrole* (*IFP*) and *Geoservices* designed the GetSMART system which aims at the detection while drilling of the main abnormal vibrations and hydraulic malfunctions. The system is based on the diagnosis

trees methodology, which allows one to take into account the empirical knowledge of the driller to analyze the signals coming from sensors or physical models and also to generate alarms.
When abnormal situations arise, the drilling operator has to quickly react to events. However, extracting only the relevant information from the huge mass of data available is not always an easy task. In addition, some diagnoses are impossible to do without an appropriate preprocessing of the data. The GetSMART system automatically warns the driller in due time of any abnormal situations, therefore, corrective actions can be done immediately.
The key to such a system is the implementation on a dedicated workstation of a special processing software extracting a corrected signal from the raw sensor output. This signal, which takes into account not only the current outputs of the sensors, but also constantly updated typical rig/well responses, will automatically trigger a drilling floor alarm [241].

8.7 Notes and References

In this chapter, some of the main experience-based techniques for reducing stick-slip vibrations were evaluated through simulations of the torsional neutral-type time-delay model. The most common operational instructions involve an increment of the rotary velocity and a decrement of the weight on bit. Alternative empirical methods suggest the manipulation of BHA characteristics such as the modification of the drilling fluid to increase the damping at the down end and the implementation of a weight on bit variation law [252].

It has been demonstrated that the stick-slip is less likely to occur when the rotary velocity is high or when the weight on bit is not excessive. Changing the dampingcharacteristics at the bottom end by modifying the drilling mud properties or by implementing energy absorbers is an alternative technique to overcome drilling oscillations.

Experts of the oil drilling industry have developed diverse empirical methods to avoid drilling vibrations. For instance, in [184], a systematic analysis of various factors, such as the structure of the bit and the geometry of drilled trajectory and formation character, is carried out to improve the drilling process. Some useful operational guidelines can be found in reports of the *Society of Petroleum Engineers* (*SPE*). For example, in [162], an integrated approach is proposed; the method includes the identification of safe operating windows for rotary speed and a regular inspection of drillstring components. In [62], a technique for monitoring drillstring vibrations to determine optimal drilling parameters is proposed. The proposal gives rise to a reduction of drillstring failures and of the amount of time spent on drilling, besides it increases bit life and drilling rate. In [193], various practical approaches are proposed to eliminate lateral vibrations once the whirling phenomenon is recognized; surface identification of BHA whirl is achieved through the measurement of drillstring vibrations. In [197], eight case studies involving actual BHA failures are analyzed; a method to predict lateral vibrations allows selecting operating speeds to avoid unwanted behaviors.

Even though operational guidelines are helpful in reducing string vibrations, they cannot guarantee an optimal drilling operation. As a case in point, a substantial increase of rotary speed may induce lateral vibrations causing drillstring whirling which entails chaotic displacements and collar-borehole wall impacts. Additionally, significant WoB reduction could reduce the penetration rate or even stop the perforation. These are the reasons why an effective elimination of drilling vibrations requires feedback control actions.

It is well known that special attention has been paid to the problem of eliminating stick-slip vibration in drilling systems. There is a wide range of control solutions developed by the scientific community; some recently proposed techniques are briefly described below:

- Adaptive PID controller [106]. This strategy solves a trajectory tracking problem which improves the dynamic behavior of the drilling rotary system and leads to the system stabilization.
- PID and lead–lag controllers in conjunction with genetic algorithms [142]. The proposed strategy to reduce stick-slip vibrations is validated through numerical simulations.
- Linear PI-type control [212]. The control goals are to eliminate the bit-sticking phenomena and drive the bit velocity to a desired value. The proposed method was developed using a lumped parameter model of three DOF, however, it can be generalized to treat a multidegree-of-freedom system [214].
- Nonlinear friction compensation [2]. This control strategy involves a combination of a feedback model-based compensation of friction and a Proportional–Integral (PI) controller which allows reducing the stick-slip oscillations.
- Robust μ-synthesis controller [143]. The stick-slip oscillation is reduced through this robust control technique that allows considering the modeling errors in terms of uncertainty weights. The considered nonlinear model has to be linearized around an operating point in order to use μ-synthesis.
- Time-varying sliding mode adaptive control [174]. This strategy to tackle torsional vibrations is based on a lumped parameter model of the drilling system; it takes into account existing changes of the model parameters and external disturbances.
- Integral high-order sliding mode control [131]. A novel sliding mode control law combined with a cascade control scheme is proposed for the suppression of stick-slip oscillations; the strategy considers uncertain parameters and external disturbances.
- Identification of the root cause of stick-slip vibration [307]. This strategy combines the drilling expertise with an advanced dynamics model to identify the root cause of torsional vibrations. The use of this model allows focusing on suitable solutions rather than trial and error approaches.
- Torque estimator-based control [225]. This approach consists of an automatically tuned active damping control scheme based on torque estimation to reduce the

stick-slip vibrations; it was experimentally verified using a hardware-in-the-loop assembly under both laboratory and field conditions.
- Sliding backstepping control technique [315]. Torsional vibrations are reduced through a sliding backstepping control method; a reduced order observer allows estimating the unmeasurable states of the system.
- ROP optimization algorithm [54]. The use of this optimization algorithm within an automated closed loop process has shown to be effective in increasing the ROP; furthermore, a reduction in downhole tool failures was observed resulting in a notable decrease in nonproductive time.
- Model-based control [248]. The proposed stick-slip control predicts the downhole vibration intensity to determine the optimal drilling parameters. Field tests were performed to evaluate the system and identify the dynamic model representing the drilling process.
- Self-adapting vibration damper [133]. This method aimed at preventing stick-slip vibration detects drilling dysfunctions and then adjusts the damping characteristics of the device.

Additional stick-slip control strategies are reviewed in [314].

Subsequent chapters address the design of stabilizing feedback controllers developed by the authors of this contribution.[10] The main advantage of our methods with respect to the existing ones is that they are based on infinite-dimensional models that consider the coupling between the axial and torsional dynamics of the system; therefore, they are able to eliminate the stick-slip and the bit-bounce phenomena, guaranteeing the system stability.

The following chapter presents some low-order control techniques to reduce drilling vibrations. Three classical solutions will be reviewed: a PI-like control technique,the well-known soft torque controller, and the torsional rectification method. Furthermore, based on the bifurcation analysis studied in Chap. 6, a pair of control strategies aimed at suppressing the stick-slip and bit-bounce are designed: a delayed proportional and a delayed PID controller.

[10] Except for the angular velocity regulation controller, the torsional rectification control and the soft torque technique reviewed in Sects. 9.1 and 9.2.

Chapter 9
Low-Order Controllers

Drillstring platforms usually operate with reduced-order simple control laws which make the drillstring rotate at a constant speed; only a few of them include controllers to tackle the vibration problem.

This chapter presents some of the most frequently used low-order controllers to regulate the angular velocity and tackle the stick-slip phenomenon.

First, a PI-like control law to maintain a constant rotary speed is presented. The controller is designed under the basis of a two DOF lumped parameter model; its gains are adjusted by means of the classic two-time-scales separation method [50].

Next, two classic solutions to counteract the stick-slip phenomenon are discussed: the soft torque and the torsional rectification controllers. The torsional rectification control constitutes an improved version of the classical PI speed controller; it allows the absorption of the energy at the top extremity to avoid the reflection of torsional waves back down to the drillstring. The soft torque is one of the most popular vibration control methods; it has the form of a standard speed controller but includes a high-pass filtered torque signal. Both control methods are evaluated in this chapter; furthermore, an analytic treatment to characterize the torsional energy reflection provided by the torsional rectification controller is developed.

Finally, a novel technique to reduce the stick-slip and bit-bounce is introduced. Based on the bifurcation analysis of the drilling system, a pair of low-order controllers aimed at eliminating axial and torsional coupled vibrations are designed: delayed proportional and delayed PID. The performances of the proposed control techniques are highlighted through simulation of the coupled system.

9.1 Angular Velocity Regulation

A simple technique to overcome the problem of angular speed regulation is proposed in [51]. The low-order controller presented below is designed on the basis of the following two DOF lumped parameter model describing the drillstring torsional dynamics:

M.B. Saldivar Márquez et al., *Analysis and Control of Oilwell Drilling Vibrations*, Advances in Industrial Control, DOI 10.1007/978-3-319-15747-4_9

$$
\begin{aligned}
I_p\ddot{\Phi}_p + c\left(\dot{\Phi}_p - \dot{\Phi}_b\right) + k\left(\Phi_p - \Phi_b\right) + d_p\dot{\Phi}_p &= u_T \\
I_b\ddot{\Phi}_b - c\left(\dot{\Phi}_p - \dot{\Phi}_b\right) - k\left(\Phi_p - \Phi_b\right) + d_b\dot{\Phi}_b &= -T(\dot{\Phi}_b),
\end{aligned}
\tag{9.1}
$$

where $\dot{\Phi}_p$ and $\dot{\Phi}_b$ denote the angular velocity at the top and bottom extremities, respectively, u_T is the motor torque provided by the rotary table and $T(\dot{\Phi}_b)$ is the frictional torque describing the bit-rock interaction. Notations I_p, I_b, d_p, d_b, c, and k defined as for model (2.1).

The angular velocity regulation control law is defined as:

$$
u_T = k_1\left(\Omega_0 - \dot{\Phi}_p\right) + k_2\int\left(\Omega_0 - \dot{\Phi}_p\right)dt - k_3\left(\dot{\Phi}_p - \dot{\Phi}_b\right). \tag{9.2}
$$

This reduced-order controller, aimed at regulating the rotational velocity to a certain reference velocity Ω_0, is inspired by the one proposed in [56].

Figures 9.1 and 9.2 show the control performance for $\Omega_0 = 5\,\mathrm{rad\,s^{-1}}$ and $\Omega_0 = 10\,\mathrm{rad\,s^{-1}}$, respectively. The numerical values of the system parameters are given in Table C.1 of Appendix C.

Notice the controller (9.2) is not able to regulate the angular velocity at the bottom end for $\Omega_0 = 5\,\mathrm{rad\,s^{-1}}$.

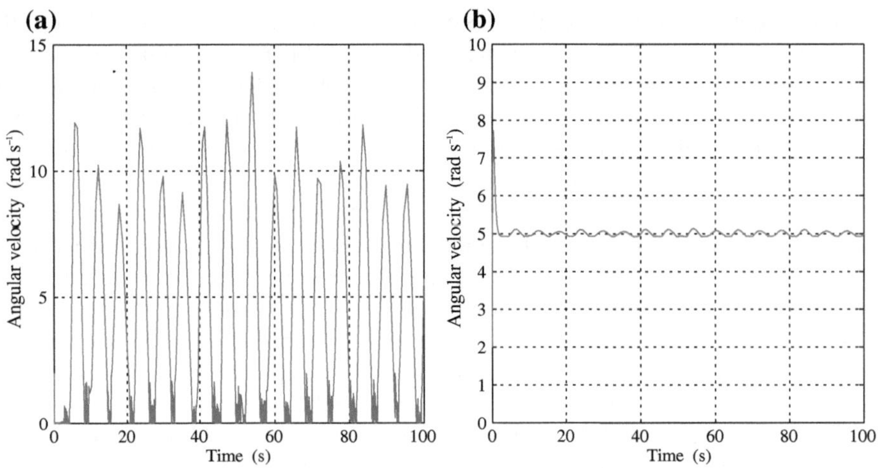

Fig. 9.1 Simulation of the lumped parameter model (9.1) coupled to the frictional torque (3.14) with the friction coefficient given in (3.15). Trajectories of the drilling system in closed loop with the speed regulation controller (9.2) with $k_1 = 15{,}725$, $k_2 = 30{,}576$, $k_3 = 194$, for a reference angular velocity $\Omega_0 = 5\,\mathrm{rad\,s^{-1}}$. **a** Angular velocity at the bottom extremity. **b** Angular velocity at the upper extremity

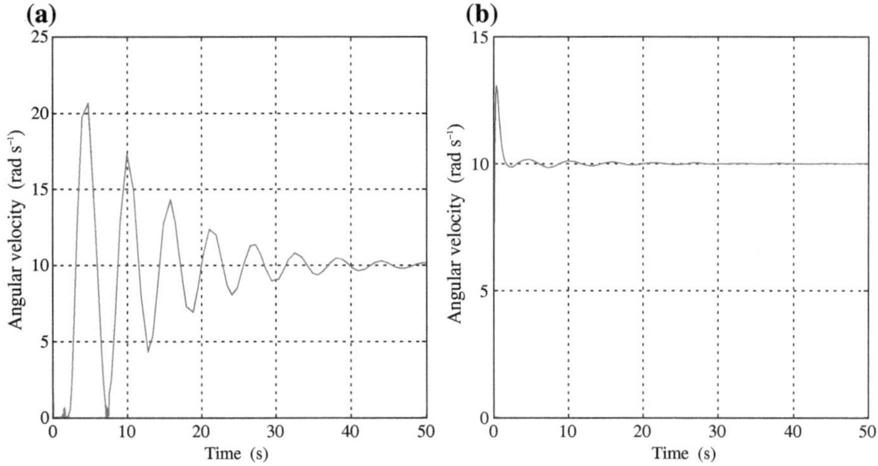

Fig. 9.2 Simulation of the lumped parameter model (9.1) coupled to the frictional torque (3.14) with the friction coefficient given in (3.15). Trajectories of the drilling system in closed loop with the speed regulation controller (9.2) with $k_1 = 15{,}725$, $k_2 = 30{,}576$, $k_3 = 194$, for a reference angular velocity $\Omega_0 = 10\,\mathrm{rad\,s^{-1}}$. **a** Angular velocity at the bottom extremity. **b** Angular velocity at the upper extremity

9.1.1 Synthesis of the Controller

There are different methods for adjusting the controller gains; one of them is the classic two-time-scales separation method [50] detailed below.

Following [51], a variable z and a constant κ defined as:

$$z = k\left(\Phi_p - \Phi_b\right), \quad \kappa^2 = \frac{1}{k},$$

are introduced.

According to the drilling model (9.1), the dynamics of Φ_p and z are described by:

$$\begin{aligned} u_T &= I_p\ddot{\Phi}_p + c\kappa^2\dot{z} + z + d_p\dot{\Phi}_p \\ \kappa^2\ddot{z} &= \ddot{\Phi}_p - \ddot{\Phi}_b = -\left(\frac{1}{I_p} + \frac{1}{I_b}\right)z - \kappa^2\left(\frac{c}{I_p} + \frac{c}{I_b}\right)\dot{z} \\ &\quad -\frac{d_p}{I_p}\dot{\Phi}_p + \frac{d_b}{I_b}\dot{\Phi}_b + \frac{1}{I_p}u_T + \frac{1}{I_b}T(\dot{\Phi}_b), \end{aligned}$$

which can be written as:

$$
\begin{aligned}
u_T &= I_p\ddot{\Phi}_p + c\kappa^2\dot{z} + z + d_p\dot{\Phi}_p \\
\kappa^2\ddot{z} &= -\frac{z}{I_{eq}} - \kappa^2\left(\frac{c}{I_{eq}} - \frac{d_b}{I_b}\right)\dot{z} + \left(\frac{d_b}{I_b} - \frac{d_p}{I_p}\right)\dot{\Phi}_p + \frac{u_T}{I_p} + \frac{T(\dot{\Phi}_b)}{I_b},
\end{aligned} \tag{9.3}
$$

with $I_{eq} = \left(\frac{1}{I_p} + \frac{1}{I_b}\right)^{-1}$.

Since the torsional stiffness takes large values, it is assumed that $\kappa^2 \simeq 0$; then it follows that

$$
u_{Ts} = I_p\ddot{\Phi}_p + z + d_p\dot{\Phi}_p \tag{9.4}
$$

$$
0 = -\frac{z}{I_{eq}} + \left(\frac{d_b}{I_b} - \frac{d_p}{I_p}\right)\dot{\Phi}_p + \frac{u_{Ts}}{I_p} + \frac{T(\dot{\Phi}_b)}{I_b}, \tag{9.5}
$$

where the rotary torque u_T is split into slow (u_{Ts}) and fast (u_{Tf}) modes:

$$
u_T = u_{Ts} + \kappa u_{Tf}. \tag{9.6}
$$

From (9.5), we have that

$$
z = I_{eq}\left[\left(\frac{d_b}{I_b} - \frac{d_p}{I_p}\right)\dot{\Phi}_p + \frac{u_{Ts}}{I_p} + \frac{T(\dot{\Phi}_b)}{I_b}\right],
$$

by substituting the above equation into (9.4) we obtain:

$$
\ddot{\Phi}_p + \left(\frac{d_b + d_p}{I_p + I_b}\right)\dot{\Phi}_p = \frac{1}{I_p + I_b}u_{Ts} - \frac{1}{I_p + I_b}T(\dot{\Phi}_b).
$$

Assume that u_{Ts} is chosen as:

$$
u_{Ts} = \left(k_1 + \frac{k_2}{s}\right)(\Omega_0 - \dot{\Phi}_p),
$$

and define the variable $\tilde{\Phi}_p$ as follows

$$
\dot{\tilde{\Phi}}_p = \Omega_0 - \dot{\Phi}_p
$$

then, $\tilde{\Phi}_p = \frac{\Omega_0 - \dot{\Phi}_p}{s}$ and $\ddot{\tilde{\Phi}}_p = -\ddot{\Phi}_p$.

For $(d_b + d_p) \ll (I_p + I_b)$, we have that

$$
\ddot{\tilde{\Phi}}_p + \left(\frac{k_1}{I_p + I_b}\right)\dot{\tilde{\Phi}}_p + \left(\frac{k_2}{I_p + I_b}\right)\tilde{\Phi}_p = \frac{1}{I_p + I_b}T(\dot{\Phi}_b).
$$

Then, by imposing a damping δ_r and natural frequency ω_n in the closed-loop dynamics, the values of k_1 and k_2 can be computed from

$$k_1 = (I_p + I_b)\left(2\delta_r\omega_n - \frac{d_b + d_p}{I_p + I_b}\right)$$
$$k_2 = (I_p + I_b)\,\omega_n^2.$$

Substituting (9.6) into (9.3) gives:

$$\kappa^2\ddot{z} = -\frac{z}{I_{eq}} - \kappa^2\left(\frac{c}{I_{eq}} - \frac{d_b}{I_b}\right)\dot{z} + \left(\frac{d_b}{I_b} - \frac{d_p}{I_p}\right)\dot{\Phi}_p + \frac{u_{Ts}}{I_p} + \frac{\kappa}{I_p}u_{Tf} + \frac{T(\dot{\Phi}_b)}{I_b}. \quad (9.7)$$

Following the two-time-scales-separation approach, it is assumed that the slow mode has reached its steady-state value, thus, Eq. (9.7) can be rewritten as:

$$\kappa^2\ddot{z} = -\frac{z}{I_{eq}} - \kappa^2\left(\frac{c}{I_{eq}} - \frac{d_b}{I_b}\right)\dot{z} + \frac{\kappa}{I_p}u_{Tf} + \frac{T(\dot{\Phi}_b)}{I_b} + \rho_p\left(\dot{\Phi}_p^*, \Phi_p^*\right), \quad (9.8)$$

where the superscript * denotes the steady-state value and

$$\rho_p\left(\dot{\Phi}_p^*, \Phi_p^*\right) = \left(\frac{d_b}{I_b} - \frac{d_p}{I_p}\right)\dot{\Phi}_p^* + \frac{u_{Ts}\left(\dot{\Phi}_p^*, \Phi_p^*\right)}{I_p}$$
$$= \frac{z^*}{I_{eq}} - \frac{T(\dot{\Phi}_b)}{I_b}.$$

Defining the fast error coordinate as $\zeta = z - z^*$, Eq. (9.8) can be rewritten as:

$$\ddot{\zeta} + \left(\frac{c}{I_{eq}} + \frac{d_b}{I_b}\right)\dot{\zeta} + \frac{1}{\kappa^2 I_{eq}}\zeta = \frac{1}{I_p\kappa}u_{Tf}.$$

By choosing $u_{Tf} = -k_3\dot{\zeta}$, we obtain:

$$\ddot{\zeta} + \left(\frac{c}{I_{eq}} + \frac{d_b}{I_b} + \frac{k_3}{I_p\kappa}\right)\dot{\zeta} + \frac{1}{\kappa^2 I_{eq}}\zeta = 0,$$

then, prescribing a damping value for the torsion dynamics δ_{tor}, the value of k_3 can be computed as follows:

$$k_3 = \frac{I_p}{\sqrt{k}}\left(2\delta_{tor}\sqrt{k}I_{eq} - \frac{c}{I_{eq}} - \frac{d_b}{I_b}\right).$$

9.2 Drilling Vibration Control

The control law (9.2) is designed to regulate the drillstring angular velocity, but it does not consider the vibration problem. This section presents a pair of reduced-order torque feedback controllers aimed at maintaining a constant rotary speed while reducing torsional vibrations.

9.2.1 Torsional Rectification Control

A torsional rectification method to suppress the stick-slip phenomenon is proposed in [284]; the strategy takes advantage of the fact that the general solution of the wave equation describing torsional drillstring vibrations allows the identification of both "up" and "down" moving components.

The underlying idea consists in maintaining the energy of the "down" traveling wave constant in the presence of the nonlinear boundary conditions describing the bit-rock frictional interface.

Without loss of generality, it is assumed that the speed of torsional waves is one; the torsional excitations of the drilling system, described by the drillstring angular displacement $\Phi(s,t)$, are modeled by the wave equation:

$$\frac{\partial^2 \Phi}{\partial t^2}(s,t) = \frac{\partial^2 \Phi}{\partial s^2}(s,t), \quad 0 \leqslant s \leqslant 1 \tag{9.9}$$

with boundary conditions

$$\frac{\partial^2 \Phi}{\partial t^2}(0,t) = F_0\left(\frac{\partial \Phi}{\partial t}(0,t), \frac{\partial \Phi}{\partial s}(0,t), \Phi(0,t), t\right)$$
$$\frac{\partial^2 \Phi}{\partial t^2}(1,t) = F_1\left(\frac{\partial \Phi}{\partial t}(1,t), \frac{\partial \Phi}{\partial s}(1,t), \Phi(1,t), t\right)$$

where $s = 0$ denotes the connection of the drillstring with the rotary table, and $s = 1$ the bottom extremity. The functions F_0 and F_1 are determined by the top drive and friction torques at the upper and bottom extremities, respectively. The general solution of the wave equation is given by

$$\Phi(s,t) = \varphi(t+s) + \psi(t-s),$$

where φ and ψ are arbitrary continuously differentiable real-valued functions, with φ representing an arbitrary up-traveling wave and ψ an arbitrary down-traveling wave. The time and spatial derivatives of $\Phi(s,t)$ are

$$\frac{\partial \Phi}{\partial t}(s,t) = \dot{\varphi}(t+s) + \dot{\psi}(t-s),$$
$$\frac{\partial \Phi}{\partial s}(s,t) = \dot{\varphi}(t+s) - \dot{\psi}(t-s),$$

respectively. Since the contact torque at any point "s" of the drillstring is proportional to $\frac{\partial \Phi}{\partial s}(s,t)$, $\dot{\psi}$ represents the transmission of torque to the BHA. In order to drive the angular velocity to a prescribed constant rotary speed, $\dot{\psi}$ must be maintained close to a constant value. To this end, the quantity $\Psi(t)$ defined by

$$\Psi(t) = \frac{\partial \Phi}{\partial t}(0,t) - \frac{\partial \Phi}{\partial s}(0,t) = 2\dot{\psi}(t) \tag{9.10}$$

must be monitored.

A Newton-type equation is chosen to describe the top boundary condition:

$$\frac{\partial^2 \Phi}{\partial t^2}(0,t) = G_{top}\frac{\partial \Phi}{\partial s}(0,t) + u_T(t) \tag{9.11}$$

where G_{top} is proportional to the torsional rigidity of the drillstring ($G_{top} = \frac{GJ}{LI_T}$ according to equation (2.17)).

A commonly used control law to maintain a constant angular velocity Ω_0 is given by:

$$u_T(t) = k_p\dot{\xi}(t) + k_i\xi(t), \tag{9.12}$$

where $k_p > 0$ and k_i are the proportional and integral gain variables, respectively, and

$$\xi(t) = \Omega_0 t - \Phi(0,t) + \xi_0$$
$$\dot{\xi}(t) = \Omega_0 - \frac{\partial \Phi}{\partial t}(0,t),$$

where ξ_0 denotes the displacement of the drillstring at the upper extremity from its reference value.

In [284], an improved control strategy is proposed; the contact torque between the drillstring and the rotary table can be monitored by introducing a compensating drive torque that rectifies the uptravelling torsional waves on the drillstring. The control law is written as:

$$u_T(t) = k_p\dot{\xi}(t) + k_i\xi(t) - \lambda\Psi(t), \tag{9.13}$$

where $\lambda \geqslant 0$ and $\Psi(t)$ is given in (9.10). The torsional rectification control law (9.13) is evaluated in [284] for different values of the control gains k_p, k_i, and λ; besides, the effect of an incident torsional harmonic wave on the rotary table is explored. By analytical and numerical analyses it is concluded that the reflected torsional energy from the upper extremity can be decreased by increasing the control parameter λ.

9.2.2 Soft Torque Control

The soft torque, introduced in [128], is a control approach widely used in the drilling industry to tackle torsional vibrations. This controller has the form of the standard speed controller (9.12) but it includes a high-pass filtered torque signal, i.e.,

$$u_T(t) = \kappa_p \dot{\tilde{\xi}}(t) + \kappa_i \tilde{\xi}(t), \tag{9.14}$$

with

$$\begin{aligned} \tilde{\xi}(t) &= \Omega_0 t - h\int T_f(t)dt - \Phi(0,t) + \xi_0, \\ \dot{\tilde{\xi}}(t) &= \Omega_0 - hT_f(t) - \frac{\partial \Phi}{\partial t}(0,t), \end{aligned} \tag{9.15}$$

where h is an additional control parameter and T_f is defined as:

$$T_f(t) \equiv T_{\text{contact}}(t) - T_c(t),$$

T_c denotes the output of a low-pass filter applied to the contact torque:

$$T_{\text{contact}}(t) = -G_{top} \frac{\partial \Phi}{\partial s}(0,t),$$

measured at the upper extremity. An AC low-pass filter is modeled by:

$$\dot{T}_c(t) = \omega_c \left(T_{\text{contact}}(t) - T_c(t)\right),$$

where ω_c is the cut-off angular frequency.

By considering that

$$\int T_f(t)\, dt = \frac{1}{\omega_c} \int \dot{T}_c(t)dt = \frac{1}{\omega_c} T_c(t),$$

Equation (9.15) can be rewritten as:

$$\tilde{\xi}(t) = \Omega_0 t - \frac{h}{\omega_c} T_c(t) - \Phi(0,t) + \xi_0.$$

9.2.3 Torsional Energy Reflection and Stick-Slip Reduction

The performance of the torsional rectification controller (9.13) can be evaluated through the analysis of the energy reflection at the top extremity. To this end, consider the following solution to the Eq. (9.9) which describes a harmonic wave:

$$\Phi(s,t) = A_{\varphi,\omega}\sin(\omega(t+s)) + A_{\psi,\omega}\sin(\omega(t-s)+\alpha_\omega) + \Omega_0 t + C_0 s$$

where ω is the angular frequency, $A_{\varphi,\omega}$ is the amplitude of the wave incident on the rotary from below, and $A_{\psi,\omega}$ is the amplitude of the reflected wave. The constants C_0, α_ω and the ratio $A_{\psi,\omega}/A_{\varphi,\omega}$ are determined from Eqs. (9.11)–(9.13) as follows:

$$\begin{aligned}
C_0 &= \frac{\lambda\Omega_0 - k_i\xi_0}{G_{top}+\lambda} \\
\alpha_\omega &= \tan^{-1}\left(\frac{2\omega\left(k_i-\omega^2\right)\left(G_{top}+\lambda\right)}{\omega^4-\left(2k_i+\left(G_{top}-k_p\right)\eta\right)\omega^2+k_i^2}\right) \\
\frac{A_{\psi,\omega}}{A_{\varphi,\omega}} &= \frac{k_i-\omega^2}{\left(k_i-\omega^2\right)\cos(\alpha_\omega)+\omega\eta\sin(\alpha_\omega)} \\
\eta &= G_{top}+k_p+2\lambda.
\end{aligned}$$

The reflection coefficient is thus defined as

$$r_\omega = \left|\frac{A_{\psi,\omega}}{A_{\varphi,\omega}}\right| = \sqrt{\frac{\omega^4+a\omega^2+k_i^2}{\omega^4+b\omega^2+k_i^2}}$$

where

$$a = (G_{top}-k_p)^2 - 2k_i, \quad b = \eta^2 - 2k_i.$$

The absorption of vibrational energy is greater when the reflection coefficient takes small values. It is easy to prove that the torsional energy reflected from the rotary table is reduced by increasing the control parameter λ. This is shown in Fig. 9.3 which depicts the shape of the reflection coefficient r_ω as a function of the angular frequency ω for different values of λ. Different datasets are considered: Fig. 9.3a uses the parameters $kp = 1.314$, $k_i = 0.08336$, $G_{top} = 0.4836$, and Fig. 9.3b the parameters $k_p = 0.3658$, $k_i = 0.1672$, $G_{top} = 0.5765$.

Notice that as the torsional rectification feedback λ is increased, the reflection coefficient r_ω uniformly decreases; this behavior corresponds to an increase of torsional vibration absorption.

A similar analysis for evaluating the performance of the soft torque controller (9.14) is presented in [284]; it is concluded that the shape of the reflection coefficient does not varies uniformly with the control parameter h. Besides, it is proven that the torsional rectification controller exhibits an improved performance compared with the soft torque one.

In order to asses the ability of the torque feedback controllers (9.12), (9.13) and (9.14) in eliminating the stick-slip phenomenon, we propose to test them by using two different modeling approaches. A lumped parameter model describing the torsional drilling behavior coupled to the simplified ToB model given in (3.10)

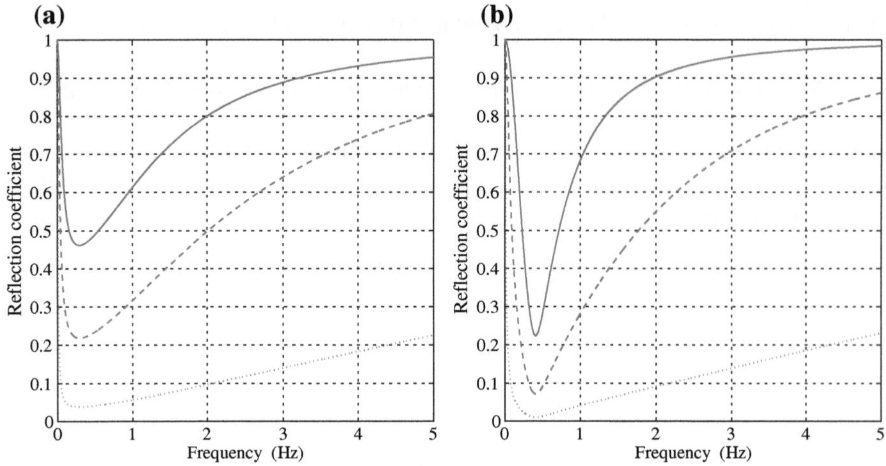

Fig. 9.3 Reflection coefficient r_ω as a function of the frequency ω with $\lambda = 0$ (*solid line*), $\lambda = 1$ (*dashed line*) and $\lambda = 10$ (*dotted line*). **a** Shape of r_ω for the parameters $kp = 1.314$, $k_i = 0.08336$, $G_{top} = 0.4836$. **b** Shape of r_ω for the parameters $k_p = 0.3658$, $k_i = 0.1672$, $G_{top} = 0.5765$

is first considered. Next, we use a neutral-type time-delay model derived from a distributed parameter representation of the coupled torsional-axial drilling dynamics (see Sects. 2.2 and 2.3), subject to the ToB model given in (3.14) with the friction coefficient given in (3.15), which is inspired by the Karnopp friction law.

According to the geometrical formulation of elasticity laws established in [10], the boundary conditions of the wave equation can be modeled as follows:

$$\frac{\partial^2 \Phi}{\partial t^2}(0,t) - G_{top}\frac{\partial \Phi}{\partial s}(0,t) - u_T(t) = 0$$
$$\frac{\partial^2 \Phi}{\partial t^2}(1,t) + G_{bit}\frac{\partial \Phi}{\partial s}(1,t) + T\left(\frac{\partial \Phi}{\partial t}(1,t)\right) = 0$$

where T describes the frictional torque at the bit, $u_T(t)$ the motor torque considered as a control input. The top drive and BHA inertias are scaled into the constants $G_{top} = \frac{GJ}{LI_T}$ and $G_{bit} = \frac{GJ}{LI_B}$.

An approximation of the drillstring torsional dynamics is obtained by ignoring the infinite-dimensional character of the system and considering it as a torsional pendulum that couples the top drive torque with the frictional torque arising from the bit-rock contact. By defining the torsional displacements at the top and bottom extremities $\Phi(0,t)$, $\Phi(1,t)$ as Φ_p and Φ_b, respectively, and by considering $\frac{\partial \Phi}{\partial s}(1,t) \simeq \Phi_b - \Phi_p$, the equations of motion are given by:

$$\ddot{\Phi}_p + G_{top}\left(\Phi_p - \Phi_b\right) - u_T(t) = 0$$
$$\ddot{\Phi}_b + G_{bit}\left(\Phi_b - \Phi_p\right) + T\left(\dot{\Phi}_b\right) = 0. \tag{9.16}$$

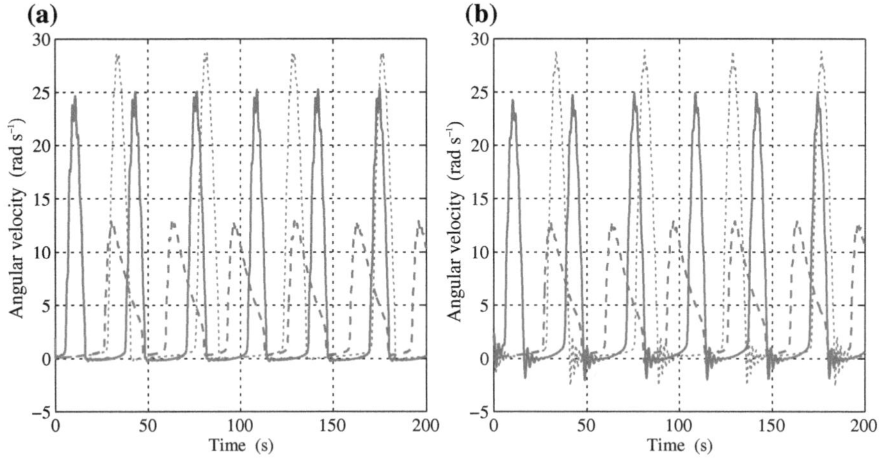

Fig. 9.4 Simulation of the simplified model (9.16) with the frictional torque given in (9.17). Trajectories of the drilling system in closed loop with the standard speed controller (9.12) (*dotted line*), the torsional rectification control law (9.13) (*dashed line*), and the soft torque control (9.14) (*solid line*) with $k_p = 0.3658$, $k_i = 0.1672$, $\lambda = 1$, for a reference angular velocity $\Omega_0 = 5\,\text{rad}\,\text{s}^{-1}$. **a** Angular velocity at the bottom extremity. **b** Angular velocity at the upper extremity

As discussed in Chap. 3, the frictional torque T can be approximated by the following function (see Eq. (3.10)):

$$T(\dot{\Phi}_b(t)) = \frac{2\bar{k}p\dot{\Phi}_b(t)}{\dot{\Phi}_b^2(t) + \bar{k}^2}, \quad \bar{k} > 0,\ p > 0. \tag{9.17}$$

Figures 9.4 and 9.5 show the system response to the torque feedback controllers: standard (9.12), torsional rectification (9.13), and soft torque (9.14) for $\Omega_0 = 5\,\text{rad}\,\text{s}^{-1}$ and $\Omega_0 = 10\,\text{rad}\,\text{s}^{-1}$, respectively. The numerical values of the system parameters are given in Table C.1 of Appendix C.

Notice that, as remarked in [284], the torsional rectification controller provides superior performance when the simplified model (9.16) with the frictional torque given in (9.17) is considered.

Consider now, the neutral-type time-delay model of the torsional-axial coupled drilling dynamics given in (2.32)–(2.34). Consider also the frictional torque on bit model, derived from the Karnopp friction law, given in (3.14) with the friction coefficient given in (3.15). Obviously, this system representation constitutes a more reliable description of the dynamics taking place during the drilling performance. Figures 9.6 and 9.7 show the system trajectories in closed loop with the controllers (9.12)–(9.14). Observe that, under this modeling approach, the soft torque control method performs better than the standard speed controller and than the torsional rectification one; in fact, these last ones are unable to eliminate the stick-slip for low angular velocities.

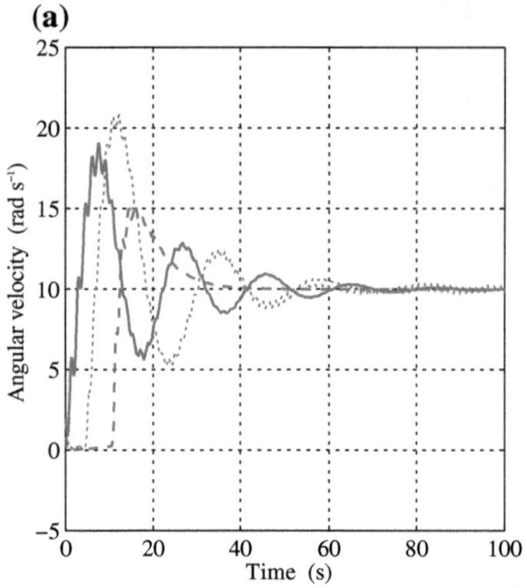

Fig. 9.5 Simulation of the simplified model (9.16) with the frictional torque given in (9.17). Trajectories of the drilling system in closed loop with the standard speed controller (9.12) (*dotted line*), the torsional rectification control law (9.13) (*dashed line*), and the soft torque control (9.14) (*solid line*) with $k_p = 0.3658$, $k_i = 0.1672$, $\lambda = 1$, for a reference angular velocity $\Omega_0 = 10\,\text{rad}\,\text{s}^{-1}$. **a** Angular velocity at the bottom extremity. **b** Angular velocity at the upper extremity

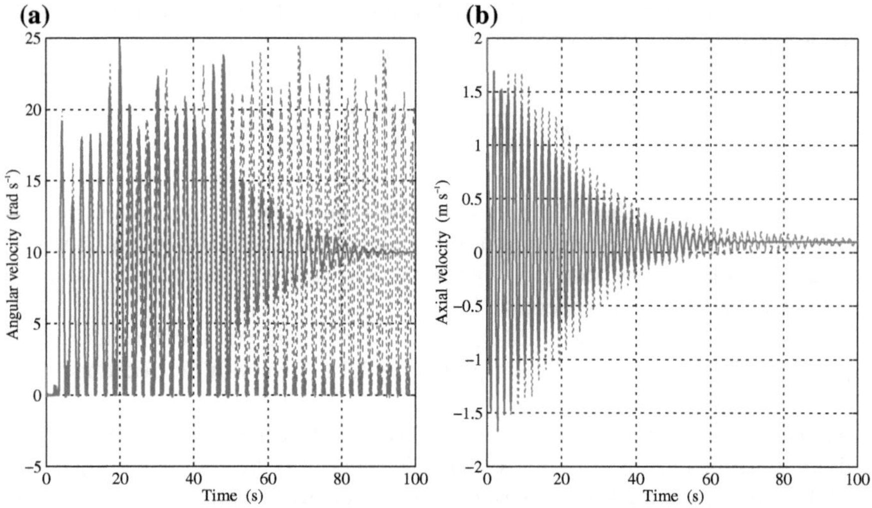

Fig. 9.6 Simulation of the neutral-type time-delay model (2.32)–(2.34) coupled to the frictional torque model (3.14) with the friction coefficient given in (3.15). Trajectories of the drilling system in closed loop with the standard speed controller (9.12) (*dotted line*), the torsional rectification control law (9.13) (*dashed line*), and the soft torque control (9.14) (*solid line*) with $k_p = 0.3658$, $k_i = 0.1672$, $\lambda = 0.01$, for a reference angular velocity $\Omega_0 = 10\,\text{rad}\,\text{s}^{-1}$. **a** Angular velocity at the bottom extremity. **b** Axial velocity at the bottom extremity

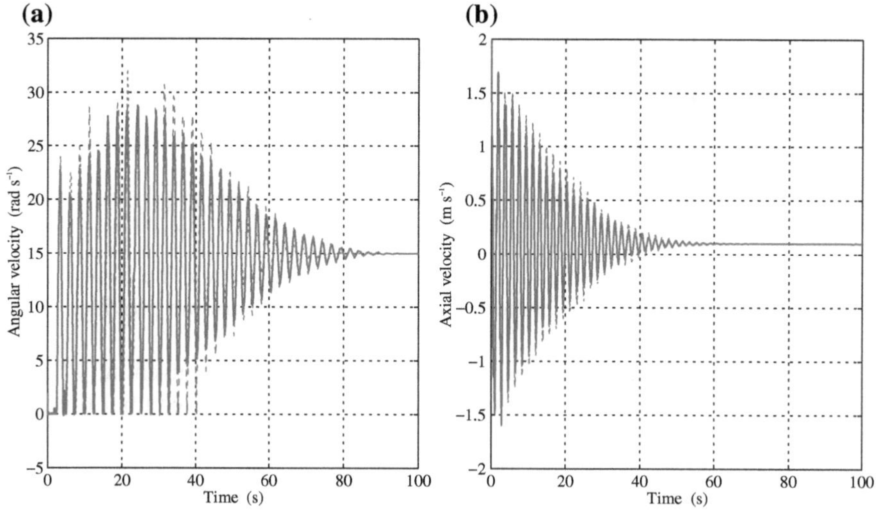

Fig. 9.7 Simulation of the neutral-type time-delay model (2.32)–(2.34) coupled to the frictional torque model (3.14) with the friction coefficient given in (3.15). Trajectories of the drilling system in closed loop with the standard speed controller (9.12) (*dotted line*), the torsional rectification control law (9.13) (*dashed line*), and the soft torque control (9.14) (*solid line*) with $k_p = 0.3658$, $k_i = 0.1672$, $\lambda = 0.01$, for a reference angular velocity $\Omega_0 = 15\,\mathrm{rad\,s^{-1}}$. **a** Angular velocity at the bottom extremity. **b** Axial velocity at the bottom extremity

9.3 Bifurcation Analysis-Based Controllers

This section presents a pair of control schemes to tackle both axial and torsional vibrations occurring along a rotary oilwell drilling system. It will be shown that the implementation of low-order delayed feedback controllers allows reducing undesired vibrations leading to an acceptable drilling performance.

Consider the following coupled torsional-axial model:

$$\begin{cases} \dfrac{\partial^2 U}{\partial s^2}(s,t) = c^2 \dfrac{\partial^2 U}{\partial t^2}(s,t), \quad c = \sqrt{\dfrac{\rho_a}{E}} \\ E\Gamma \dfrac{\partial U}{\partial s}(0,t) = \alpha \dfrac{\partial U}{\partial t}(0,t) - u_H(t) \\ E\Gamma \dfrac{\partial U}{\partial s}(L,t) = -M_B \dfrac{\partial^2 U}{\partial t^2}(L,t) + F\left(\dfrac{\partial U}{\partial t}(L,t)\right) \end{cases} \tag{9.18}$$

and

$$\begin{cases} \dfrac{\partial^2 \Phi}{\partial s^2}(s,t) = \tilde{c}^2 \dfrac{\partial^2 \Phi}{\partial t^2}(s,t), \quad \tilde{c} = \sqrt{\dfrac{\rho_a}{G}} \\ GJ \dfrac{\partial \Phi}{\partial s}(0,t) = \beta \dfrac{\partial \Phi}{\partial t}(0,t) - u_T(t) \\ GJ \dfrac{\partial \Phi}{\partial s}(L,t) = -I_B \dfrac{\partial^2 \Phi}{\partial t^2}(L,t) - \tilde{F}\left(\dfrac{\partial U}{\partial t}(L,t)\right) \end{cases} \tag{9.19}$$

where U and Φ stand for the axial and the torsional position, and the control inputs u_H and u_T correspond to the brake motor control and the torque provided by the rotary table.

By using the d'Alembert transformation, the system of PDE given in (9.18) and (9.19) is reduced to the following system of Neutral Delay Differential Equations (NDDE):

$$\begin{cases} \ddot{U}_b(t) - \tilde{\Upsilon}_n \ddot{U}_b(t-2) = -\tilde{\Psi}_n \dot{U}_b(t) - \tilde{\Upsilon}_n \tilde{\Psi}_n \dot{U}_b(t-2) + \dfrac{1}{M_B} F\big(\dot{U}_b(t)\big) \\ \qquad - \dfrac{1}{M_B} \tilde{\Upsilon}_n F\big(\dot{U}_b(t-2)\big) + \tilde{\Pi}_n u_H(t-1), \\ \ddot{\Phi}_b(t) - \Upsilon_n \ddot{\Phi}_b(t-2\tau) = -\Psi_n \dot{\Phi}_b(t) - \Upsilon_n \Psi_n \dot{\Phi}_b(t-2\tau) + \dfrac{1}{I_B} \tilde{F}\big(\dot{U}_b(t)\big) \\ \qquad - \dfrac{1}{I_B} \Upsilon_n \tilde{F}\big(\dot{U}_b(t-2\tau)\big) + \Pi_n u_T(t-\tau), \end{cases} \tag{9.20}$$

with

$$\tilde{\Pi}_n = \frac{2\tilde{\Psi}_n}{\alpha+1}, \quad \tilde{\Upsilon}_n = \frac{\alpha-1}{\alpha+1}, \quad \tilde{\Psi}_n = \frac{1}{M_B},$$
$$\Pi_n = \frac{2\tilde{c}GJ}{I_B\,(cE\Gamma\beta + \tilde{c}GJ)}, \quad \Upsilon_n = \frac{cE\Gamma\beta - \tilde{c}GJ}{cE\Gamma\beta + \tilde{c}GJ}, \quad \Psi_n = \frac{\tilde{c}GJ}{cE\Gamma I_B},$$

where τ is the ratio of the speeds $\tau = \frac{\tilde{c}}{c}$, and U_b and Φ_b are, respectively, the axial and torsional bit positions, $U_b(t) = U(L,t)$, $\Phi_b(t) = \Phi(L,t)$. For the sake of simplicity, we consider functions F and $\tilde{F}$ of the form: $z \mapsto p\bar{k}\,z/(\bar{k}^2 z^2 + \zeta)$ where the parameters p, $\bar{k}$, and ζ are positive constants related to the sharpness of the top angle of the friction force, satisfying $0 < \zeta \ll 1$ and $0 < \bar{k} < 1$, and the constant p provides the amplitude of the friction force.

Following the ideas introduced in Chap. 6, the qualitative analysis based on the center manifold theorem [52] and normal forms theory [121] is applied to reduce the model to a singularly perturbed system of ordinary differential equations (ODE) by means of a spectral projection.

9.4 Delayed Proportional Feedback Controller

Setting $x = (x_1, x_2, x_3, x_4)^T$, $x_1 = U_b$, $x_2 = \Phi_b$, $x_3 = \dot{U}_b$, and $x_4 = \dot{\Phi}_b$, and defining:

$$\begin{aligned} u_H(t) &= p_H\, U_b(t-1) \\ u_T(t) &= p_\Omega\, \Phi_b(t-\tau), \end{aligned} \tag{9.21}$$

the drilling model can be rewritten as:

$$\begin{cases} \dot{x}(t) = D_1\,\dot{x}(t-2) + D_2\,\dot{x}(t-2\tilde{c}/c) + A_0\,x(t) + A_1\,x(t-2) \\ \qquad + A_2\,x(t-2\tilde{c}/c) + \mathscr{F}(x(t), x(t-2), x(t-2\tilde{c}/c)) \end{cases} \tag{9.22}$$

where $\mathscr{F}$ is the nonlinear part of system (9.20) with the matrices defined below:

$$D_1 = \begin{bmatrix} 0_2 & 0_2 \\ 0_2 & \tilde{d}_1 \end{bmatrix}, \quad D_2 = \begin{bmatrix} 0_2 & 0_2 \\ 0_2 & \tilde{d}_2 \end{bmatrix}, \quad A_0 = \begin{bmatrix} 0_2 & I_2 \\ 0_2 & \tilde{a}_0 \end{bmatrix},$$
$$A_1 = \begin{bmatrix} 0_2 & 0_2 \\ \tilde{p}_H & \tilde{a}_1 \end{bmatrix}, \quad A_2 = \begin{bmatrix} 0_2 & 0_2 \\ \tilde{p}_\Omega & \tilde{a}_2 \end{bmatrix},$$

where 0_2 and I_2 are the zero and the identity 2×2 matrices, and

$$\tilde{d}_1 = \begin{bmatrix} \tilde{\Upsilon}_n & 0 \\ 0 & 0 \end{bmatrix}, \quad \tilde{d}_2 = \begin{bmatrix} 0 & 0 \\ 0 & \Upsilon_n \end{bmatrix}, \quad \tilde{p}_H = \begin{bmatrix} p_H & 0 \\ 0 & 0 \end{bmatrix}, \quad \tilde{p}_\Omega = \begin{bmatrix} 0 & 0 \\ 0 & p_\Omega \end{bmatrix},$$

$$\tilde{a}_0 = \begin{bmatrix} -\frac{\tilde{\Psi}_n(p\bar{k}+\zeta)}{\zeta} & 0 \\ \frac{pk}{J\zeta} & -\Psi_n \end{bmatrix}, \quad \tilde{a}_1 = \begin{bmatrix} -\frac{\tilde{\Upsilon}_n\tilde{\Psi}_n(p\bar{k}+\zeta)}{\zeta} & 0 \\ 0 & 0 \end{bmatrix}, \quad \tilde{a}_2 = \begin{bmatrix} 0 & 0 \\ -\frac{\Upsilon_n p\bar{k}}{\zeta I_B} & -\Upsilon_n\Psi_n \end{bmatrix}.$$

Setting the numerical values of the physical parameters given in Table C.1 of Appendix C, the following proposition is established.

Proposition 9.4.1 *The following properties are established for system* (9.22), *depending on the values of the gains* p_H *and* p_Ω *in* (9.21).
For $p_\Omega = p_H = 0$*:*

- *Zero is the only eigenvalue with zero real part and the remaining eigenvalues are with negative real parts. Moreover, zero is an eigenvalue of algebraic multiplicity 2 and of geometric multiplicity 1, that is, the zero eigenvalue is non-semisimple and the singularity is of Bogdanov-Takens-type.*
- *The system* (9.22) *is formally stable but not asymptotically stable* (*although there are no characteristic roots with positive real parts*).

For $p_H = -24\delta\, r$ *and* $p_\Omega = \frac{45\mu r^2}{10}$ *for a small enough* r:

- *The dynamics of* (9.22) *are reduced on a cubic center manifold to*

$$\begin{cases} \dot{z}_1 = z_2 \\ \dot{z}_2 = \delta z_1 + \mu z_2 - 3 z_2 z_1^2 \end{cases} \tag{9.23}$$

for which the function

$$I(z) = \frac{1}{2} {z_1}^2 - \frac{1}{2}\frac{{z_2}^2}{\delta} - \frac{1}{2}\delta\, {z_1}^2 {z_2}^2 + \frac{1}{4}\delta^2 {z_1}^4 + \frac{1}{4} {z_2}^4$$

is a Lyapunov function. For $\delta < 0$ *and* $\mu < 0$, *the system is globally asymptotically stable.*

Sketch of the proof: The first assertion is obtained by establishing a linear analysis of the system. Indeed, it can be easily verified by computing the associated characteristic equation and substituting the physical values. Numerical tools as the *Quasi-Polynomial Mapping Based Rootfinder QPMR* [300] are useful for locating the spectral values.

Next, we address the second part of the proposition concerning the nonlinear analysis. Following the approach described in [127], which considers a singular delay system linearly dependent on one parameter, and in the same spirit of the decomposition established in [87], with the goal of computing the normal form of delay systems, we extend the scheme of computing the center manifold to the case of NDDE. System (9.22) is regarded as a perturbation of

$$\frac{d}{dt}\mathscr{D}\, x_t = \mathscr{L}_0\, x_t, \quad \text{where} \quad \mathscr{L}_0 = \mathscr{L}|_{\{p_H=0,\, p_\Omega=0\}}, \tag{9.24}$$

Indeed, system (9.22) can be written as

$$\begin{aligned} \frac{d}{dt}\mathscr{D}\, x_t :&= \mathscr{L}_0\, x_t + \tilde{\mathscr{F}}(x_t) \\ &= \mathscr{L}_0\, x_t + (\mathscr{L} - \mathscr{L}_0)\, x_t + \mathscr{F}(x_t), \end{aligned} \tag{9.25}$$

such that

$$\mathscr{F}_{\mu,p} = \begin{bmatrix} -0.0405\, x_1^3(t) + 0.0377\, x_3^3(t-2) + p_H\, x_1(t-2) \\ -1.875\, p x_1^3(t) + 1.874998\, p x_1^3(t - 1.264911064) \end{bmatrix}.$$

Next, following the theoretical schemes presented in [48], the computations steps for obtaining the evolution equation of system's (9.24) solutions on the center manifold are given.
First, we compute the basis of the generalized eigenspace corresponding to the double eigenvalue $\lambda_0 = 0$.

$$\Phi(\theta)^T = \begin{bmatrix} 1+\theta & 1+2\theta & 1 & 2 \\ 1 & 2 & 0 & 0 \end{bmatrix},$$

where $\theta \in [-2, 0]$. Recall that the adjoint linear equation associated to (9.22) is

$$\begin{cases} \dot{u}(t) = D_1\dot{u}(t+2) + D_2\dot{u}(t+2\tilde{c}/c) \\ \qquad - A_0u(t) - A_1u(t+2) - A_2u(t+2\tilde{c}/c). \end{cases} \tag{9.26}$$

A basis for the generalized eigenspace associated to the double eigenvalue zero can be defined as

$$\Psi(\theta) = \begin{bmatrix} -1 & -3 & -7 & -13 \\ \xi+1 & 3\xi+2 & 7\xi+3 & 13\xi+4 \end{bmatrix}.$$

Let us consider the bilinear form, see [126]

$$\begin{aligned} (\psi, \varphi) = {} & \psi(0)(\varphi(0) - D_1\varphi(-2) - D_2\varphi(-1.264911)) \\ & + \int_{-2}^{0} \psi(\xi+2)A_1\varphi(\xi)d\xi \\ & + \int_{-1.264911}^{0} \psi(\xi+1.264911)A_2\varphi(\xi)d\xi \\ & - \int_{-2}^{0} \psi'(\xi+2)D_1\varphi(\xi)d\xi \\ & - \int_{-1.264911}^{0} \psi'(\xi+1.264911)D_2\varphi(\xi)d\xi. \end{aligned} \tag{9.27}$$

By using (9.27), we can easily normalize Ψ such that $(\Psi, \Phi) = I_d$, thus the space C can be decomposed as $C = P \bigoplus Q$, where $P = \{\varphi = \Phi z;\ z \in \mathbb{R}^2\}$ and $Q = \{\varphi \in C;\ (\Psi, \varphi) = 0\}$. Recall that each of these subspaces is invariant under the semigroup $T(t)$ and that the matrix B (concerned by the theoretical settings), satisfying $\mathscr{A}\Phi = \Phi B$, is given by

$$B = \begin{bmatrix} 0 & 0 \\ 1 & 0 \end{bmatrix}. \tag{9.28}$$

Let us first set the following decomposition $x_t = \Phi y(t) + z(t)$, where $z(t) \in Q$, $y(t) \in \mathbb{R}^2$, $z(t) = h(y(t))$ and h is some analytic function $h : P \to Q$. Following [48], the explicit solution on the center manifold is given by

$$\dot{y}(t) = By(t) + \Psi(0)\mathscr{F}[\Phi(\theta)y(t) + h(\theta, y(t))], \tag{9.29}$$

$$\frac{\partial h}{dy}\{By + \Psi(0)\mathscr{F}[\Phi(\theta)y + h]\} + \Phi(\theta)\Psi(0)\mathscr{F}[\Phi(\theta)y + h]$$
$$= \begin{cases} \dfrac{\partial h}{d\theta}, & -2 \leqslant \theta \leqslant 0, \\ \mathscr{L}(h(\theta, y)) + \mathscr{F}[\Phi(\theta)y + h(\theta, y)], & \theta = 0, \end{cases} \tag{9.30}$$

where $h = h(\theta, y)$ and $\tilde{\mathscr{F}}$ is defined in (9.25).

Since our aim is to establish the parameter values of p_H and p_Ω guaranteeing an asymptotic suppression of the vibrations after some fixed time t_0, we investigate the parameter bifurcations. The computation of the evolution equation on the center manifold of the solutions of system (9.25) is required.

The next step consists in introducing a small scaling parameter r in order to zoom the neighborhood of the singularity. Consider the following change of coordinates:

$$p_H = -24\delta\, r, \quad p_\Omega = \frac{45\mu r^2}{10}, \quad y_1 = r z_1, \quad y_2 = r^2 z_2.$$

A time scaling defined by $t_{old} = r\, t_{new}$, leads to the following cubic normal form reduction of (9.22),

$$\begin{cases} \dot{z}_1 = z_2, \\ \dot{z}_2 = \delta z_1 + \mu z_2 - z_1^3, \end{cases}$$

for which a normal form is given in (9.23).

If μ is a positive function beyond (0, 0) and $\delta < 0$, then,

$$\dot{l}(z_1, z_2) = -\frac{{z_2}^2\mu}{\delta} + 3\,\frac{{z_2}^2{z_1}^2}{\delta} - \delta\, {z_1}^2{z_2}^2\mu + 3\,\delta\, {z_1}^4{z_2}^2 + {z_2}^4\mu - 3\,{z_2}^4{z_1}^2$$

which is always negative. Thus the system is globally asymptotically stable and the undesired vibrations are suppressed. The proposed scheme offset the computation of a Lyapunov function for a system of PDE with nonlinear boundary conditions.

Figures 9.8 and 9.9 show a substantial reduction of drilling vibrations through the delayed feedback controllers:

$$u_H(t) = -24\delta r U_b(t-1) \tag{9.31}$$

$$u_T(t) = 4.5\mu r^2 \Phi_b(t-\tau) \tag{9.32}$$

for $r = 0.1$, $\delta = -1{,}000$, and $\mu = -10$.

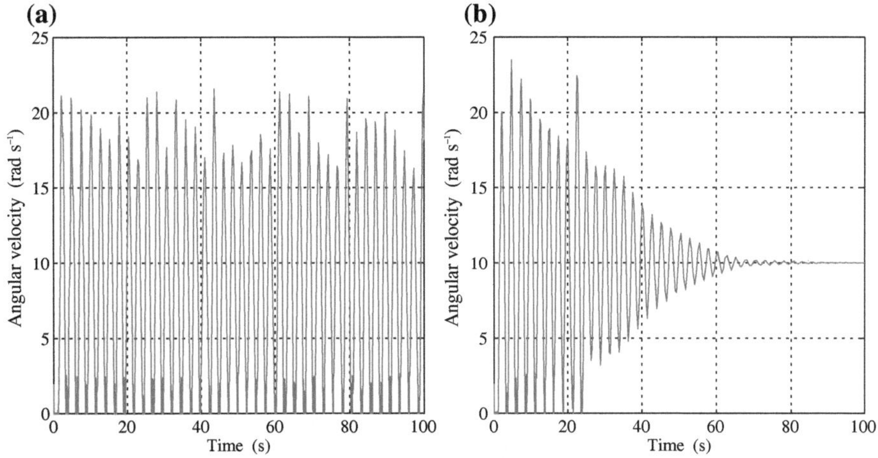

Fig. 9.8 Simulation of the coupled torsional-axial model (9.20). Angular velocity at the bottom extremity for a reference velocity of $10\,\mathrm{rad\,s^{-1}}$. **a** Trajectory without control actions (stick-slip). **b** Reduction of stick-slip oscillations by means of the delayed feedback controllers (9.31) and (9.32)

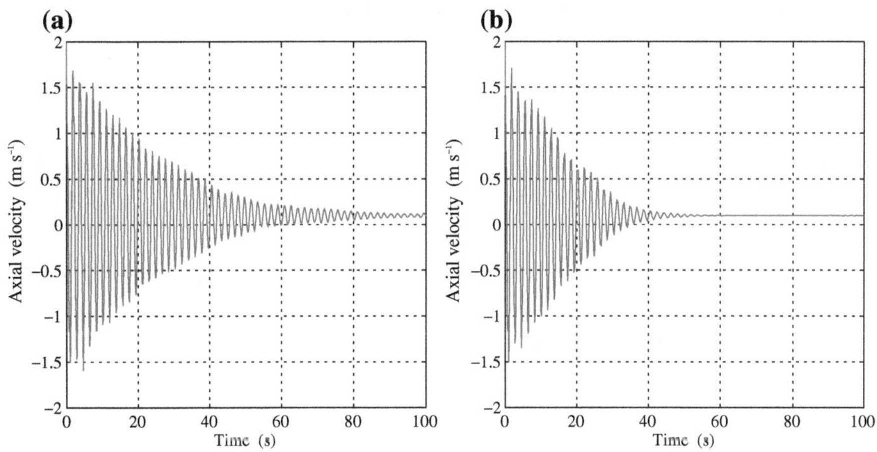

Fig. 9.9 Simulation of the coupled torsional-axial model (9.20). Axial velocity at the bottom extremity for a reference velocity of $0.1\,\mathrm{m\,s^{-1}}$. **a** Trajectory without control actions (bit-bounce). **b** Reduction of bit-bounce oscillations by means of the delayed feedback controllers (9.31) and (9.32)

9.5 Delayed PID Controller

Setting $x = (x_1, x_2, x_3, x_4, x_5, x_6)^T$, $x_3 = U_b$, $x_4 = \Phi_b$, $x_5 = \dot{U}_b$, $x_6 = \dot{\Phi}_b$, $x_1 = \int_0^t x_3(s)\,ds$, and $x_2 = \int_0^t x_4(s)\,ds$ and considering:

$$\begin{aligned} u_H(t) &= H_p\, U_b(t-1) + H_d\, \dot{U}_b(t-1) \\ u_T(t) &= \Omega_p \Phi_b(t-\tau) + \Omega_i \int_0^{t-\tau} \Phi_b(s)ds, \end{aligned} \tag{9.33}$$

the matrix representation of the system is given by:

$$\begin{cases} \dot{x}(t) = \hat{D}_1\, \dot{x}(t-2) + \hat{D}_2\, \dot{x}(t-2\tilde{c}/c) + \hat{A}_0\, x(t) + \hat{A}_1\, x(t-2) \\ \qquad + \hat{A}_2\, x(t-2\tilde{c}/c) + \mathscr{F}(x(t), x(t-2), x(t-2\tilde{c}/c)) \end{cases} \tag{9.34}$$

where $\mathscr{F}$ is the nonlinear part of system (9.20) with the matrices defined below:

$$\hat{D}_1 = \begin{bmatrix} 0_{4\times4} & 0_{4\times2} \\ 0_{2\times4} & \tilde{d}_1 \end{bmatrix}, \quad \hat{D}_2 = \begin{bmatrix} 0_{4\times4} & 0_{4\times2} \\ 0_{2\times4} & \tilde{d}_2 \end{bmatrix}, \quad \hat{A}_0 = \begin{bmatrix} 0_2 & I_2 & 0_2 \\ 0_2 & 0_2 & I_2 \\ \tilde{\Omega}_i & \tilde{H}_p & \tilde{a}_0 + \tilde{H}_d \end{bmatrix},$$

$$\hat{A}_1 = \begin{bmatrix} 0_{4\times4} & 0_{4\times2} \\ 0_{2\times4} & \tilde{a}_1 \end{bmatrix}, \quad \hat{A}_2 = \begin{bmatrix} 0_{4\times4} & 0_{4\times2} \\ 0_{2\times4} & \tilde{a}_2 \end{bmatrix},$$

where $0_{m\times n}$ is the zero matrix of $m \times n$ dimension, and

$$\tilde{\Omega}_i = \begin{bmatrix} 0 & 0 \\ 0 & \Omega_i \end{bmatrix}, \quad \tilde{H}_p = \begin{bmatrix} H_p & 0 \\ 0 & \Omega_p \end{bmatrix}, \quad \tilde{H}_d = \begin{bmatrix} H_d & 0 \\ 0 & 0 \end{bmatrix}.$$

Proposition 9.5.1 *The following properties are established for system* (9.34), *depending on the values of the gains H_p, H_d, Ω_p, and Ω_i in* (9.33).
For $\Omega_p = \Omega_i = H_p = H_d = 0$:

- *Zero is the only eigenvalue with zero real part and the remaining eigenvalues have negative real parts. Moreover, zero is an eigenvalue of algebraic multiplicity 4 and of geometric multiplicity 1, i.e., there is a generalized Bogdanov-Takens singularity.*
- *The system* (9.34) *is formally stable but not asymptotically stable (although there are no characteristic roots with positive real parts).*

For $H_d = 13.27\, r^{11}\delta_3$, $H_p = 26.55\, r^{12}\delta_2$, $\Omega_i = -11.47\, r^{13}\delta_1$, $\Omega_p = 11.47\, r^{10}\delta_4$ and a small parameter r:

- *The dynamics of* (9.34) *reduces on a degree 15 center manifold to*

$$\begin{cases} \dot{z}_1 = z_2, \ \dot{z}_2 = z_3, \ \dot{z}_3 = z_4 \\ \dot{z}_4 = \delta_1 z_1 + \delta_2 z_2 + \delta_3 z_3 + \delta_4 z_4 - z_4^3 \end{cases} \tag{9.35}$$

for which the linear part is written as a companion matrix:

$$A = \begin{bmatrix} 0 & 1 & 0 & 0 \\ 0 & 0 & 1 & 0 \\ 0 & 0 & 0 & 1 \\ \delta_1 & \delta_2 & \delta_3 & \delta_4 \end{bmatrix}$$

thus there exist values for δ_1, δ_2, δ_3, and δ_4 such that A is Hurwitz, which guaranteeing local asymptotic stability.

Sketch of the proof: Similar to the case of the delayed feedback controller, the spectral projection methodology is applied. The singularity here is zero with algebraic multiplicity 4 and geometric multiplicity 1. A basis for the generalized eigenspace $\mathcal{M}_0$ can be defined by

$$\Phi(\theta) = \begin{bmatrix} \frac{1}{6}\theta^3 + \theta^2 + \theta - 1 & \frac{1}{2}\theta^2 + 2\theta + 1 & \theta + 2 & 1 \\ -\frac{1}{6}\theta^3 + \frac{1}{2}\theta^2 - \theta + 2 & -\frac{1}{2}\theta^2 + \theta - 1 & -\theta + 1 & -1 \\ \theta^2 + \theta & 2\theta + 1 & 2 & 0 \\ \theta + 1 & 1 & 0 & 0 \\ 4\theta & 4 & 0 & 0 \\ 1 & 0 & 0 & 0 \end{bmatrix}.$$

The matrix B, satisfying $\mathcal{A}\Phi = B\Phi$, is given by

$$B = \begin{bmatrix} 0 & 1 & 0 & 0 \\ 0 & 0 & 1 & 0 \\ 0 & 0 & 0 & 1 \\ 0 & 0 & 0 & 0 \end{bmatrix}.$$

A basis Ψ of the adjoint space satisfying $(\Psi, \Phi) = I_d$, evaluated at zero is given by:

$$\Psi(0) = \begin{bmatrix} -0.00082 & -0.00082 & 0.00124 & 0.12511 & 0.01883 & 0.08717 \\ 0.01066 & 0.01066 & -0.016 & -0.14123 & -0.00384 & -0.17785 \\ 0.33261 & 0.33261 & 0.00107 & -0.69201 & -0.7673 & -0.28423 \\ 0.25042 & -0.74957 & 0.12435 & 3.20018 & 0.52271 & 2.56715 \end{bmatrix}.$$

The change of coordinates given in the second assertion of Proposition 9.4.1 leads to the reduced system (9.35).

Figure 9.10 shows the performance of the delayed PID feedback controllers:

$$u_H(t) = 26.55\, r^{12}\delta_2\, U_b(t-1) + 13.27\, r^{11}\delta_3\, \dot{U}_b(t-1) \tag{9.36}$$

$$u_T(t) = 11.47\, r^{10}\delta_4 \Phi_b(t-\tau) - 11.47\, r^{13}\delta_1 \int_0^{t-\tau} \Phi_b(s)ds \tag{9.37}$$

for $r^{10} = 1$, $r^{11} = 0.0001$, $r^{12} = 10$, $r^{13} = 1$, $\delta_1 = -10$, $\delta_2 = -100$, $\delta_3 = -24$ and $\delta_4 = -10$.

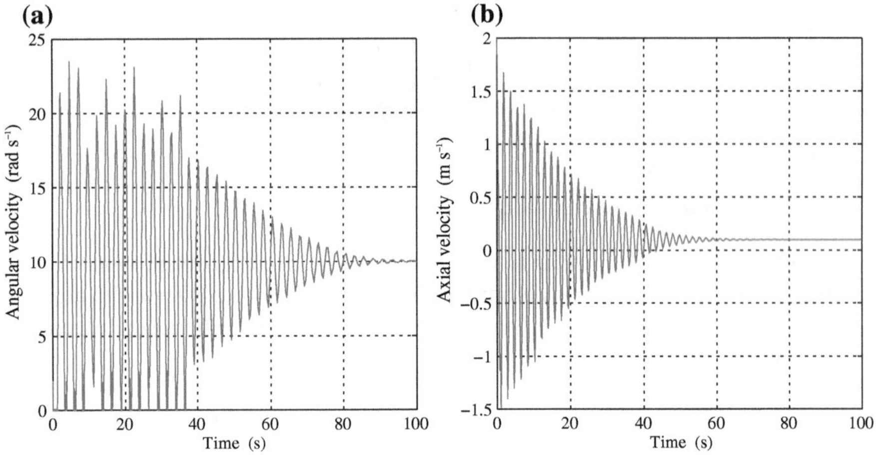

Fig. 9.10 Simulation of the coupled axial-torsional model (9.20). Reduction of stick-slip and bit-bounce oscillations by means of the delayed PID feedback controllers (9.36) and (9.37). **a** Angular velocity at the bottom extremity $\dot{\Phi}_b(t)$ for a reference velocity of $10\,\mathrm{rad\,s^{-1}}$. **b** Axial velocity at the bottom extremity $\dot{U}_b(t)$ for a reference velocity of $0.1\,\mathrm{m\,s^{-1}}$

9.6 Notes and References

The torsional rectification method reviewed in this chapter is based on the identification of both "up" and "down" moving components of the general solution of the wave equation describing torsional drillstring vibrations. A similar approach is proposed in [161]; the suggested method allows reducing torsional vibrations by exactly decomposing the drill string dynamics into two traveling waves: those traveling in the direction of the top extremity and those traveling in the direction of the bit. The decomposition is achieved with two sensors that can be placed at the top drive and at a short distance below the top drive. The velocity of the top drive is controlled in order to absorb the wave traveling in the direction of the top drive, thus achieving a reflection coefficient of zero for the frequency range of the undesired torsional vibrations. The performance of the control algorithm was evaluated through a small-scale experimental setup.

The design of low-order controllers presented in Sect. 9.3 was conducted by an analytic treatment of the NFDE describing the coupled axial-torsional drilling dynamics. Via spectral projection, a finite-dimensional approximation of the PDE model allowing preserving essential dynamics of the original system is obtained; then, the Lyapunov function guaranteeing the global stability of the reduced model ensures the local stability of the infinite-dimensional system. The model transformation procedure is explained in detail in [39].

Drilling system analysis led to the identification of a local bifurcation of Bogdanov-Takens type (double zero eigenvalues) which has not been studied for NDDE, except for [39], where an analytical study of the uncontrolled drilling vibrations is developed. A similar analysis approach is presented in [127], where a physiological system described by a DDE with double-zero eigenvalue singularity is investigated. In [305], explicit conditions for a Hopf bifurcation to occur are derived; using center manifold reduction and normal form theory the stability coefficient of the periodic orbit on the center manifold is explicitly determined.

There are some computational techniques to determine center manifolds. For instance, [6] presents an algorithm to establish the center manifold of a neutral functional differential equation; Bogdanov-Takens and Hopf singularities are considered. In [48], an overview on the theoretical settings for calculating center manifolds is provided; furthermore, it is shown how the computations may be implemented in the symbolic algebra package *Maple*.

So far we have studied some simple control techniques that help reduce unwanted drilling oscillations. It has been shown that its performance is acceptable; however, it is natural to think that more sophisticated control methods could produce better results. The following chapter concerns the design of a control strategy that exploits an intrinsic characteristic of the system: its flatness property. As will be seen later, this property is useful to solve trajectory tracking problems; based on this idea, we design a pair of controllers aimed at steering the system trajectories to prescribed paths, achieving in this way the elimination of axial and torsional vibrations.

Chapter 10
Flatness-Based Control of Drilling Vibrations

In nonlinear systems theory, the *flatness property* refers to the ability for dynamical systems of being exactly linearized via endogenous feedback [90]. A system satisfying the flatness property is called a differentially flat system. The main attribute of flat systems is that the state and input variables can be directly expressed without integrating any differential equation, in terms of one particular set of variables called a flat output (or linearizing output) and a finite number of its derivatives [90]. The flat terminology is due to the fact that the linearizing output plays an analogous role to the flat coordinates in the differential geometric approach of the Frobenius theorem (see, e.g., [218]). A considerable amount of realistic models is indeed flat.

Through the d'Alembert method, the differential flatness property of the drilling system described by a pair of wave equations subject to Newton-like nonlinear boundary conditions is proved in this chapter. It is worthy of mention that the flatness property of a nonlinear dynamical system is useful to deal with trajectory tracking problems. Based on the idea that the elimination of drilling vibrations requires the angular and axial velocities of the drilling bit to follow a constant reference path, we design a pair of controllers aimed at tackling the steering problem. Two configurations are considered:

- **Feedforward control**. Since in differentially flat systems the input variables can be directly expressed through the parameterization provided by the flat output, an open-loop controller is directly obtained by prescribing reference trajectories for the angular and axial bit velocities.
- **Feedback control**. Feedback controllers are intended to drive the errors between the reference and the actual bit velocities exponentially toward zero.

Simulation results show that the flatness-based feedback controllers, allowing an exponential convergence of the system trajectories, accurately eliminate the stick-slip and bit-bounce drilling vibrations.

M.B. Saldivar Márquez et al., *Analysis and Control of Oilwell Drilling Vibrations*,
Advances in Industrial Control, DOI 10.1007/978-3-319-15747-4_10

10.1 Differential Flatness Concept

The flatness concept was first introduced by Fliess et al. [90] within the framework of differential algebra. In the formalism of differential algebra, any system is considered as a differential field generated by a set of variables (states and inputs). The system is called flat if it is possible to find a set of variables (flat outputs) such that the system is (nondifferentially) algebraic over the differential field generated by the set of flat outputs. In general terms, a system is flat if there is a set of outputs such that all states and inputs can be determined from these outputs without integration.

Originally, the notion of flatness was defined for finite-dimensional models; however, its basic idea can be generalized to the infinite-dimensional case. The extension of the flatness concept to the infinite-dimensional linear case was introduced in [92].

The main difference between the finite and the infinite-dimensional case lies in the character of the relation between the flat output and the system variables: while in the first case, this relation involves only finite-order derivatives of the flat output, it may comprise derivatives of arbitrary order or delays and predictions, in the latter case depending on the type of the underlying partial differential equation.

This chapter aims at designing a control strategy to eliminate the coupled axial–torsional drilling vibrations. The underlying idea is to steer the system trajectories toward prescribed trajectories, i.e., in order to avoid the velocity fluctuations characterizing the stick-slip and bit-bounce, the angular and the axial bit velocities must follow constant reference paths.

In order to describe the drillstring dynamics with high accuracy, the distributed parameter model involving the coupled wave equations is considered. The design the vibration control scheme will be conducted as follows:

- The first step is to determine the flatness property of the system, i.e., we prove that the control inputs can be directly characterized through the system state and a finite number of its derivatives. To this end, we use the general solution of the wave equation provided by the d'Alembert formula.
- The second step is to define a feedforward controller aimed at solving the steering problem. Since the flatness analysis of the drilling system yields the exact characterization of the controllers trajectories, it suffices to set suitable trajectories for the flat outputs and calculate the control inputs from them.
- Finally, through algebraic manipulations, we design a feedback control scheme which guarantees that the error between actual and prescribed trajectories converges to zero.

10.2 Flatness of Finite-Dimensional Systems

The notion of differential flatness for nonlinear finite-dimensional systems was introduced in [90]. This concept is outlined as follows. Consider a differential system of the form:

$$\dot{x}_i = f_i(x, u), \quad i = 1, \dots, n, \tag{10.1}$$

where x_i are the state variables and $u = (u_i, \dots, u_m)$ is the control vector. System (10.1) is called flat if there exist some flat outputs $y = h(x, u, \dots, u^{(j)})$ with $j \in \mathbb{N}$ such that the components of y and all their derivatives are differentially independent and such that we can parametrize every solution (x, u) of (10.1) by means of the flat output y and its derivatives up to a finite order i:

$$x = F\left(y, \dots y^{(i)}\right),$$
$$u = G\left(y, \dots y^{(i)}\right).$$

The flat output of a system is not unique; it is generally chosen according to the system physical properties that it expresses.

The concept of flatness can be seen as a nonlinear generalization of Kalman's controllability and of the Brunovsky decomposition. Hence, every linear controllable system is flat [220].

10.2.1 A Simple Finite-Dimensional Example

Consider again the model (2.1), where, for the sake of simplicity, we have neglected viscous and torsional damping:

$$\begin{cases} I_p \ddot{\Phi}_p + k\left(\Phi_p - \Phi_b\right) = u_T, \\ I_b \ddot{\Phi}_b - k\left(\Phi_p - \Phi_b\right) = -T(\dot{\Phi}_b), \end{cases} \tag{10.2}$$

where Φ_p and Φ_b are the angular displacements of the rotary table and of the BHA, respectively. The control signal u_T is the drive torque coming from the rotary table transmission box used to regulate the rotary angular velocity $\dot{\Phi}_b$. The frictional torque T represents the torque on bit and the nonlinear frictional forces along the drill collars. The inertial masses are I_p and I_b, and are connected one to each other by a linear spring with torsional stiffness k.

A flat output of this model is Φ_b. Indeed, the second equation above yields Φ_p as a function of the first and second derivative of Φ_b; then, u_T is given through the first equation:

$$\begin{cases} \Phi_p = \Phi_b + \frac{1}{k}\left(I_b \ddot{\Phi}_b + T(\dot{\Phi}_b)\right) \\ u_T = (I_p + I_b)\ddot{\Phi}_b + T(\dot{\Phi}_b) + \frac{I_p}{k}\left(I_b \Phi_b^{(4)} + \frac{d^2}{dt^2}\left(T(\dot{\Phi}_b)\right)\right). \end{cases} \tag{10.3}$$

10.3 Flatness of Infinite-Dimensional Delay Systems

The notion of differential flatness for nonlinear infinite-dimensional delay systems was introduced in [203]. This concept is outlined as follows: A (nonlinear) system with delays $\delta_1, \ldots, \delta_r$ is a system described by a (finite) set of differential–difference equations

$$F_l(z, \ldots, z^{(i)}(t - s_1\tau_1 - \cdots - s_r\tau_r), \ldots, z^{(\nu_l)}(t - \underset{1}{\sigma}\tau_1 - \cdots - \underset{r}{\sigma}\tau_r)) = 0, \quad l = 1, \ldots, N.$$

We use the multi-index notation

$$\delta^s = \Pi_{j=1,\ldots,r}\delta_j^{s_j} = \delta_1^{s_1} \cdots \delta_r^{s_r},$$

with $(s_1, \ldots, s_r)$ any r-tuple of nonnegative integers and $\delta = (\delta_1, \ldots, \delta_r)$, the delay operators of finite amplitude which map $f(t)$ on $\delta_i f(t) = f(t - \tau_i)$, $i = 1, \ldots, r$.

With this, the above system equations read

$$F_l(z, \ldots, \delta^s z^{(i)}, \ldots, \delta^\sigma z^{(\nu_l)}) = 0, \qquad l = 1, \ldots, N.$$

Broadly speaking, the notion of flatness is extended to systems with delays in the following way (see [203] for formal definitions): A system with delays $\delta_1, \ldots, \delta_r$ is called *δ-flat* if there exists a collection $y = (y_1, \ldots, y_m)$ of functions, called a *δ-flat output*, with the following three properties:

1. The components of y can be expressed in terms of the system variables z via difference–differential relations of the type

$$y_i = P_i(z, \ldots, \delta^s \delta^{-\bar{s}} z^{(k)}, \ldots, \delta^\theta \delta^{-\bar{\theta}} z^{(\rho_i)})$$

   for $i = 1, \ldots, m$.
2. The components of y are difference–differentially independent, i.e., they are not related by any (nontrivial) difference–differential equation

$$Q(y, \ldots, \delta^s \delta^{-\bar{s}} y^{(k)}, \ldots, \delta^\beta \delta^{-\bar{\beta}} y^{(\alpha)}) = 0.$$

3. Every variable z_i used to describe the system, for instance states or inputs—and with them all their derivatives, the delayed variables, and all functions of these variables—can be calculated from y using only differentiations, delays, and advances $\delta_1^{-1}, \ldots, \delta_r^{-1}$. In other words, any such z_i satisfies a relation of the type

$$z_i = R(y, \ldots, \delta^s \delta^{-\bar{s}} y^{(k)}, \ldots, \delta^\varepsilon \delta^{-\bar{\varepsilon}} y^{(\gamma)}).$$

10.3.1 An Infinite-Dimensional System Example

The following example provides a better understanding of the flatness notion in distributed parameter systems.

Consider a dynamical system described by the following PDE:

$$\begin{cases} \dfrac{\partial^2 \theta}{\partial t^2}(x,t) = \dfrac{\partial^2 \theta}{\partial x^2}(x,t), \quad x \in [0,1] \\ \dfrac{\partial \theta}{\partial x}(0,t) = -u(t) \\ \dfrac{\partial \theta}{\partial x}(1,t) = F\left(\dfrac{\partial \theta}{\partial t}(1,t)\right) \end{cases} \tag{10.4}$$

where $u(t)$ is the control input. It will be shown that all system variables can be parameterized by a flat output, i.e., once a trajectory for the flat output has been prescribed, the trajectories of all system variables can be computed from it.

The general solution of the wave equation $\frac{\partial^2 \theta}{\partial t^2}(x,t) = \frac{\partial^2 \theta}{\partial x^2}(x,t)$ is given by the d'Alembert formula:

$$\theta(x,t) = \varphi(t+x) + \psi(t-x),$$

where the arbitrary functions φ and ψ correspond to traveling waves of speed -1 and 1, respectively. Substituting it into the boundary conditions yields:

$$\begin{aligned} -u(t) &= \varphi'(t) - \psi'(t) \\ F(\dot{y}(t)) &= \varphi'(t+1) - \psi'(t-1), \end{aligned}$$

where $'$ denotes derivation with respect to x and the variable y is defined as $y(t) = \theta(1,t) = \varphi(t+1) + \psi(t-1)$. The above equations can be solved as follows:

$$\begin{aligned} u(t) &= \frac{1}{2}\left(\dot{y}(t+1) - F\left(\dot{y}(t+1)\right)\right) - \frac{1}{2}\left(\dot{y}(t-1) + F\left(\dot{y}(t-1)\right)\right) \\ \varphi'(t) &= \frac{1}{2}\left(\dot{y}(t-1) + F\left(\dot{y}(t-1)\right)\right) \\ \psi'(t) &= \frac{1}{2}\left(\dot{y}(t+1) - F\left(\dot{y}(t+1)\right)\right). \end{aligned}$$

Integrating the above equations we obtain:

$$\varphi(t) = a + \frac{1}{2}\left(y(t-1) + \int_0^{t-1} F\left(\dot{y}(\tau)\right) d\tau\right)$$
$$\psi(t) = b + \frac{1}{2}\left(y(t+1) - \int_0^{t+1} F\left(\dot{y}(\tau)\right) d\tau\right),$$

where the constants a and b are such that $a+b=0$ since $y(t) = \varphi(t+1) + \psi(t-1)$.
The following parametrization is thus obtained:

$$2\theta(x,t) = y(t+x-1) + y(t-x-1) - \int_{t-1+x}^{t+1-x} F\left(\dot{y}(\tau)\right) d\tau.$$

Since this parametrization is explicit with respect to any arbitrary function $y(t)$, system (10.4) is flat and y is a flat output, i.e., there is a one-to-one correspondence between the solutions of (10.4) and $y(t)$.

10.4 Differential Flatness of the Drillstring Model

As reviewed in Chap. 2, the coupled torsional–axial dynamics of a drilling system can be modeled by a pair of wave equations of the form (2.13) with Newton-like boundary conditions (2.17). Consider the normalized rod length $\sigma = s/L$, the model described by the rotary angle $\Phi(\sigma,t)$ and the longitudinal position $U(\sigma,t)$ is given by:

$$\frac{\partial^2 \Phi}{\partial \sigma^2}(\sigma,t) = \tau^2 \frac{\partial^2 \Phi}{\partial t^2}(\sigma,t), \quad \tau = L\tilde{c}, \quad \sigma \in (0,1), \tag{10.5a}$$
$$I_T \frac{\partial^2 \Phi}{\partial t^2}(0,t) = \bar{G}\frac{\partial \Phi}{\partial \sigma}(0,t) + u_T(t), \quad \bar{G} = GJ/L, \tag{10.5b}$$
$$I_B \frac{\partial^2 \Phi}{\partial t^2}(1,t) = -\bar{G}\frac{\partial \Phi}{\partial \sigma}(1,t) - \tilde{p}F\left(\frac{\partial \Phi}{\partial t}(1,t)\right), \tag{10.5c}$$

and

$$\frac{\partial^2 U}{\partial \sigma^2}(\sigma,t) = \tilde{\tau}^2 \frac{\partial^2 U}{\partial t^2}(\sigma,t), \quad \tilde{\tau} = Lc, \quad \sigma \in (0,1), \tag{10.6a}$$
$$M_T \frac{\partial^2 U}{\partial t^2}(0,t) = \bar{E}\frac{\partial U}{\partial \sigma}(0,t) + u_H(t), \quad \bar{E} = E\Gamma/L, \tag{10.6b}$$
$$M_B \frac{\partial^2 U}{\partial t^2}(1,t) = -\bar{E}\frac{\partial U}{\partial \sigma}(1,t) - pF\left(\frac{\partial \Phi}{\partial t}(1,t)\right). \tag{10.6c}$$

The spatial variable σ is chosen such that $\sigma = 0$ denotes the top of the drill string and $\sigma = 1$ the bottom extremity. As explained in Sect. 2.2, the speeds of propagation $v_U = c^{-1}$ and $v_\Phi = \tilde{c}^{-1}$ can be computed from the density ρ_a and the shear and Young modulus G, E, as: $\tilde{c} = \sqrt{\rho_a/G}$ and $c = \sqrt{\rho_a/E}$. The total length of the drillstring, its cross section and its second moment of area are denoted by L, Γ and J, respectively. M_T and M_B stand for the top and bottom drive masses and I_T and I_B for the inertias. The terms $\tilde{p}F(\frac{\partial \Phi}{\partial t}(1,t))$ and $pF(\frac{\partial \Phi}{\partial t}(1,t))$ account for the frictional torque resulting from the interaction between the drill bit and the rock. The system is controlled by the boundary torque u_T coming from the motor drive at the surface and the boundary force u_H provided by the lifting hook at the drilling platform.

In what follows, a flatness-based parameterization of the drillstring model will be derived. Firstly, this will be done for the subsystem (10.5a–10.5c) describing the torsional vibrations. Because of the structural similarity of both subsystems, the considerations made in this first step will greatly simplify the analysis of the entire system.

10.4.1 Flatness of the Torsional Subsystem

Typically, the unactuated boundary of a distributed system corresponds to a flat output. For the case under consideration, the flatness property of $\Phi(1,\cdot)$ will be proven.

To this end, it will be shown that all the system variables appearing in the torsional subsystem, i.e., the distributed variable Φ and the control input u_T can be parameterized by the boundary values of Φ and $\frac{\partial \Phi}{\partial \sigma}$ at $\sigma = 1$. Afterwards, it will be proven that these values and, thus, the whole solution can be computed from the flat output $y_\Phi = \Phi(1,\cdot)$.

The parameterization of the torsional displacement $\Phi(\sigma,t)$ is obtained by solving the partial differential equation (10.5a). This can be done, for example, by means of the classical d'Alembert solution, the method of characteristics or operational calculus, yielding:

$$\begin{aligned} 2\Phi(\sigma,t) = {} & \Phi(1, t+\tau(1-\sigma)) + \Phi(1, t-\tau(1-\sigma)) \\ & + \frac{1}{\tau}\int_{t-\tau(1-\sigma)}^{t+\tau(1-\sigma)} \frac{\partial \Phi}{\partial \sigma}(1,\xi)d\xi. \end{aligned} \tag{10.7}$$

It becomes obvious that the solution depends on predicted and delayed boundary values at $\sigma = 1$.

For the sake of notation simplicity, let us denote as $y_\Phi(t)$ the rotational angle at the bottom end $\Phi(1,t)$, and as $y_U(t)$ the axial displacement at the bottom extremity of the drillstring $U(1,t)$.

The relation between $\frac{\partial \Phi}{\partial \sigma}(1, t)$ and $y_\Phi(t)$ can be established from the boundary condition (10.5c):

$$\begin{aligned}\frac{\partial \Phi}{\partial \sigma}(1, t) &= -\frac{I_B}{\bar{G}}\frac{\partial^2 \Phi}{\partial t^2}(1, t) - \frac{\tilde{p}}{\bar{G}}F\left(\frac{\partial \Phi}{\partial t}(1, t)\right) \\ &= -\frac{I_B}{\bar{G}}\ddot{y}_\Phi(t) - \frac{\tilde{p}}{\bar{G}}F\left(\dot{y}_\Phi(t)\right).\end{aligned} \tag{10.8}$$

Substituting (10.7) into the boundary condition (10.5b) yields the parameterization of the torque $u_T(t)$:

$$\begin{aligned}2u_T(t) = {} & I_T\ddot{y}_\Phi(t+\tau) - \frac{\tilde{p}I_T}{\tau\bar{G}}\frac{d}{dt}F\left(\dot{y}_\Phi(t+\tau)\right) - \frac{I_BI_T}{\tau\bar{G}}\dddot{y}_\Phi(t+\tau) \\ & + I_B\ddot{y}_\Phi(t+\tau) + I_T\ddot{y}_\Phi(t-\tau) + \frac{\tilde{p}I_T}{\tau\bar{G}}\frac{d}{dt}F\left(\dot{y}_\Phi(t-\tau)\right) \\ & + \frac{I_BI_T}{\tau\bar{G}}\dddot{y}_\Phi(t-\tau) + I_B\ddot{y}_\Phi(t-\tau) \\ & + \frac{\tilde{p}}{\tau}\left(F\left(\dot{y}_\Phi(t+\tau)\right) + F\left(\dot{y}_\Phi(t-\tau)\right)\right).\end{aligned} \tag{10.9}$$

It has been shown that all involved torsional variables can be parameterized by the flat output y_Φ and a finite number of its derivatives.

10.4.2 Flatness of the Entire System

It can be observed from the model equations that the torsional and the axial subsystems are structurally similar. Thus, the previous results suggest that the boundary value $U(1, \cdot)$ might be a flat output of the axial system, if the torsional displacement Φ is considered known. This would amount to a flat output of the entire system comprising both distributed system variables evaluated at $\sigma = 1$, i.e., $\boldsymbol{y} = (y_\Phi, y_U) = (\Phi(1, \cdot), U(1, \cdot))$. Next, it will be shown that $\boldsymbol{y}$ is indeed a flat output of the entire system. As it has already been shown, Φ and u_T can be parameterized by the first component y_Φ of $\boldsymbol{y}$; the same statement will be proven for U and u_H.

Again, the solution of the governing partial differential equation (10.6a) is parameterized in terms of the boundary values at $\sigma = 1$:

$$\begin{aligned}2U(\sigma, t) = {} & U(1, t+\tilde{\tau}(1-\sigma)) + U(1, t-\tilde{\tau}(1-\sigma)) \\ & + \frac{1}{\tilde{\tau}}\int_{t-\tilde{\tau}(1-\sigma)}^{t+\tilde{\tau}(1-\sigma)} \frac{\partial U}{\partial \sigma}(1, \xi)d\xi.\end{aligned} \tag{10.10}$$

The relation between $\frac{\partial U}{\partial \sigma}(1,t)$ and $y_U(t)$ can be determined by the boundary condition (10.6c) as follows:

$$\frac{\partial U}{\partial \sigma}(1,t) = -\frac{M_B}{\bar{E}}\frac{\partial^2 U}{\partial t^2}(1,t) - \frac{p}{\bar{E}}F\left(\frac{\partial \Phi}{\partial t}(1,t)\right), \qquad (10.11)$$
$$= -\frac{M_B}{\bar{E}}\ddot{y}_U(t) - \frac{p}{\bar{E}}F\left(\dot{y}_\Phi(t)\right).$$

The parameterization of the control input $u_H(t)$ is obtained by substituting (10.10) into the boundary condition (10.6b):

$$\begin{aligned} 2u_H(t) &= M_T\ddot{y}_U(t+\tilde{\tau}) - \frac{pM_T}{\tilde{\tau}\bar{E}}\frac{d}{dt}F\left(\dot{y}_\Phi(t+\tau)\right) - \frac{M_B M_T}{\tilde{\tau}\bar{E}}\dddot{y}_U(t+\tilde{\tau}) \\ &+ M_B\ddot{y}_U(t+\tilde{\tau}) + M_T\ddot{y}_U(t-\tilde{\tau}) + \frac{pM_T}{\tilde{\tau}\bar{E}}\frac{d}{dt}F\left(\dot{y}_\Phi(t-\tau)\right) \\ &+ \frac{M_B M_T}{\tilde{\tau}\bar{E}}\dddot{y}_U(t-\tilde{\tau}) + M_B\ddot{y}_U(t-\tilde{\tau}) \qquad (10.12) \\ &+ \frac{p}{\tilde{\tau}}\left(F\left(\dot{y}_\Phi(t+\tau)\right) + F\left(\dot{y}_\Phi(t-\tau)\right)\right). \end{aligned}$$

We conclude that the entire drilling system is flat, i.e., its solutions can be parameterized by the flat output $\boldsymbol{y} = (y_\Phi, y_U)$.

10.5 Control Design: Tracking Problem

Based on the flatness-based parameterization previously analyzed, the trajectory tracking problem can be addressed. Two different control schemes are proposed: feedforward and feedback.

10.5.1 Feedforward Control

For the feedforward scheme, it suffices to prescribe appropriate trajectories $\dot{y}_{\Phi r}, \dot{y}_{Ur}$ for the flat output $\boldsymbol{y}$ and compute the required control inputs from it (cf. [202]).

According to (10.7) and (10.8), the trajectories of the torsional subsystem can be parameterized as follows:

$$\begin{aligned} 2\Phi(\sigma,t) &= y_\Phi(t+\tau(1-\sigma)) + y_\Phi(t-\tau(1-\sigma)) \qquad (10.13) \\ &- \frac{1}{\tau\bar{G}}\int_{t-\tau(1-\sigma)}^{t+\tau(1-\sigma)}\left[I_B\ddot{y}_\Phi(t) + \tilde{p}F\left(\dot{y}_\Phi(t)\right)\right]dt. \end{aligned}$$

Taking the time derivative of (10.13) and evaluating in $\sigma = 0$ yields

$$\begin{aligned} 2\frac{\partial \Phi}{\partial t}(0,t) &= \dot{y}_\Phi(t+\tau) + \dot{y}_\Phi(t-\tau) \\ &\quad -\frac{I_B}{\tau\bar{G}}\left[\ddot{y}_\Phi(t+\tau) - \ddot{y}_\Phi(t-\tau)\right] \\ &\quad -\frac{\tilde{p}}{\tau\bar{G}}\left[F\left(\dot{y}_\Phi(t+\tau)\right) - F\left(\dot{y}_\Phi(t-\tau)\right)\right], \end{aligned} \tag{10.14}$$

which also gives:

$$\begin{aligned} -\frac{I_B}{\tau\bar{G}}\left[\dddot{y}_\Phi(t+\tau) - \dddot{y}_\Phi(t-\tau)\right] &= 2\frac{\partial^2 \Phi}{\partial t^2}(0,t) - \ddot{y}_\Phi(t+\tau) - \ddot{y}_\Phi(t-\tau) \\ &\quad +\frac{\tilde{p}}{\tau\bar{G}}\frac{d}{dt}\left[F\left(\dot{y}_\Phi(t+\tau)\right) - F\left(\dot{y}_\Phi(t-\tau)\right)\right]. \end{aligned}$$

Then, (10.9) can be rewritten as:

$$\begin{aligned} 2u_T(t) &= I_T\ddot{y}_\Phi(t+\tau) - \frac{\tilde{p}I_T}{\tau\bar{G}}\frac{d}{dt}F\left(\dot{y}_\Phi(t+\tau)\right) \\ &\quad +I_B\ddot{y}_\Phi(t+\tau) + I_T\ddot{y}_\Phi(t-\tau) + \frac{\tilde{p}I_T}{\tau\bar{G}}\frac{d}{dt}F\left(\dot{y}_\Phi(t-\tau)\right) \\ &\quad +I_B\ddot{y}_\Phi(t-\tau) + \frac{\tilde{p}}{\tau}\left(F\left(\dot{y}_\Phi(t+\tau)\right) + F\left(\dot{y}_\Phi(t-\tau)\right)\right) \\ &\quad +2I_T\frac{\partial^2 \Phi}{\partial t^2}(0,t) - I_T\ddot{y}_\Phi(t+\tau) - I_T\ddot{y}_\Phi(t-\tau) \\ &\quad +\frac{\tilde{p}I_T}{\tau\bar{G}}\frac{d}{dt}\left[F\left(\dot{y}_\Phi(t+\tau)\right) - F\left(\dot{y}_\Phi(t-\tau)\right)\right]. \end{aligned} \tag{10.15}$$

Similarly, from (10.10) and (10.11), it follows that

$$\begin{aligned} 2U(\sigma,t) &= y_U(t+\tilde{\tau}(1-\sigma)) + y_U(t-\tilde{\tau}(1-\sigma)) \\ &\quad -\frac{1}{\tilde{\tau}\bar{E}}\int_{t-\tilde{\tau}(1-\sigma)}^{t+\tilde{\tau}(1-\sigma)} \left[M_B\ddot{y}_U(t) + pF\left(\dot{y}_\Phi(t)\right)\right]dt, \end{aligned}$$

and the following equality is obtained,

$$\begin{aligned} -\frac{M_B}{\tilde{\tau}\bar{E}}\left[\dddot{y}_U(t+\tilde{\tau}) - \dddot{y}_U(t-\tilde{\tau})\right] &= 2\frac{\partial^2 U}{\partial t^2}(0,t) - \ddot{y}_U(t+\tilde{\tau}) - \ddot{y}_U(t-\tilde{\tau}) \\ &\quad +\frac{p}{\tilde{\tau}\bar{E}}\frac{d}{dt}\left[F\left(\dot{y}_\Phi(t+\tau)\right) - F\left(\dot{y}_\Phi(t-\tau)\right)\right]. \end{aligned}$$

Then, (10.12) can be rewritten as:

$$\begin{aligned}2u_H(t) &= M_T\ddot{y}_U(t+\tilde{\tau}) - \frac{pM_T}{\tilde{\tau}\bar{E}}\frac{d}{dt}F\left(\dot{y}_\Phi(t+\tau)\right)\\&+M_B\ddot{y}_U(t+\tilde{\tau}) + M_T\ddot{y}_U(t-\tilde{\tau}) + \frac{pM_T}{\tilde{\tau}\bar{E}}\frac{d}{dt}F\left(\dot{y}_\Phi(t-\tau)\right)\\&+M_B\ddot{y}_U(t-\tilde{\tau}) + \frac{p}{\tilde{\tau}}\left(F\left(\dot{y}_\Phi(t+\tau)\right) + F\left(\dot{y}_\Phi(t-\tau)\right)\right)\\&+2M_T\frac{\partial^2 U}{\partial t^2}(0,t) - M_T\ddot{y}_U(t+\tilde{\tau}) - M_T\ddot{y}_U(t-\tilde{\tau})\\&+\frac{pM_T}{\tilde{\tau}\bar{E}}\frac{d}{dt}\left[F\left(\dot{y}_\Phi(t+\tau)\right) - F\left(\dot{y}_\Phi(t-\tau)\right)\right].\end{aligned} \tag{10.16}$$

In view of (10.15) and (10.16), the feedforward controllers are then defined as:

$$\begin{aligned}2u_T(t) &= 2I_T\frac{\partial^2 \Phi}{\partial t^2}(0,t) + I_B\ddot{y}_{\Phi r}(t+\tau) + I_B\ddot{y}_{\Phi r}(t-\tau)\\&+\frac{\tilde{p}}{\tau}\left(F\left(\dot{y}_{\Phi r}(t+\tau)\right) + F\left(\dot{y}_{\Phi r}(t-\tau)\right)\right)\end{aligned} \tag{10.17}$$

and

$$\begin{aligned}2u_H(t) &= 2M_T\frac{\partial^2 U}{\partial t^2}(0,t) + M_B\ddot{y}_{Ur}(t+\tilde{\tau}) + M_B\ddot{y}_{Ur}(t-\tilde{\tau})\\&+\frac{p}{\tilde{\tau}}\left(F\left(\dot{y}_{\Phi r}(t+\tau)\right) + F\left(\dot{y}_{\Phi r}(t-\tau)\right)\right).\end{aligned} \tag{10.18}$$

10.5.2 Feedback Control

The open-loop control laws (10.17) and (10.18) are designed under the supposition that the model under consideration is perfect, which, due to the uncertainties, model mismatch and poorly known initial conditions, is not the case. In order to overcome this problem, closed-loop controllers ensuring the system stabilization around the reference trajectories must be designed. The underlying idea for the design of feedback controllers is to compute the control inputs such that the errors between actual and reference trajectories $e_\Phi := \dot{y}_\Phi - \dot{y}_{\Phi r}$ and $e_U := \dot{y}_U - \dot{y}_{Ur}$ satisfy the stable dynamics: $\dot{e}_\Phi = -\lambda e_\Phi$, $\dot{e}_U = -\bar{\lambda} e_U$.

The result on the stabilization of the drilling system regarding its torsional and axial dynamics is stated as follows:

Theorem 10.5.1 *The controllers*

$$\begin{aligned} u_T(t) &= I_T\frac{\partial^2\Phi}{\partial t^2}(0,t)+\tau\bar{G}\frac{\partial\Phi}{\partial t}(0,t)-\tau\bar{G}\left[\dot{y}_\Phi(t-\tau)+\frac{1}{2}\gamma(t)\right]\\ &\quad +I_B v(t)+\tilde{p}F\left(\dot{y}_\Phi(t-\tau)+\gamma(t)\right),\\ v(t) &= \frac{\bar{G}\chi}{\lambda}\ddot{y}_{\Phi r}(t+\tau)-\bar{G}\chi I+\frac{I_B\chi}{\tau}\ddot{y}_\Phi(t-\tau)\\ &\quad -\frac{\tilde{p}\chi}{\tau}\left[F\left(\dot{y}_\Phi(t-\tau)+\gamma(t)\right)-F\left(\dot{y}_\Phi(t-\tau)\right)\right],\\ I &= 2\frac{\partial\Phi}{\partial t}(0,t)-\dot{y}_\Phi(t-\tau)-\dot{y}_{\Phi r}(t+\tau),\\ \gamma(t) &= \int_{t-2\tau}^{t} v(\xi)d\xi,\quad \chi=\frac{\lambda\tau}{\tau\bar{G}+\lambda I_B}, \end{aligned}$$

and

$$\begin{aligned} u_H(t) &= M_T\frac{\partial^2 U}{\partial t^2}(0,t)+\tilde{\tau}\bar{E}\frac{\partial U}{\partial t}(0,t)-\tilde{\tau}\bar{E}\left[\dot{y}_U(t-\tilde{\tau})+\frac{1}{2}\bar{\gamma}(t)\right]\\ &\quad +M_B\bar{v}(t)+pF\left(\dot{y}_\Phi(t-\tau)+\gamma(t)\right),\\ \bar{v}(t) &= \frac{\bar{E}\bar{\chi}}{\bar{\lambda}}\ddot{y}_{Ur}(t+\tilde{\tau})-\bar{E}\bar{\chi}\bar{I}+\frac{M_B\bar{\chi}}{\tilde{\tau}}\ddot{y}_U(t-\tilde{\tau})\\ &\quad -\frac{p\bar{\chi}}{\tilde{\tau}}\left[F\left(\dot{y}_\Phi(t-\tau)+\gamma(t)\right)-F\left(\dot{y}_\Phi(t-\tau)\right)\right],\\ \bar{I} &= 2\frac{\partial U}{\partial t}(0,t)-\dot{y}_U(t-\tilde{\tau})-\dot{y}_{Ur}(t+\tilde{\tau}),\\ \bar{\gamma}(t) &= \int_{t-2\tilde{\tau}}^{t} \bar{v}(\xi)d\xi,\quad \bar{\chi}=\frac{\bar{\lambda}\tilde{\tau}}{\tilde{\tau}\bar{E}+\bar{\lambda}M_B}, \end{aligned}$$

lead to an exponential convergence of the torsional and axial trajectories $\dot{y}_\Phi(t)=\frac{\partial\Phi}{\partial t}(1,t)$ and $\dot{y}_U(t)=\frac{\partial U}{\partial t}(1,t)$ to the reference velocities $\dot{y}_{\Phi r}(t)$ and $\dot{y}_{Ur}(t)$.

Proof In view of (10.9) and (10.14), we can write

$$\begin{aligned} &I_T\frac{\partial^2\Phi}{\partial t^2}(0,t)+\tau\bar{G}\frac{\partial\Phi}{\partial t}(0,t)-u_T(t)-\frac{\tau\bar{G}}{2}\dot{y}_\Phi(t-\tau)= \\ &-I_B\ddot{y}_\Phi(t+\tau)-\tilde{p}F\left(\dot{y}_\Phi(t+\tau)\right)+\frac{\tau\bar{G}}{2}\dot{y}_\Phi(t+\tau). \end{aligned} \tag{10.19}$$

The introduction of a new variable $v(t)$ defined as $v(t) := \ddot{y}_\Phi(t+\tau)$ implies

$$\dot{y}_\Phi(t+\tau)=\dot{y}_\Phi(t-\tau)+\int_{t-2\tau}^{t} v(\xi)d\xi. \tag{10.20}$$

In view of (10.19) and (10.20), the controller $u_T(t)$ can be written as:

$$u_T(t) = I_T \frac{\partial^2 \Phi}{\partial t^2}(0,t) + \tau \bar{G} \frac{\partial \Phi}{\partial t}(0,t) - \tau \bar{G} \left[\dot{y}_\Phi(t-\tau) + \frac{1}{2}\gamma(t) \right]$$
$$+ I_B v(t) + \tilde{p} F\left(\dot{y}_\Phi(t-\tau) + \gamma(t)\right),$$

where $\gamma(t) = \int_{t-2\tau}^{t} v(\xi)d\xi$. Now, by substituting (10.20) into (10.14), the prediction term is written as follows:

$$\dot{y}_\Phi(t+\tau) = 2\frac{\partial \Phi}{\partial t}(0,t) - \dot{y}_\Phi(t-\tau) + \frac{I_B}{\tau \bar{G}}\left[v(t) - \ddot{y}_\Phi(t-\tau)\right] \quad (10.21)$$
$$+ \frac{\tilde{p}}{\tau \bar{G}}\left[F\left(\dot{y}_\Phi(t-\tau) + \gamma(t)\right) - F\left(\dot{y}_\Phi(t-\tau)\right)\right].$$

Regarding the velocity tracking problem under consideration, the error is defined as $e_\Phi := \dot{y}_\Phi(t+\tau) - \dot{y}_{\Phi r}(t+\tau)$. The controller must guarantee stable closed-loop error dynamics ($\dot{e_\Phi} = -\lambda e_\Phi$), to this end we set

$$v(t) = \ddot{y}_{\Phi r}(t+\tau) - \lambda\left[\dot{y}_\Phi(t+\tau) - \dot{y}_{\Phi r}(t+\tau)\right],$$

which, in view of (10.21), is written as:

$$v(t) = \frac{\tau \bar{G}}{\tau \bar{G} + \lambda I_B}\ddot{y}_{\Phi r}(t+\tau) + \frac{\lambda I_B}{\tau \bar{G} + \lambda I_B}\ddot{y}_\Phi(t-\tau) - \frac{\tau \bar{G}\lambda}{\tau \bar{G} + \lambda I_B} I$$
$$- \frac{\lambda \tilde{p}}{\tau \bar{G} + \lambda I_B}\left[F\left(\dot{y}_\Phi(t-\tau) + \gamma(t)\right) - F\left(\dot{y}_\Phi(t-\tau)\right)\right],$$

with

$$I = 2\frac{\partial \Phi}{\partial t}(0,t) - \dot{y}_\Phi(t-\tau) - \dot{y}_{\Phi r}(t+\tau).$$

Similarly, for the axial subsystem, we define the variable $\bar{v}(t)$ as $\bar{v}(t) = \ddot{y}_U(t+\tilde{\tau})$, which implies

$$\dot{y}_U(t+\tilde{\tau}) = \dot{y}_U(t-\tilde{\tau}) + \int_{t-2\tilde{\tau}}^{t} \bar{v}(\xi)d\xi, \quad (10.22)$$

and

$$u_H(t) = M_T \frac{\partial^2 U}{\partial t^2}(0,t) + \tilde{\tau}\bar{E}\frac{\partial U}{\partial t}(0,t) - \tilde{\tau}\bar{E}\left[\dot{y}_U(t-\tilde{\tau}) + \frac{1}{2}\bar{\gamma}(t)\right]$$
$$+ M_B \bar{v}(t) + pF\left(\dot{y}_\Phi(t-\tau) + \gamma(t)\right),$$

where $\bar{\gamma}(t) = \int_{t-2\tilde{\tau}}^{t} \bar{v}(\xi)d\xi$. As we have shown before, the flatness property allows the following parameterization:

$$\begin{aligned} 2U(\sigma,t) &= y_U(t+\tilde{\tau}(1-\sigma)) + y_U(t-\tilde{\tau}(1-\sigma)) \\ &\quad - \frac{1}{\tilde{\tau}\bar{E}} \int_{t-\tilde{\tau}(1-\sigma)}^{t+\tilde{\tau}(1-\sigma)} \left[M_B \ddot{y}_U(t) + pF(\dot{y}_\Phi(t))\right] d\xi. \end{aligned} \tag{10.23}$$

Substituting (10.22) into the time derivative of (10.23) at $\sigma = 0$ yields:

$$\begin{aligned} \dot{y}_U(t+\tilde{\tau}) &= 2\frac{\partial U}{\partial t}(0,t) - \dot{y}_U(t-\tilde{\tau}) + \frac{M_B}{\tilde{\tau}\bar{E}}\left[\bar{v}(t) - \ddot{y}_U(t-\tilde{\tau})\right] \\ &\quad + \frac{p}{\tilde{\tau}\bar{E}}\left[F(\dot{y}_\Phi(t-\tau) + \gamma(t)) - F(\dot{y}_\Phi(t-\tau))\right]. \end{aligned} \tag{10.24}$$

The error related to the axial dynamics e_U is defined as $e_U := \dot{y}_U(t+\tilde{\tau}) - \dot{y}_{Ur}(t+\tilde{\tau})$. In order to ensure stable closed-loop error dynamics, we set

$$\bar{v}(t) = \ddot{y}_{Ur}(t+\tilde{\tau}) - \bar{\lambda}\left[\dot{y}_U(t+\tilde{\tau}) - \dot{y}_{Ur}(t+\tilde{\tau})\right],$$

which, in view of (10.24), is written as:

$$\begin{aligned} \bar{v}(t) = &\frac{\tilde{\tau}\bar{E}}{\tilde{\tau}\bar{E} + \bar{\lambda}M_B}\ddot{y}_{Ur}(t+\tilde{\tau}) + \frac{\bar{\lambda}M_B}{\tilde{\tau}\bar{E} + \bar{\lambda}M_B}\ddot{y}_U(t-\tilde{\tau}) - \frac{\tilde{\tau}\bar{E}\bar{\lambda}}{\tilde{\tau}\bar{E} + \bar{\lambda}M_B}\bar{I} \\ &- \frac{\bar{\lambda}p}{\tilde{\tau}\bar{E} + \bar{\lambda}M_B}\left[F(\dot{y}_\Phi(t-\tau) + \gamma(t)) - F(\dot{y}_\Phi(t-\tau))\right], \end{aligned}$$

with

$$\bar{I} = 2\frac{\partial U}{\partial t}(1,t) - \dot{y}_U(t-\tilde{\tau}) - \dot{y}_{Ur}(t+\tilde{\tau}).$$

Remark 10.1 Note that the previous tracking feedback control laws of Theorem 10.5.1 enable *tracking with stability* of system (2.13) with boundary conditions (2.17), which corresponds to a *non formally stable neutral delay system*, this class of systems being notoriously difficult to control.

10.5.3 Numerical Simulations

The effectiveness of the proposed control approach is highlighted through simulations results. The numerical values of the physical parameters used in the following are given in Table C. 1 of Appendix C. The friction at the rock-bit interface is approximated by the model (3.10).

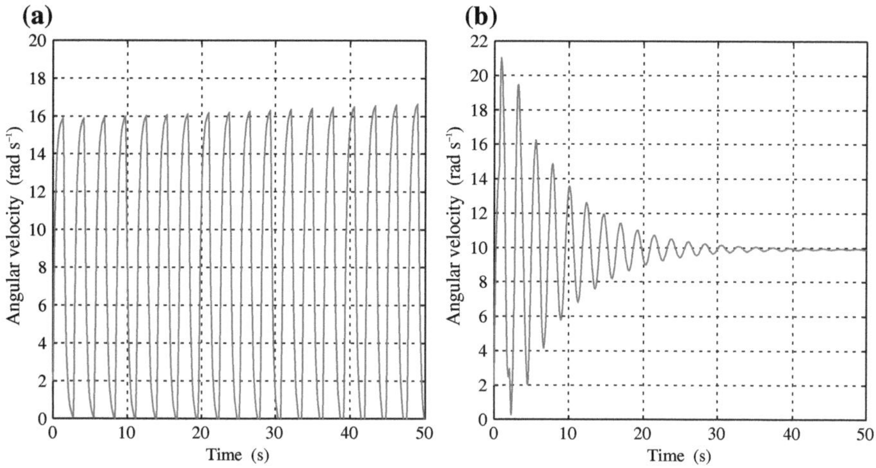

Fig. 10.1 Simulation of the coupled torsional–axial model given in (10.5a–10.5c) and (10.6a–10.6c). Angular velocity at the bottom extremity for a reference velocity of $10\,\mathrm{rad\,s^{-1}}$. **a** Trajectory without control actions (stick-slip). **b** Elimination of stick-slip oscillations by means of the flatness-based controllers stated in Theorem 10.5.1

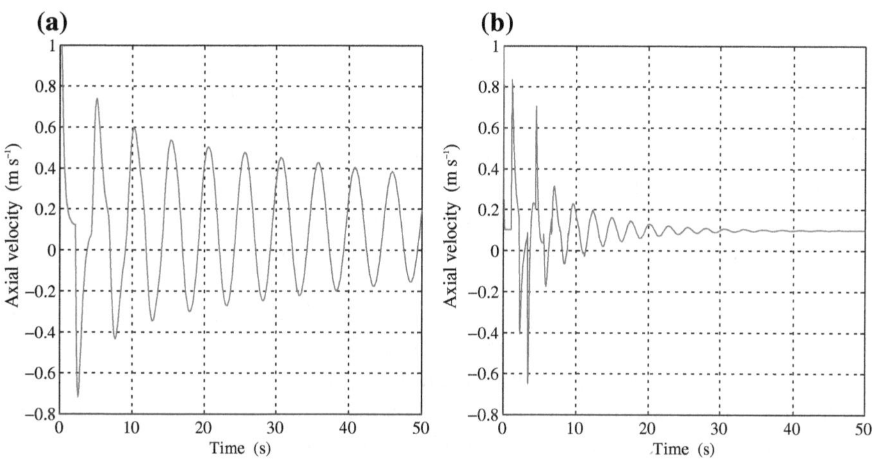

Fig. 10.2 Simulation of the coupled torsional–axial model given in (10.5a–10.5c) and (10.6a–10.6c). Axial velocity at the bottom extremity for a reference velocity of $0.1\,\mathrm{m\,s^{-1}}$. **a** Trajectory without control actions (bit-bounce). **b** Elimination of bit-bounce oscillations by means of the flatness-based controllers stated in Theorem 10.5.1

System trajectories corresponding to the rotational and longitudinal velocities of the drilling rod at the bottom end $(\dot{y}_{\Phi}(t), \dot{y}_U(t))$ without feedback control actions are shown in Figs. 10.1a and 10.2a.

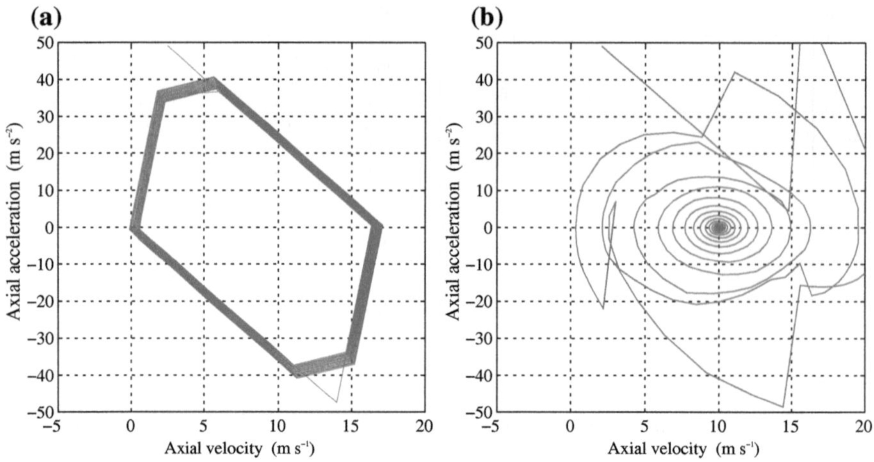

Fig. 10.3 Simulation of the coupled torsional–axial model given in (10.5a–10.5c) and (10.6a–10.6c). Phase portrait of the drillstring torsional trajectories. **a** System trajectories without control actions converging to a stick-slip limit cycle. **b** System trajectories under the F controllers stated in Theorem 10.5.1 converging to the stationary solution

As explained above, the flatness approach allows the design of feedback controllers to track prescribed reference trajectories ensuring stable error dynamics and consequently the drilling vibration elimination.

Figures 10.1b and 10.2b show the closed-loop response of the drilling system subject to the proposed flatness-based control approach. Angular and axial velocities at the bottom end of the drillstring follow the references $\dot{y}_{\Phi r} = 10\,\mathrm{rad\,s^{-1}}$ and $\dot{y}_{Ur} = 0.1\,\mathrm{m\,s^{-1}}$. The considered exponential decay rates are $\lambda = \bar{\lambda} = 2.5$.

In absence of control actions, and under certain parametric conditions, the drilling system exhibits oscillatory characteristics leading to stick-slip. Figures 10.3a and 10.4a show the torsional trajectories of the drilling system in a two-dimensional phase with the relative variable $\dot{y}_{\Phi}(t)$; the bit motion converges to a limit cycle.[1] Such behavior, frequently occurring in mechanical oscillatory systems, is generally assumed to be caused by friction forces. There are some works devoted to the analysis of limit cycles driven by friction, for instance in [221], the influence of stiction is investigated, and in [130], the effect of the static switch and LuGre-type frictions on limit cycles are studied.

The proposed feedback scheme based on the differential flatness property of the drilling system ensures the elimination of undesirable oscillations, consequently leading to the elimination of the limit cycle. The closed-loop system trajectories leave the limit cycle and converge to the stationary solution, see Figs. 10.3b and 10.4b.

[1] A limit cycle is defined as a closed trajectory which has the property that at least one other trajectory spirals into it as time approaches infinity.

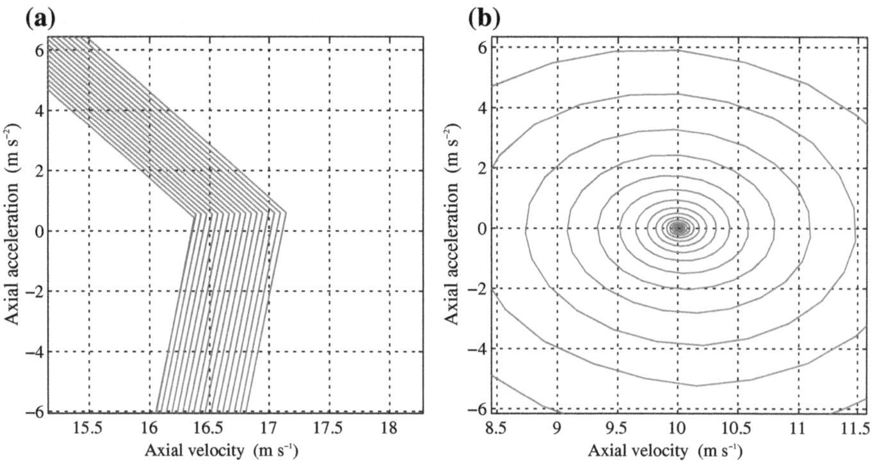

Fig. 10.4 Zomm-in of Fig. 10.3

10.6 Notes and References

Motion planning (tracking problem), i.e., steering a system from one state to another, is a basic control issue. For a certain class of systems with the flatness property, motion planning admits simple and explicit solutions. The trajectories of this class of systems can be explicitly described by an arbitrary function, called a flat output, and a finite number of its time derivatives.

The flatness property is also useful to solve inverse dynamics problems and many other control issues. There is a wide range of engineering problems that have been addressed by exploiting their flatness properties. Indeed, many classes of systems commonly used in nonlinear control theory are flat; some of them are listed below:

- **Mechanical systems**: aircrafts [74], cranes [156], towed cable systems [206], satellites [188], mobile robots [294], batch processes [120]. Other examples of flat mechanical systems are given in [205].
- **Electromechanical systems:** DC to DC converters [109], magnetic bearings [172], induction motors [194].
- **Chemical systems**: Continuous stirred tank reactors [246], polymerization reactors [229], chemical reactors with time delays [202].

In this chapter, the differential flatness of the drilling system described by a pair of wave equations with nonlinear coupled boundary conditions was investigated. This property allows us to design a pair of feedback controllers to tackle a trajectory tracking problem leading to the suppression of the stick-slip and bit-bounce vibrations. The analysis was carried out by considering Newton-type boundary conditions (see Sect. 2.2). Flatness-based controllers for the drilling system subject to a speed difference at the top extremity are determined in [260]. Further insights on the differential flatness of the drilling system can be found in [155].

In [250], the flatness property of the torsional drilling model is used to tackle the stick-slip phenomenon. The proposed feedback control is based on a linearized PDE model which neglects the BHA inertia, diminishing the model reliability. Besides the flatness approach, [250] addresses the design of a stabilizing control law to eliminate the stick-slip based on the backstepping technique introduced in [274]. The backstepping-based control method uses a drilling model of the form:

$$\begin{cases} \dfrac{\partial^2 \Phi}{\partial t^2}(s,t) = \dfrac{\partial^2 \Phi}{\partial s^2}(s,t) - \gamma \dfrac{\partial \Phi}{\partial t}(s,t), \quad s \in [0,1], \ t > 0 \\ \dfrac{\partial \Phi}{\partial s}(0,t) = u_T(t) \\ \dfrac{\partial^2 \Phi}{\partial t^2}(1,t) = a \dfrac{\partial \Phi}{\partial s}(1,t) + ab \dfrac{\partial \Phi}{\partial t}(1,t). \end{cases}$$

The change of variables given by

$$v(s,t) = \frac{\partial\left(\Phi - \bar{\Phi}\right)}{\partial s}(s,t), \quad y(t) = \frac{\partial \Phi}{\partial t}(1,t), \quad u(t) = u_T(t) - \bar{u}_T(t),$$

where $\bar{\Phi}$ and $\bar{u}_T$ correspond to reference angular position and control input, respectively, gives rise to the following PDE–ODE cascade system:

$$\begin{cases} \dfrac{\partial^2 v}{\partial t^2}(s,t) = \dfrac{\partial^2 v}{\partial s^2}(s,t) - \gamma \dfrac{\partial v}{\partial t}(s,t), \quad s \in [0,1] \\ v(0,t) = u(t) \\ \dfrac{\partial v}{\partial s}(1,t) = av(1,t) + (\gamma + ab)\, y(t) \\ \dot{y}(t) = aby(t) + av(1,t). \end{cases} \tag{10.25}$$

The underlying idea for eliminating the stick-slip consists in designing a control law that maps system (10.25) to a target system with desirable stability properties. The proposed controller is given by:

$$u(t) = \int_0^1 k(0,\xi)\, v(\xi,t) d\xi + \int_0^1 s(0,\xi)\, v(\xi,t) d\xi + \lambda(0) y(t).$$

A suitable choice of the variables $k(0,\xi)$, $s(0,\xi)$ and $\lambda(0)$ allows the mapping between system (10.25) and the target system defined as:

$$\begin{cases} \dfrac{\partial^2 w}{\partial t^2}(s,t) = \dfrac{\partial^2 w}{\partial s^2}(s,t) - \gamma \dfrac{\partial w}{\partial t}(s,t), \quad s \in [0,1] \\ w(0,t) = 0 \\ \dfrac{\partial w}{\partial s}(1,t) = c \dfrac{\partial w}{\partial t}(1,t) \\ \qquad \dot{y}(t) = -\delta y(t) + a w(1,t), \quad c > 0, \ \delta > 0, \end{cases}$$

which is exponentially stable in the sense of an appropriate system norm.

A major drawback of the backstepping-based control strategy is that it uses full-state feedback, which is not realistic in practice. Furthermore, a boundary observer is required to implement it on a real plant.

A different strategy to stabilize a coupled PDE–ODE system of the form (10.25), is proposed in [26]. A Lyapunov-based stability analysis of a cascade model of a wave equation with a nonlinear ODE, which describes the torsional drilling dynamics is presented. Furthermore, the design methodology is extended to nonlinear systems with actuator dynamics that are governed by a wave PDE with antidamping on the uncontrolled boundary.

The problem of eliminating oilwell drillstring vibrations is addressed from a different perspective in the following chapter. Based on the Lyapunov stability theory two different control strategies to tackle the stick-slip phenomenon are developed. Different representations of the drilling system modeled by the nonlinear neutral-type time-delay equation given in Sect. 2.3 are used. For the first control method, the system is represented by a switching-type equation in which the switching rule is autonomous and depends on the bit angular velocity. The second method uses a linear polytopic approximation of the nonlinear NFDE describing the torsional dynamics. Both strategies guarantee the exponential stability of the closed-loop system.

Chapter 11
Stick-Slip Control: Lyapunov-Based Approach

Based on the neutral-type time-delay model of the drillstring torsional dynamics, this chapter addresses the design of stabilizing controllers aimed at eliminating the stick-slip phenomenon. Within the framework of Lyapunov theory, two control approaches based on different system representations are proposed:

- **Switched model-based control**. As explained in Chap. 3, the frictional torque on bit is usually modeled by a nonlinear function subject to the "sgn" function, the torsional drilling model can thus be regarded as a nonlinear autonomous state-dependent switching system. The drilling system stabilization is addressed within the framework of switched systems theory.
- **Multimodel representation-based control**. This strategy is based on a linear approximation of the neutral-type torsional drilling model describing torsional drilling vibrations. Stabilization conditions are derived from the descriptor approach [97] and a proposal of a Lyapunov-Krasovskii functional.

Both strategies lead to LMI-type conditions (cf. [41]) guaranteeing an exponential convergence of the system trajectories. Numerical simulations illustrate effective stick-slip suppression.

11.1 Stability Analysis: Switched Systems Approach

Switched systems are a particular branch of hybrid systems consisting of a finite number of subsystems and a logical rule which orchestrates switching between these subsystems. Such systems are useful for modeling various real-world systems such as chemical processes [82], communication networks [147], traffic control [176], manufacturing system control [37], among many others.

Since the frictional torque on bit is usually modeled by a nonlinear function which is discontinuous at zero velocity (i.e., in sticking phase), the oil well drilling system can be regarded as a particular class of autonomous switching systems. More precisely, the system switches between a submodel describing the stick phase

M.B. Saldivar Márquez et al., *Analysis and Control of Oilwell Drilling Vibrations*,
Advances in Industrial Control, DOI 10.1007/978-3-319-15747-4_11

and a submodel characterizing the slip phase. The switching is assumed to be state-dependent since it depends on the bit velocity.

In this section we derive asymptotic and exponential stability conditions for a general class of nonlinear neutral-type time-delay systems subject to autonomous state-dependent switching.

11.1.1 Asymptotic Stability Conditions

Consider a nonlinear neutral-type time-delay system with state-dependent switching of the form:

$$\begin{cases} \dot{x}(t) - C_\sigma \dot{x}(t-\tau_1) = A_\sigma x(t) + B_\sigma x(t-\tau_1) + D_\sigma u(t) + f_{1\sigma}(t, x(t)) \\ \qquad + f_{2\sigma}(t, x(t-\tau_1)) \\ x(t_0+\theta) = \varphi(\theta), \quad \forall \theta \in [-\tau_1, 0], \end{cases} \tag{11.1}$$

where $x(t) \in \mathbb{R}^n$ is the state vector, $u(t) \in \mathbb{R}^m$ is the control input, τ_1 is a positive constant time-delay, φ is a continuously differentiable initial function, and $\sigma \in \{1, 2, \ldots, N\}$ is a piecewise constant switching signal. The matrices $(A_\sigma, B_\sigma, C_\sigma)$ are allowed to take values, at an arbitrary time, in the finite set $(A_\sigma, B_\sigma, C_\sigma) \in \{(A_1, B_1, C_1), \ldots, (A_N, B_N, C_N)\}$.

We consider that the nonlinear functions $f_{1\sigma}(t, x(t)), f_{2\sigma}(t, x(t-\tau_1))$ are bounded in magnitude, i.e., there exist positive constants $\alpha_{1\sigma}, \alpha_{2\sigma}$ such that

$$\begin{aligned} \|f_{1\sigma}(t, x(t))\| &\leqslant \alpha_{1\sigma} \|x(t)\|, \quad \forall t \geqslant 0, \\ \|f_{2\sigma}(t, x(t-\tau_1))\| &\leqslant \alpha_{2\sigma} \|x(t-\tau_1)\|, \ \sigma \in \{1, \ldots, N\}. \end{aligned} \tag{11.2}$$

Let $u(t)$ be a state-feedback controller in the form $u(t) = Kx(t-\tau_1)$. Substituting this control law into (11.1), we obtain the closed-loop system:

$$\begin{aligned} \dot{x}(t) - C_\sigma \dot{x}(t-\tau_1) =& A_\sigma x(t) + \bar{B}_\sigma x(t-\tau_1) + f_{1\sigma}(t, x_2(t)) \\ &+ f_{2\sigma}(t, x_2(t-\tau_1)) \\ x(t_0+\theta) =& \varphi(\theta), \quad \forall \theta \in [-\tau, 0], \end{aligned} \tag{11.3}$$

where $\bar{B}_\sigma = B_\sigma + D_\sigma K$.

In [311], asymptotic stability of switched neutral systems is analyzed. The following theorem provides exponential stability conditions for nonlinear switched neutral-type time-delay systems.

Theorem 11.1.1 (Asymptotic stability) *Given a gain matrix K, the switched neutral system* (11.3) *with $f_{1\sigma}(t, x(t))$ and $f_{2\sigma}(t, x(t-\tau_1))$ satisfying* (11.2) *is asymptotically stable if there are symmetric positive definite matrices P, Q_1, Q_2, R_1 such that the LMI*

$$\Psi_i = \begin{pmatrix} \Psi_{i11} & \sqrt{2}P & \sqrt{2}\alpha_{1i}W & \Psi_{i14} & 0 & \Psi_{i16} \\ * & -I & 0 & 0 & 0 & 0 \\ * & * & -I & 0 & 0 & 0 \\ * & * & * & \Psi_{i44} & \alpha_{2i}W & \Psi_{i46} \\ * & * & * & * & -I & 0 \\ * & * & * & * & * & \Psi_{i66} \end{pmatrix} < 0, \tag{11.4}$$

with

$$\begin{aligned} \Psi_{i11} &= PA_i + A_i^T P + Q_1 + A_i^T WA_i - R_1 + \alpha_{1i}^2 I + \alpha_{1i}^2 W + 2A_i^T A_i \\ \Psi_{i14} &= P\bar{B}_i + A_i^T W\bar{B}_i + R_1 + 2A_i^T \bar{B}_i \\ \Psi_{i16} &= PC_i + A_i^T WC_i + 2A_i^T C_i \\ \Psi_{i44} &= -Q_1 + \bar{B}_i^T W\bar{B}_i - R_1 + 2\alpha_{2i}^2 I + \alpha_{2i}^2 W + 2\bar{B}_i^T \bar{B}_i \\ \Psi_{i46} &= \bar{B}_i^T WC_i + 2\bar{B}_i^T C_i \\ \Psi_{i66} &= -Q_2 + C_i^T WC_i + 2C_i^T C_i \\ \bar{B}_i &= B_i + D_i K \\ W &= Q_2 + \tau_1^2 R_1 \end{aligned}$$

is feasible, for all $i = 1, \ldots, N$.

Proof As in [311], we consider the energy functional

$$V(x_t) = x^T(t)Px(t) + \int_{t-\tau_1}^{t} x^T(s)Q_1 x(s)ds + \int_{t-\tau_1}^{t} \dot{x}^T(s)Q_2\dot{x}(s)ds + \tau_1 \int_{-\tau_1}^{0}\int_{t+\theta}^{t} \dot{x}^T(s)R_1\dot{x}(s)dsd\theta. \tag{11.5}$$

Taking the derivative of $V(x_t)$ along the trajectories of any subsystem of (11.3), we have

$$\begin{aligned} \dot{V}(x_t) &= 2x^T(t)P\dot{x}(t) - x^T(t-\tau_1)Q_1x(t-\tau_1) + x^T(t)Q_1x(t) \\ &\quad - \dot{x}^T(t-\tau_1)Q_2\dot{x}(t-\tau_1) + \dot{x}^T(t)\left(Q_2 + \tau_1^2 R_1\right)\dot{x}(t) \\ &\quad - \tau_1 \int_{t-\tau_1}^{t} \dot{x}^T(s)R_1\dot{x}(s)ds. \end{aligned} \tag{11.6}$$

Using the Jensen's inequality we obtain

$$\begin{aligned} -\tau_1 \int_{t-\tau_1}^{t} \dot{x}^T(s)R_1\dot{x}(s)ds &\leqslant -\int_{t-\tau_1}^{t} \dot{x}^T(s)dsR_1 \int_{t-\tau_1}^{t} \dot{x}(s)ds \\ &= -(x(t) - x(t-\tau_1))^T R_1(x(t) - x(t-\tau_1)). \end{aligned} \tag{11.7}$$

Then, substituting (11.7) into (11.6) gives

$$
\begin{aligned}
\dot{V}(x_t) \leqslant\ & 2x^T(t)P\left[C_i\dot{x}(t-\tau_1)+A_ix(t)+\bar{B}_ix(t-\tau_1)\right]-x^T(t-\tau_1)Q_1x(t-\tau_1)\\
&+x^T(t)Q_1x(t)-\dot{x}^T(t-\tau_1)Q_2\dot{x}(t-\tau_1)\\
&-(x(t)-x(t-\tau_1))^TR_1(x(t)-x(t-\tau_1))+F_i\\
&+\left[C_i\dot{x}(t-\tau_1)+A_ix(t)+\bar{B}_ix(t-\tau_1)\right]^T\\
&\times W\left[C_i\dot{x}(t-\tau_1)+A_ix(t)+\bar{B}_ix(t-\tau_1)\right],
\end{aligned}
$$

where $W := Q_2 + \tau_1^2 R_1$, and

$$
\begin{aligned}
F_i = F_i(x_t, f_i) = & 2x^T(t)P\left[f_{1i}(\cdot)+f_{2i}(\cdot)\right]+G_i^TW\left[f_{1i}(\cdot)+f_{2i}(\cdot)\right]\\
&+\left[f_{1i}(\cdot)+f_{2i}(\cdot)\right]^TWG_i+\left[f_{1i}(\cdot)+f_{2i}(\cdot)\right]^TW\left[f_{1i}(\cdot)+f_{2i}(\cdot)\right],
\end{aligned}
\tag{11.8}
$$

where

$$
G_i = G_i(x_t) := \left[C_i\dot{x}(t-\tau_1)+A_ix(t)+\bar{B}_ix(t-\tau_1)\right].
$$

We look for an upper bound on F_i. Considering that, for any vectors $a, b \in \mathbb{R}^n$, the inequality $2a^Tb \leqslant a^Ta + b^Tb$ is satisfied, and taking into account the bounds (11.2), we obtain

$$
2x^T(t)Pf_{1i}(\cdot) \leqslant x^T(t)PPx(t)+f_{1i}(\cdot)^Tf_{1i}(\cdot) \leqslant x^T(t)PPx(t)+\alpha_{1i}^2x^T(t)x(t),
$$

and

$$
\begin{aligned}
2x^T(t)Pf_2(\cdot) &\leqslant x^T(t)PPx(t)+f_{2i}(\cdot)^Tf_{2i}(\cdot)\\
&\leqslant x^T(t)PPx(t)+\alpha_{2i}^2x^T(t-\tau_1)x(t-\tau_1).
\end{aligned}
$$

Similarly,

$$
G_i^TWf_{1i}(\cdot)+f_{1i}^T(\cdot)WG \leqslant G_i^TG_i+f_{1i}^T(\cdot)WWf_{1i}(\cdot) \leqslant G_i^TG_i+\alpha_{1i}^2x^T(t)WWx(t),
$$

$$
\begin{aligned}
G_i^TWf_{2i}(\cdot)+f_{2i}^T(\cdot)WG_i &\leqslant G_i^TG_i+f_{2i}^T(\cdot)WWf_{2i}(\cdot)\\
&\leqslant G_i^TG_i+\alpha_{2i}^2x^T(t-\tau_1)WWx(t-\tau_1),
\end{aligned}
$$

and

$$\begin{aligned}\left[f_{1i}(\cdot)+f_{2i}(\cdot)\right]^T W\left[f_{1i}(\cdot)+f_{2i}(\cdot)\right] &= f_{1i}(\cdot)^T Wf_{1i}(\cdot)+f_{2i}^T(\cdot)Wf_{2i}(\cdot)\\ &\quad +f_{1i}(\cdot)^T Wf_{2i}(\cdot)+f_{2i}(\cdot)^T Wf_{1i}(\cdot)\\ &\leqslant \alpha_{1i}^2 x^T(t)Wx(t)\\ &\quad +\alpha_{2i}^2 x^T(t-\tau_1)Wx(t-\tau_1)\\ &\quad +\alpha_{1i}^2 x^T(t)WWx(t)\\ &\quad +\alpha_{2i}^2 x^T(t-\tau_1)x(t-\tau_1).\end{aligned}$$

Substituting the above inequalities into (11.8) yields

$$\begin{aligned}F_i &\leqslant 2x^T(t)PPx(t)+\alpha_{1i}^2 x^T(t)x(t)\\ &\quad +2\alpha_{2i}^2 x^T(t-\tau_1)x(t-\tau_1)+2G_i^T G_i\\ &\quad +2\alpha_{1i}^2 x^T(t)WWx(t)+\alpha_{2i}^2 x^T(t-\tau_1)WWx(t-\tau_1)\\ &\quad +\alpha_{1i}^2 x^T(t)Wx(t)+\alpha_{2i}^2 x^T(t-\tau_1)Wx(t-\tau_1).\end{aligned}$$

Then, the derivative of $V(x_t)$ along the trajectories of any subsystem ith of (11.3) satisfies

$$\begin{aligned}\dot V(x_t) &\leqslant 2x^T(t)PG_i - x^T(t-\tau_1)Q_1x(t-\tau_1)+x^T(t)Q_1x(t)\\ &\quad -\dot x^T(t-\tau_1)Q_2\dot x(t-\tau_1)+G_i^T WG_i\\ &\quad -(x(t)-x(t-\tau_1))^T R_1(x(t)-x(t-\tau_1))\\ &\quad +2x^T(t)PPx(t)+\alpha_{1i}^2 x^T(t)x(t)\\ &\quad +2\alpha_{2i}^2 x^T(t-\tau_1)x(t-\tau_1)+2G_i^T G_i+2\alpha_{1i}^2 x^T(t)WWx(t)\\ &\quad +\alpha_{2i}^2 x^T(t-\tau_1)WWx(t-\tau_1)\\ &\quad +\alpha_{1i}^2 x^T(t)Wx(t)+\alpha_{2i}^2 x^T(t-\tau_1)Wx(t-\tau_1).\end{aligned}$$

Setting $\xi(t)=(x^T(t)\ \ x^T(t-\tau_1)\ \ \dot x^T(t-\tau_1))$, the above inequality is written as

$$\dot V(x_t)\leqslant \xi(t)\Phi_i(P,Q_1,Q_2,R_1)\xi^T(t), \tag{11.9}$$

where

$$\Phi_i=\begin{pmatrix}\Phi_{i11} & \Phi_{i12} & \Phi_{i13}\\ * & \Phi_{i22} & \Phi_{i23}\\ * & * & \Phi_{i33}\end{pmatrix}, \tag{11.10}$$

$$
\begin{aligned}
\Phi_{i11} &= PA_i + A_i^T P + Q_1 + A_i^T W A_i - R_1 + 2PP + \alpha_{1i}^2 I \\
&\quad + 2\alpha_{1i}^2 WW + \alpha_{1i}^2 W + 2A_i^T A_i \\
\Phi_{i12} &= P\bar{B}_i + A_i^T W \bar{B}_i + R_1 + 2A_i^T \bar{B}_i \\
\Phi_{i13} &= PC_i + A_i^T W C_i + 2A_i^T C_i \\
\Phi_{i22} &= -Q_1 + \bar{B}_i^T W \bar{B}_i - R_1 + 2\alpha_{2i}^2 I + \alpha_{2i}^2 WW + \alpha_{2i}^2 W + 2\bar{B}_i^T \bar{B}_i \\
\Phi_{i23} &= \bar{B}_i^T W C_i + 2\bar{B}_i^T C_i \\
\Phi_{i33} &= -Q_2 + C_i^T W C_i + 2C_i^T C_i \\
\bar{B}_i &= B_i + D_i K \\
W &= Q_2 + \tau_1^2 R_1 .
\end{aligned}
$$

By Schur's complements, it follows that $\Phi_i < 0$ in (11.10) is equivalent to $\Psi_i < 0$ in (11.4) and the result follows.

11.1.2 Exponential Stability Conditions

The closed-loop system (11.3) is said to be exponentially stable or α-stable with decay rate α if there exists a scalar $\kappa \geqslant 1$ such that, for any continuously differentiable initial condition φ, the solution $x(t, t_0, \varphi)$, satisfies:

$$
|x(t, t_0, \varphi)| \leqslant \kappa \, |\varphi| \, \mathrm{e}^{-\alpha(t-t_0)}.
$$

Using the change of variable:

$$
x^*(t) := \mathrm{e}^{\alpha t} x(t),
$$

we can rewrite the system (11.3) as

$$
\begin{cases}
\dot{x}^*(t) = C_\sigma \mathrm{e}^{\alpha\tau_1} \dot{x}^*(t-\tau_1) + \bar{A}_\sigma x^*(t) + \mathrm{e}^{\alpha\tau_1} \bar{B}_\sigma x^*(t-\tau_1) \\
\qquad\qquad -\alpha \mathrm{e}^{\alpha\tau_1} C_\sigma x^*(t-\tau_1) + f_{1\sigma}(t, x^*(t)) + f_{2\sigma}(t, x^*(t-\tau_1)), \\
x(t_0+\theta) = \varphi(\theta), \quad \forall \theta \in [-\tau, 0],
\end{cases}
\tag{11.11}
$$

where $\bar{A}_\sigma = A_\sigma + \alpha I$ and $\bar{B}_\sigma = B_\sigma + D_\sigma K$. Notice that the condition (11.2) imply

$$
\begin{aligned}
\left\| f_{1\sigma}(t, \dot{\Phi}_b(t)) \right\| &\leqslant \alpha_{1\sigma} \left\| x^*(t) \right\|, \quad \forall t \geqslant 0, \\
\left\| f_{2\sigma}(t, \dot{\Phi}_b(t-\tau_1)) \right\| &\leqslant \alpha_{2\sigma} \left\| x^*(t-\tau_1) \right\|, \quad \sigma \in \{1, \ldots, N\}.
\end{aligned}
$$

Our proposal is to find conditions for which the solution $x^* = 0$ of the transformed system (11.11) is stable. Clearly, these conditions will ensure the exponential stability of the system (11.3). The following result follows from Theorem 11.1.1.

Theorem 11.1.2 (Exponential stability) *Given a gain matrix* K, *the switched neutral system* (11.3) *with* $f_{1\sigma}(t, x(t))$ *and* $f_{2\sigma}(t, x(t-\tau_1))$ *satisfying* (11.2) *is exponentially stable if there are symmetric positive definite matrices* P, Q_1, Q_2, R_1 *such that the LMI*

$$\Psi_i = \begin{pmatrix} \Psi_{i11} & \sqrt{2}P & \sqrt{2}\alpha_{1i}W & \Psi_{i14} & 0 & \Psi_{i16} \\ * & -I & 0 & 0 & 0 & 0 \\ * & * & -I & 0 & 0 & 0 \\ * & * & * & \Psi_{i44} & \alpha_{2i}W & \Psi_{i46} \\ * & * & * & * & -I & 0 \\ * & * & * & * & * & \Psi_{i66} \end{pmatrix} < 0, \tag{11.12}$$

with

$$\begin{aligned}
\Psi_{i11} &= P\bar{A}_i + \bar{A}_i^T P + Q_1 + \bar{A}_i^T W \bar{A}_i - R_1 + \alpha_{1i}^2 I + \alpha_{1i}^2 W + 2\bar{A}_i^T \bar{A}_i \\
\Psi_{i14} &= e^{\alpha\tau_1} P\left(\bar{B}_i - \alpha C_i\right) + e^{\alpha\tau_1}\bar{A}_i^T W\left(\bar{B}_i - \alpha C_i\right) + R_1 + 2e^{\alpha\tau_1}\bar{A}_i^T\left(\bar{B}_i - \alpha C_i\right) \\
\Psi_{i16} &= e^{\alpha\tau_1} P C_i + e^{\alpha\tau_1}\bar{A}_i^T W C_i + 2e^{\alpha\tau_1}\bar{A}_i^T C_i \\
\Psi_{i44} &= -Q_1 + e^{2\alpha\tau_1}\left(\bar{B}_i - \alpha C_i\right)^T W\left(\bar{B}_i - \alpha C_i\right) - R_1 + 2\alpha_{2i}^2 I + \alpha_{2i}^2 W \\
&\quad + 2e^{2\alpha\tau_1}\left(\bar{B}_i - \alpha C_i\right)^T\left(\bar{B}_i - \alpha C_i\right) \\
\Psi_{i46} &= e^{2\alpha\tau_1}\left(\bar{B}_i - \alpha C_i\right)^T W C_i + 2e^{2\alpha\tau_1}\left(\bar{B}_i - \alpha C_i\right)^T C_i \\
\Psi_{i66} &= -Q_2 + e^{2\alpha\tau_1} C_i^T W C_i + 2e^{2\alpha\tau_1} C_i^T C_i \\
\bar{A}_i &= A_i + \alpha I \\
\bar{B}_i &= B_i + D_i K \\
W &= Q_2 + \tau_1^2 R_1
\end{aligned}$$

is feasible, for all $i = 1, \ldots, N$.

11.1.3 Stability Analysis of the Drilling System

The torsional drilling model (2.32) is represented as an autonomous state-dependent switching system as follows,

$$\ddot{\Phi}_b(t) - \Upsilon \ddot{\Phi}_b(t-2\tau) = \left(-\Psi - \frac{c_b}{I_B}\right)\dot{\Phi}_b(t) + \left(\frac{\Upsilon c_b}{I_B} - \Upsilon\Psi\right)\dot{\Phi}_b(t-2\tau) \tag{11.13}$$
$$+ \Pi\Omega(t-\tau) + f_{1\sigma}\left(t, \dot{\Phi}_b(t)\right) + f_{2\sigma}\left(t, \dot{\Phi}_b(t-2\tau)\right),$$

$$\Pi = \frac{2\Psi\beta}{\beta + \tilde{c}GJ}, \quad \Upsilon = \frac{\beta - \tilde{c}GJ}{\beta + \tilde{c}GJ}, \quad \Psi = \frac{\tilde{c}GJ}{I_B}, \quad \tau = \tilde{c}L, \quad \tilde{c} = \sqrt{\frac{I}{GJ}}.$$

where $\dot{\Phi}_b$ is the bit angular velocity and Ω is the angular velocity coming from the rotary table, taken as the control input.

The functions $f_{1\sigma}, f_{2\sigma}$, $\sigma = 1, 2$ of system (11.13) are switched according to the following autonomous state-dependent rule:

$$\begin{cases} \text{for } \dot{\Phi}_b(t) = 0: \\ f_{11}\left(t, \dot{\Phi}_b(t)\right) = f_{21}\left(t, \dot{\Phi}_b(t-2\tau)\right) = 0 \\ \\ \text{for } \dot{\Phi}_b(t) > 0: \\ f_{12}\left(t, \dot{\Phi}_b(t)\right) = -c_1 - c_2 e^{-\gamma_b \dot{\Phi}_b(t)} \\ f_{22}\left(t, \dot{\Phi}_b(t-2\tau)\right) = c_1\Upsilon + c_2\Upsilon e^{-\gamma_b \dot{\Phi}_b(t-2\tau)}, \end{cases} \tag{11.14}$$

with $c_1 = \frac{W_{ob}R_b}{I_B}\mu_{cb}$, and $c_2 = \frac{W_{ob}R_b}{I_B}(\mu_{sb} - \mu_{cb})$.

Notice that the switching depends on the angular velocity at the bottom end of the drillstring $\dot{\Phi}_b(t)$. If we approximate the switching rule (11.14) by the following one,

$$\begin{cases} \text{for } 0 \leqslant \dot{\Phi}_b(t) < 0.1: \\ f_{11}\left(t, \dot{\Phi}_b(t)\right) = f_{21}\left(t, \dot{\Phi}_b(t-2\tau)\right) = 0 \\ \\ \text{for } \dot{\Phi}_b(t) > 0.1: \\ f_{12}\left(t, \dot{\Phi}_b(t)\right) = -c_1 - c_2 e^{-\gamma_b \dot{\Phi}_b(t)} \\ f_{22}\left(t, \dot{\Phi}_b(t-2\tau)\right) = c_1\Upsilon + c_2\Upsilon e^{-\gamma_b \dot{\Phi}_b(t-2\tau)}, \end{cases} \tag{11.15}$$

then, the conditions (11.2) on $f_{12}\left(t, \dot{\Phi}_b(t)\right)$ and $f_{22}\left(t, \dot{\Phi}_b(t-2\tau)\right)$ are satisfied for some relatively small constants α_1, α_2. The switching law (11.15) means that for small values of the angular velocity at the bottom end ($\dot{\Phi}_b < 0.1\,\text{rad s}^{-1}$) the non-linear part of the torque on bit has no effect (this actually happens when $\dot{\Phi}_b = 0$).

According to (11.15) we have that for $0 \leqslant \dot{\Phi}_b(t) < 0.1$:

$$\begin{aligned} \left\| f_{11}(t, \dot{\Phi}_b(t)) \right\| &= 0 \leqslant \alpha_1 \left\| \dot{\Phi}_b(t) \right\|, \\ \left\| f_{21}(t, \dot{\Phi}_b(t-2\tau)) \right\| &= 0 \leqslant \alpha_2 \left\| \dot{\Phi}_b(t-2\tau) \right\|, \end{aligned} \tag{11.16}$$

and for $\dot{\Phi}_b(t) \geqslant 0.1$:

$$\begin{aligned} \left\| f_{12}(t, \dot{\Phi}_b(t)) \right\| &= \left\| -c_1 - c_2 \mathrm{e}^{-\gamma_b \dot{\Phi}_b(t)} \right\| \leqslant \alpha_1 \left\| \dot{\Phi}_b(t) \right\|, \\ \left\| f_{22}(t, \dot{\Phi}_b(t-2\tau)) \right\| &= \left\| c_1 \Upsilon + c_2 \Upsilon \mathrm{e}^{-\gamma_b \dot{\Phi}_b(t-2\tau)} \right\| \leqslant \alpha_2 \left\| \dot{\Phi}_b(t-2\tau) \right\|. \end{aligned} \tag{11.17}$$

Using the model parameters given in Table C.1 of Appendix C, we obtain:

$$\tau = 0.3719, \quad c_1 = 85.0829, \quad c_2 = 51.0498.$$

The conditions (11.16) and (11.17) are satisfied, for all $\alpha_1 > 1317.1$, $\alpha_2 > 974.3$.

Figure 11.1a shows a simulation of the trajectory $\dot{\Phi}_b(t)$ of the drilling system (11.13) under the rule (11.14) without control actions for $\Omega(t) = 15\,\text{rad s}^{-1}$.

Here, we propose a stabilizing control law that ensures the exponential convergence of the trajectory $x_1(t) = \dot{\Phi}_b(t) - \Omega_0$ and consequently the suppression of the stick-slip phenomenon. For stability issues the velocity at the bottom end must track the angular velocity at the upper part.

In order to achieve the velocity tracking we propose the control law:

$$\Omega(t-\tau) = -\lambda_0 \dot{x}_1(t-2\tau) - \lambda_1 x_1(t-2\tau), \tag{11.18}$$

which can be written in terms of the bit speed as:

$$\Omega(t-\tau) = -\lambda_0 \ddot{\Phi}_b(t-2\tau) - \lambda_1 \dot{\Phi}_b(t-2\tau) + \lambda_1 \Omega_0. \tag{11.19}$$

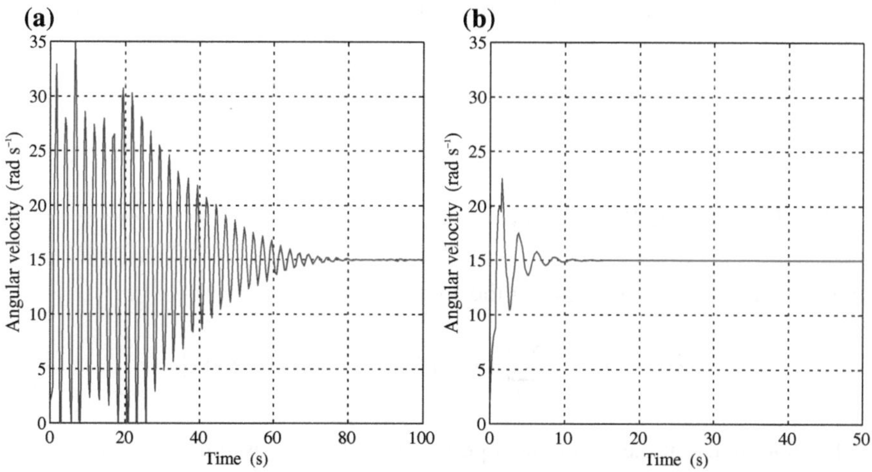

Fig. 11.1 Simulation of the torsional switching model (11.13) under the rule (11.14). Angular velocity at the bottom extremity $\dot{\Phi}_b(t)$ for $\Omega_0 = 15\,\text{rad s}^{-1}$. **a** Trajectory without control actions (stick-slip). **b** Elimination of stick-slip oscillations by means of the control law (11.19)

The result of Theorem 11.1.2 allows analyzing the exponential stability of the drilling system subject to the switching rule (11.15) in closed loop with the controller (11.18). Notice that α_1 and α_2 satisfy (11.16) and (11.17).

After computing the LMIs stated in Theorem 11.1.2 with $\lambda_0 = 0.05$, $\lambda_1 = 0.36$, $\alpha_1 = 1{,}320$, $\alpha_2 = 975$, and $\alpha = 0.6$, the exponential stability of the closed-loop drilling system is established.

The simulation result of Fig. 11.1b shows the stick-slip elimination through the application of the controller (11.19).

11.2 Multimodel Representation-Based Control

The control method developed in this section is based on Lyapunov-Krasovskii techniques; it is applicable to the class of nonlinear neutral-type time-delay systems that can be transformed into a multimodel system, which is the case of the torsional dynamics model of the oil well drilling system introduced in Sect. 2.3. The strategy, introduced in [255], yields exponential stabilization of the closed-loop system.

A multimodel system is a set of linear models nonlinearly weighted that can be represented as follows:

$$\dot{x}(t) - D\dot{x}(t-\tau_1) = \sum_{i\in I^r} h_i(x_t)\left[A_i x(t) + A_{i\tau_1} x(t-\tau_1) + Bu(t)\right], \tag{11.20}$$

where $\tau_1 > 0$ is a constant time-delay, the set I^r is the set of integers $\{1, \ldots, r\}$, where r is the number of subsystems required to describe the multimodel system. The functions $h_i(\cdot)$ are scalar weighting functions satisfying the convexity conditions:

$$\sum_{i\in I^r} h_i(x_t) = 1 \quad \forall i = 1, \ldots, r, \quad h_i(x_t) \geqslant 0. \tag{11.21}$$

11.2.1 Exponential Stability

First, we analyze the α-stability of the autonomous system:

$$\dot{x}(t) - D\dot{x}(t-\tau_1) = \sum_{i\in I^r} h_i(x_t)\left[A_i x(t) + A_{i\tau_1} x(t-\tau_1)\right]. \tag{11.22}$$

To guarantee that the difference operator is stable, we assume $|D| < 1$.

The change of variable $x_\alpha(t) = \mathrm{e}^{\alpha t} x(t)$, transforms the system (11.22) into:

$$\dot{x}_\alpha(t) D - \mathrm{e}^{\alpha\tau_1} \dot{x}_\alpha(t-\tau_1) = \sum_{i \in I^r} h_i(x_t)\big[(A_i + \alpha I_n)\, x_\alpha(t) + \mathrm{e}^{\alpha\tau_1} \left(A_{i\tau_1} - \alpha D\right) x_\alpha(t-\tau_1)\big]. \tag{11.23}$$

The proposal is to find conditions for which the solution $x_\alpha = 0$ of the transformed system (11.23) is asymptotically stable. Clearly, these conditions will guarantee the exponential stability of system (11.22). The following theorem provides these stability conditions.

Theorem 11.2.1 *The solution* $x(t) = 0$ *of the system* (11.22) *is* α*-stable if there exist matrices* $0 < P_1 = P_1^T$, P_2, P_3, $Q = Q^T$, *and* $R = R^T$, *such that for all* $i \in I^r$ *the following LMI is satisfied*

$$\begin{pmatrix} \Psi_i & P^T \begin{pmatrix} 0 \\ e^{\alpha\tau_1}\left(A_{i\tau_1} - \alpha D\right) \end{pmatrix} & P^T \begin{pmatrix} 0 \\ e^{\alpha\tau_1} D \end{pmatrix} \\ * & -R/\tau_1 & 0 \\ * & * & -Q \end{pmatrix} < 0, \tag{11.24}$$

where

$$\begin{aligned} P &:= \begin{pmatrix} P_1 & 0 \\ P_2 & P_3 \end{pmatrix}, \quad P_1 = P_1^T > 0, \\ \Psi_i &:= P^T \begin{pmatrix} 0 & I_n \\ \Lambda_i & -I_n \end{pmatrix} + \begin{pmatrix} 0 & I_n \\ \Lambda_i & -I_n \end{pmatrix}^T P + \begin{pmatrix} 0 & 0 \\ 0 & \tau_1 R + Q \end{pmatrix}, \\ \Lambda_i &:= A_i + \alpha I_n + e^{\alpha\tau_1}\left(A_{i\tau_1} - \alpha D\right). \end{aligned}$$

Proof According to the Leibniz formula,

$$x_\alpha(t-\tau_1) = x_\alpha(t) - \int_{t-\tau_1}^{t} \dot{x}_\alpha(s)ds,$$

then, we can write the system (11.23) as:

$$\begin{aligned} \dot{x}_\alpha(t) - D\mathrm{e}^{\alpha\tau_1} \dot{x}_\alpha(t-\tau_1) = &\sum_{i \in I^r} h_i(x_t)\big[\left(A_i + \alpha I_n + \mathrm{e}^{\alpha\tau_1}\left(A_{i\tau_1} - \alpha D\right)\right) x_\alpha(t)\big] \\ &- \sum_{i \in I^r} h_i(x_t)\Big[\mathrm{e}^{\alpha\tau_1}\left[A_{i\tau_1} - \alpha D\right] \int_{t-\tau_1}^{t} \dot{x}_\alpha(s)ds\Big]. \end{aligned}$$

Using the descriptor form introduced in [97], we have:

$$\begin{aligned}\dot{x}_\alpha(t) &= y(t),\\ y(t) &= \sum_{i\in I^r} h_i(x_t)\left[De^{\alpha\tau_1}y(t-\tau_1) + \Lambda_i x_\alpha(t)\right]\\ &\quad - \sum_{i\in I^r} h_i(x_t)\left[e^{\alpha\tau_1}\left(A_{i\tau_1} - \alpha D\right)\int_{t-\tau_1}^{t} y(s)ds\right],\end{aligned}$$

where

$$\Lambda_i := A_i + \alpha I_n + e^{\alpha\tau_1}\left(A_{i\tau_1} - \alpha D\right).$$

Then, we can write

$$E\begin{pmatrix}\dot{x}_\alpha(t)\\ \dot{y}(t)\end{pmatrix} = \begin{pmatrix} y(t)\\ \sum_{i\in I^r} h_i(x_t)\lambda\end{pmatrix},$$

where

$$\begin{aligned}E &= diag\left\{I_n, 0\right\},\\ \lambda &= -y(t) + De^{\alpha\tau_1}y(t-\tau_1) + \Lambda_i x_\alpha(t) - e^{\alpha\tau_1}\left(A_{i\tau_1} - \alpha D\right)\int_{t-\tau_1}^{t} y(s)ds.\end{aligned}$$

Following [97], we use the Lyapunov-Krasovskii functional

$$\begin{aligned}V_\alpha(t) &= \left(x_\alpha^T(t)\; y^T(t)\right)EP\begin{pmatrix}x_\alpha(t)\\ y(t)\end{pmatrix} + \int_{-\tau_1}^{0}\int_{t+\theta}^{t} y^T(s)Ry(s)dsd\theta\\ &\quad + \int_{t-\tau_1}^{t} y^T(s)Qy(s)ds,\end{aligned} \tag{11.25}$$

where

$$P = \begin{pmatrix}P_1 & 0\\ P_2 & P_3\end{pmatrix},\quad P_1 = P_1^T > 0,\quad R > 0,\quad Q > 0.$$

The functional $V_\alpha(t)$ is positive definite since

$$\left(x_\alpha^T(t)\; y^T(t)\right)EP\begin{pmatrix}x_\alpha(t)\\ y(t)\end{pmatrix} = x_\alpha^T(t)P_1x_\alpha(t).$$

Notice that $EP = P^TE$. Taking the derivative in "t" of $V_\alpha(t)$, we obtain:

$$\begin{aligned}\dot{V}_\alpha(t) &= 2\left(x_\alpha^T(t)\; y^T(t)\right)P^T\begin{pmatrix} y(t)\\ \sum_{i\in I^r} h_i(x_t)\lambda\end{pmatrix} + \tau_1 y^T(t)Ry(t)\\ &\quad - \int_{t-\tau_1}^{t} y^T(s)Ry(s)ds + y^T(t)Qy(t) - y^T(t-\tau_1)Qy(t-\tau_1).\end{aligned}$$

Setting $\xi = \begin{pmatrix} x_\alpha(t) & y(t) & y(t-\tau_1) \end{pmatrix}$, $\dot{V}_\alpha(t)$ can be rewritten as:

$$\dot{V}_\alpha(t) = \xi^T \begin{pmatrix} \tilde{\Psi}_i & P^T \begin{pmatrix} 0 \\ e^{\alpha\tau_1} D \end{pmatrix} \\ \begin{pmatrix} 0 & e^{\alpha\tau_1} D^T \end{pmatrix} P & -Q \end{pmatrix} \xi + \Xi$$

$$- \int_{t-\tau_1}^{t} y^T(s) R y(s) ds, \tag{11.26}$$

where

$$\Xi = -2 \int_{t-\tau_1}^{t} \begin{pmatrix} x_\alpha^T(t) & y^T(t) \end{pmatrix} P^T \begin{pmatrix} 0 \\ e^{\alpha\tau_1} \left(A_{i\tau_1} - \alpha D\right) \end{pmatrix} y(s) ds,$$

$$\tilde{\Psi}_i = \sum_{i \in I^r} h_i(x_t) \left\{ P^T \begin{pmatrix} 0 & I_n \\ \Lambda_i & -I_n \end{pmatrix} + \begin{pmatrix} 0 & I_n \\ \Lambda_i & -I_n \end{pmatrix}^T P + \begin{pmatrix} 0 & 0 \\ 0 & \tau_1 R + Q \end{pmatrix} \right\}.$$

An upper bound on Ξ is given by[1]:

$$\Xi \leqslant \begin{pmatrix} x_\alpha^T(t) & y^T(t) \end{pmatrix} P^T \begin{pmatrix} 0 \\ e^{\alpha\tau_1} \left(A_{i\tau_1} - \alpha D\right) \end{pmatrix} \tau_1 R^{-1}$$

$$\begin{pmatrix} 0 & e^{\alpha\tau_1} \left(A_{i\tau_1} - \alpha D\right)^T \end{pmatrix} P \begin{pmatrix} x_\alpha(t) \\ y(t) \end{pmatrix} \tag{11.27}$$

$$+ \int_{t-\tau_1}^{t} y^T(s) R y(s) ds.$$

From (11.26) and (11.27), it yields

$$\dot{V}_\alpha(t) \leqslant \xi^T \begin{pmatrix} \tilde{\Psi}_i & P^T \begin{pmatrix} 0 \\ e^{\alpha\tau_1} D \end{pmatrix} \\ \begin{pmatrix} 0 & e^{\alpha\tau_1} D^T \end{pmatrix} P & -Q \end{pmatrix} \xi$$

$$+ \begin{pmatrix} x_\alpha^T(t) & y^T(t) \end{pmatrix} P^T \begin{pmatrix} 0 \\ e^{\alpha\tau_1} \left(A_{i\tau_1} - \alpha D\right) \end{pmatrix} \left(\tau_1 R^{-1}\right)$$

$$\begin{pmatrix} 0 & e^{\alpha\tau_1} \left(A_{i\tau_1} - \alpha D\right)^T \end{pmatrix} P \begin{pmatrix} x_\alpha(t) \\ y(t) \end{pmatrix}.$$

[1] Inequality (11.27) is obtained by using the following property. For all vectors a, $b \in \mathbb{R}^n$ and a positive definite matrix $R \in \mathbb{R}^{n \times n}$, the inequality $\pm 2a^T b \leqslant a^T R^{-1} a + b^T R b$ is satisfied.

Finally, using Schur complements, the system (11.22) is asymptotically stable if every matrix,

$$\begin{pmatrix} \Psi_i & P^T \begin{pmatrix} 0 \\ e^{\alpha\tau_1}\left(A_{i\tau_1} - \alpha D\right) \end{pmatrix} & P^T \begin{pmatrix} 0 \\ e^{\alpha\tau_1} D \end{pmatrix} \\ * & -R/\tau_1 & 0 \\ * & * & -Q \end{pmatrix},$$

$i \in I^r$, is negative definite, i.e., if the LMI condition (11.24) is satisfied.

11.2.2 Exponential Stabilization

Having determined the criteria for exponential stability of the open-loop system (11.22), the next step is to define an algorithm that allows the synthesis of a gain K such that the feedback control law

$$u(t) = Kx(t - \tau_1), \tag{11.28}$$

exponentially stabilizes the closed-loop system

$$\dot{x}(t) - D\dot{x}(t - \tau_1) = \sum_{i \in I^r} h_i(x_t)\left[A_i x(t) + \left(A_{i\tau_1} + BK\right) x(t - \tau_1)\right], \tag{11.29}$$

with a guaranteed decay rate α.

By replacing the matrix $A_{i\tau_1}$, by the matrix $A_{i\tau_1} + BK$ in Theorem 11.2.1, we obtain that the solution $x(t) = 0$ of the system (11.29) is α-stable if there exist matrices $0 < P_1 = P_1^T$, P_2, P_3, $Q = Q^T$, $R = R^T$ such that for all $i \in I^r$ the following BMIs are satisfied:

$$\begin{pmatrix} \Psi_i & P^T \begin{pmatrix} 0 \\ e^{\alpha\tau_1} \chi \end{pmatrix} & P^T \begin{pmatrix} 0 \\ e^{\alpha\tau_1} D \end{pmatrix} \\ * & -R/\tau_1 & 0 \\ * & * & -Q \end{pmatrix} < 0, \tag{11.30}$$

where

$$
\begin{aligned}
P &:= \begin{pmatrix} P_1 & 0 \\ P_2 & P_3 \end{pmatrix}, \quad P_1 = P_1^T > 0, \\
\Psi_i &:= P^T \begin{pmatrix} 0 & I_n \\ \Lambda_i & -I_n \end{pmatrix} + \begin{pmatrix} 0 & I_n \\ \Lambda_i & -I_n \end{pmatrix}^T P + \begin{pmatrix} 0 & 0 \\ 0 & \tau_1 R + Q \end{pmatrix}, \\
\Lambda_i &:= A_i + \alpha I_n + \mathrm{e}^{\alpha\tau_1}\left(A_{i\tau_1} + BK - \alpha D\right), \\
\chi &:= \left(A_{i\tau_1} + BK - \alpha D\right).
\end{aligned}
$$

A well known synthesis gain technique which overcome the *bilinearity* of the conditions (11.30) was introduced by [280]. It consists of set $P_3 = \varepsilon P_2$, $\varepsilon \in \mathbb{R}$, where P_2 is a nonsingular matrix, and $\bar{P} = P_2^{-1}$.

Define $\bar{P}_1 = \bar{P}^T P_1 \bar{P}$, $\bar{R} = \bar{P}^T R \bar{P}$, and $Y = K\bar{P}$. Multiplying the right side of (11.30) by $\Delta_3 = diag\left\{\bar{P}, \bar{P}, \bar{P}\right\}$ and the left side by Δ_3^T, we obtain the LMI stabilization condition stated in the following theorem.

Theorem 11.2.2 *The system* (11.29) *is α-stabilizable if there exist a positive constant ε and $n \times n$ matrices $\bar{P}_1 > 0$, $\bar{P}$, $\bar{Q} = \bar{Q}^T$, $\bar{R} = \bar{R}^T$, and Y such that for all $i \in I^r$ the following LMI is satisfied:*

$$
\begin{pmatrix}
\Phi_i & \begin{pmatrix} e^{\alpha\tau_1}\vartheta \\ \varepsilon e^{\alpha\tau_1}\vartheta \end{pmatrix} & \begin{pmatrix} e^{\alpha\tau_1} D\bar{P} \\ \varepsilon e^{\alpha\tau_1} D\bar{P} \end{pmatrix} \\
* & -\bar{R}/\tau_1 & 0 \\
* & * & -\bar{Q}
\end{pmatrix} < 0,
$$

where

$$
\begin{aligned}
\vartheta &= \left(A_{i\tau_1} - \alpha D\right)\bar{P} + BY, \\
\Phi_i &= \begin{pmatrix} \Phi_{11} & \Phi_{12} \\ * & \Phi_{22} \end{pmatrix},
\end{aligned}
$$

$$
\begin{aligned}
\Phi_{11} &= \left(A_i + \alpha I_n + e^{\alpha\tau_1}\left(A_{i\tau_1} - \alpha D\right)\right)\bar{P} \\
&\quad + \bar{P}^T\left(A_i + \alpha I_n + e^{\alpha\tau_1}\left(A_{i\tau_1} - \alpha D\right)\right)^T + BY + Y^T B^T, \\
\Phi_{12} &= \bar{P}_1^T - \bar{P} + \varepsilon\bar{P}^T\left(A_i + \alpha I_n + e^{\alpha\tau_1}\left(A_{i\tau_1} - \alpha D\right)\right)^T + \varepsilon Y^T B^T, \\
\Phi_{22} &= -\varepsilon\left(\bar{P} + \bar{P}^T\right) + \tau_1\bar{R} + \bar{Q}.
\end{aligned}
$$

Moreover, the feedback gain is given by

$$
K = Y\bar{P}^{-1}.
$$

11.2.3 Drilling System Stabilization

The presented exponential stabilization results are applied to the drilling torsional model (2.32) coupled to the frictional torque (3.14) with the friction coefficient given in (3.15).

In order to shift the operating point, we introduce the new variable x_1 defined as:

$$x_1(t) = \dot{\Phi}_b(t) - \Omega_0. \tag{11.31}$$

The torsional behavior of the drilling system can be described by the following nonlinear neutral-type equation:

$$\begin{aligned}\dot{x}_1(t) + d\dot{x}_1(t-2\tau) = {} & a_0 x_1(t) + a_1 x_1(t-2\tau) + bu(t-\tau) \\ & - c_2 \mathrm{e}^{-\gamma_b (x_1(t)+\Omega_0)} \mathrm{sgn}\,(x_1(t) + \Omega_0) \\ & + c_2 \Upsilon \mathrm{e}^{-\gamma_b (x_1(t-2\tau)+\Omega_0)} \mathrm{sgn}\,(x_1(t-2\tau) + \Omega_0),\end{aligned} \tag{11.32}$$

where

$$\begin{array}{ll} a_0 = -\Psi - \frac{c_b}{I_B} & a_1 = \frac{\Upsilon c_b}{I_B} - \Upsilon \Psi \\ c_2 = \frac{W_{ob} R_b \mu_{sb}}{I_B} & b = \Pi \\ d = -\Upsilon & |d| < 1. \end{array}$$

Next, a polytopic representation of the nonlinear neutral-type model (11.32) is obtained. Consider the following change of variables:

$$\begin{cases} \varkappa_1(t) = x_1(t), \\ \varkappa_2(t) = \mathrm{e}^{-\gamma_b \varkappa_1(t) + \Omega_0}, \end{cases}$$

therefore,

$$\begin{cases} \dot{\varkappa}_1(t) = \dot{x}_1(t), \\ \dot{\varkappa}_2(t) = -\gamma_b \dot{\varkappa}_1(t) \mathrm{e}^{-\gamma_b \varkappa_1(t) + \Omega_0} = -\gamma_b \dot{\varkappa}_1(t) \varkappa_2(t). \end{cases}$$

System (11.32) can be written as

$$\dot{x}(t) - D\dot{x}(t-2\tau) = A(x)x(t) + A_{2\tau}(x)x(t-2\tau) + B_\tau u(t-\tau), \tag{11.33}$$

where $x(t)$ is defined as $x(t) = [\varkappa_1(t) \;\; \varkappa_2(t)]^T$, the control input $u(t)$ corresponds to the angular velocity provided by the rotary table and the involved matrices are given by:

$$D = \begin{pmatrix} \Upsilon & 0 \\ 0 & 0 \end{pmatrix}, \quad B_\tau = \begin{pmatrix} \Pi \\ 0 \end{pmatrix},$$

$$A_{2\tau}(x) = \begin{pmatrix} \Upsilon\left(\frac{c_b}{I_B} - \Psi\right) & c_2\Upsilon \operatorname{sgn}(\varkappa_1(t-2\tau) + \Omega_0) \\ 0 & 0 \end{pmatrix},$$

$$A(x) = \begin{pmatrix} -\left(\Psi + \frac{c_b}{I_B}\right) & -c_2 \operatorname{sgn}(\varkappa_1(t) + \Omega_0) \\ 0 & -\gamma_b \dot{\varkappa}_1(t) \end{pmatrix}.$$

Notice that the entries of the matrices D and B_τ are constant, and the entry $c_2\Upsilon \operatorname{sgn}(\varkappa_1(t-2\tau)+\Omega_0)$ of the matrix $A_{2\tau}(x)$ is bounded. If we consider that $\dot{\varkappa}_1(t)$ is a bounded variable then, so is the matrix $A(x)$. Then, we can obtain a polytopic representation of the matrices $A(x)$, $A_{2\tau}(x)$ as:

$$A(x)x(t) + A_{2\tau}(x)x(t-2\tau) = \sum_{i \in I^r} h_i(x_t)\big[A_i x(t) + A_{i2\tau}x(t-2\tau)\big], \tag{11.34}$$

where A_i, $A_{i2\tau}$ only have constant coefficients [277]. The functions $h_i(x_t)$, $i \in I^r$ are scalar not necessarily known weighting functions satisfying the convexity property (11.21).

Clearly, the nonlinear drilling system (11.32) can be written in the polytopic form (11.20).

Using parameters given in Table C.1 of Appendix C, the matrices $A(x)$, $A_{2\tau}(x)$, B_τ, and D of the oil well drilling model (11.33) take the following values:

$$D = \begin{pmatrix} 0.7396 & 0 \\ 0 & 0 \end{pmatrix}, \quad B_\tau = \begin{pmatrix} 5.8523 \\ 0 \end{pmatrix},$$

$$A(x) = \begin{pmatrix} -3.3645 & -136.1327\operatorname{sgn}(\varkappa_1(t) + \Omega_0) \\ 0 & -0.9\dot{\varkappa}_1(t) \end{pmatrix},$$

$$A_{2\tau}(x) = \begin{pmatrix} -2.4878 & 100.6802\operatorname{sgn}(\varkappa_1(t-2\tau) + \Omega_0) \\ 0 & 0 \end{pmatrix}.$$

According to [268], the polytopic representation of the system requires $A(x)$ and $A_{2\tau}(x)$ to be bounded functions.

Remark 11.2.3 Notice that there are three independent functions involved: $\dot{\varkappa}_1(t)$, $\operatorname{sgn}(\varkappa_1(t) + \Omega_0)$, and $\operatorname{sgn}(\varkappa_1(t-2\tau) + \Omega_0)$. In view of (11.31), $\dot{\varkappa}_1(t) = \dot{x}_1(t)$ represents the angular acceleration at the bottom end of the drillstring $\ddot{\Phi}_b$, which is clearly a bounded variable in real applications. As the functions $\operatorname{sgn}(\varkappa_1(t) + \Omega_0)$ and $\operatorname{sgn}(\varkappa_1(t-2\tau)+\Omega_0)$ are also bounded variables, the matrices $A(x)$ and $A_{2\tau}(x)$ are assumed to be bounded.

The polytopic representation (11.34) is obtained with $i \in I^r = 2^3 = 8$. The matrices $A_i(x)$ and $A_{i2\tau}(x)$ are given by:

$$A_i(x) = \begin{pmatrix} -3.3645 & a^i_{23}(x) \\ 0 & a^i_{33}(x) \end{pmatrix},$$

with

$$-136.1327 = a^1_{23} \leqslant a^i_{23}(x) \leqslant a^2_{23} = 0,$$
$$-0.9Acc_{\max} = a^1_{33} \leqslant a^i_{33}(x) \leqslant a^2_{33} = -0.9Dec_{\max},$$

where $Acc_{\max}$ and $Dec_{\max}$ stand for the maximum acceleration and deceleration, respectively, and

$$A_{i2\tau}(x) = \begin{pmatrix} -2.4878 & a^i_{2\tau 23}(x) \\ 0 & 0 \end{pmatrix},$$

with

$$0 = a^1_{2\tau 23}(x) \leqslant a^i_{2\tau 23}(x) \leqslant a^2_{2\tau 23}(x) = 100.6802.$$

Numerical results of Theorem 11.2.2 for the torsional drilling model in closed loop with the control law (11.28) yields:

$$K = Y\bar{P}^{-1} = \begin{pmatrix} 0.44 & -4.25 \end{pmatrix}.$$

Then, the stabilizing control law for the drilling system (11.32) is given by

$$u(t) = 0.44x_1(t-\tau) - 4.25\mathrm{e}^{-\gamma_b x_1(t-\tau)+\Omega_0},$$

which, according to (11.31), can be written in terms of the bit speed as:

$$u(t) = 0.44\dot{\Phi}_b(t-\tau) - 4.25\mathrm{e}^{-\gamma_b \dot{\Phi}_b(t-\tau)} - 0.44\Omega_0. \tag{11.35}$$

The simulation result of Fig. 11.2 shows an effective elimination of the stick-slip phenomenon by means of the application of the controller (11.35).

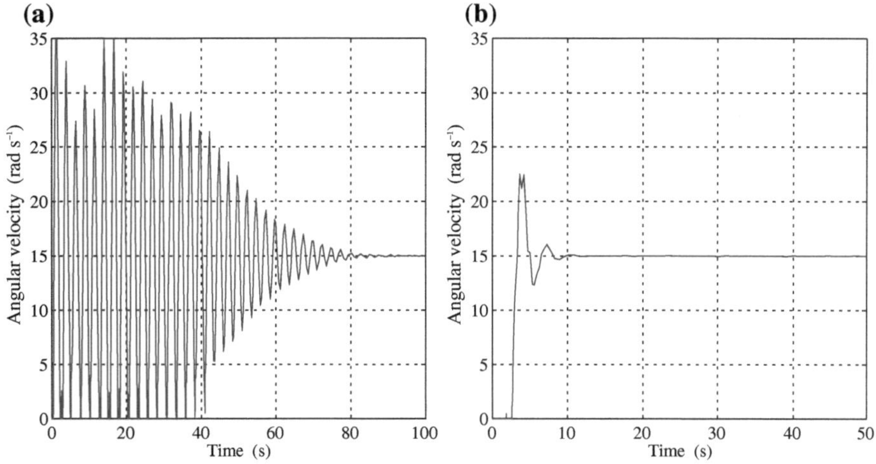

Fig. 11.2 Simulation of the torsional model (2.32) coupled to the frictional torque (3.14) with the friction coefficient given in (3.15). Angular velocity at the bottom extremity $\dot{\Phi}_b(t)$ for $\Omega_0 = 15\,\text{rad s}^{-1}$. **a** Trajectory without control actions (stick-slip). **b** Elimination of stick-slip oscillations by means of the multimodel representation-based controller (11.35)

11.3 Notes and References

Two control strategies to tackle torsional drilling vibrations were presented. The first one was developed in the framework of switching systems theory. Asymptotic and exponential stability conditions for a class of neutral switching systems were determined; the obtained results were applied to analyze the drilling system subject to stick-slip vibrations. Further details on this topic can be found in [255].

Extensive research has been devoted to the theoretical and practical understanding of switching systems, however, only a few contributions address the presence of delays in switching systems. For instance, in [153], a common Lyapunov function method switching rule is proposed to stabilize a system composed of a finite number of linear delay differential equations; in [302], LMI-type stabilization conditions for linear switched time-varying delay systems are derived from a piecewise quadratic Lyapunov function; the problem of $\mathscr{H}_\infty$ control design for such switching systems is addressed in [175]. Concerning neutral-type time-delay systems, in [179], the delay-dependent robust stability of a class of uncertain switched system with mixed delay and time-varying structure is investigated.

The second part of this chapter has addressed the control design for a class of neutral-type time-delay systems that admit a multimodel representation [255]. The proposed technique can be regarded as an extension of the method employed in [267] for the stabilization of a class of nonlinear retarded-type time-delay systems. The polytopic representation technique was introduced in [309] to investigate the robust stability and stabilization of linear systems subject to convex polytopic uncertainties.

See also [310], where improved results on the robust stabilization technique are presented.

Both control methodologies presented in this chapter give rise to the suppression of torsional oscillations; however the axial drilling dynamics leading to the occurrence of the bit-bounce phenomenon were not considered. Following the same line of analysis, the following chapter addresses the design of a pair of Lyapunov-based control strategies allowing the elimination of coupled torsional–axial vibrations. Two different modeling strategies are considered: a coupled PDE–ODE system and a coupled NDDE–ODE model. The practical stabilization of the system is achieved with the proposed control approaches.

Chapter 12
Practical Stabilization of the Drilling System

As discussed in Chap. 7, when dynamical systems are subject to external perturbations, it is not possible to establish exponential stability; nevertheless, from an engineering point of view, the system response may be considered acceptable. This idea gives rise to the notion of ultimate boundedness or practical stability, which allows characterizing the transient behavior of a perturbed system.

Under the assumption that the drilling system is subject to external disturbances, and that certain dynamics are frequently disregarded in the models, it is impractical to design control laws aimed at forcing the system response to reach a particular state; instead, we seek to drive the system trajectories into a given domain guaranteeing an acceptable system performance.

This chapter concerns the practical stabilization of the drilling system based on two different modeling approaches:

- **Coupled wave–ODE model**. As reviewed in Chap. 2, the wave equation coupled to an ODE model allows describing torsional–axial drilling oscillations. Based on this modeling approach, the practical stabilization of the system is addressed via Lyapunov techniques allowing the design of stabilizing controllers to suppress the stick-slip and the bit-bounce.
- **Coupled NDDE–ODE model**. A different approach considers a neutral-type time-delay model describing torsional drillstring oscillations coupled to a simplified model approximating axial dynamics. A pair of feedback controllers is designed based on the attractive ellipsoid method which combines Lyapunov strategies and the principle of attractive sets [215].

12.1 Coupled Wave-ODE Model-Based Control

As reviewed in Sect. 2.2, torsional excitations of a drillstring of length L, described by the rotary angle $\Phi(s, t)$ can be modeled by a wave equation of the form:

M.B. Saldivar Márquez et al., *Analysis and Control of Oilwell Drilling Vibrations*, Advances in Industrial Control, DOI 10.1007/978-3-319-15747-4_12

$$GJ\frac{\partial^2\Phi}{\partial s^2}(s,t) - I\frac{\partial^2\Phi}{\partial t^2}(s,t) - \gamma\frac{\partial\Phi}{\partial t}(s,t) = 0,$$

$$s \in (0,L), \quad t > 0,$$

with the boundary conditions:

$$GJ\frac{\partial\Phi}{\partial s}(0,t) = \beta\left(\frac{\partial\Phi}{\partial t}(0,t) - \Omega(t)\right),$$

$$GJ\frac{\partial\Phi}{\partial s}(L,t) + I_B\frac{\partial^2\Phi}{\partial t^2}(L,t) = -T\left(\frac{\partial\Phi}{\partial t}(L,t)\right).$$

A simplified model of the drillstring axial dynamics is given below, see Sect. 2.1,

$$m_0\ddot{v} + c_0\left(\dot{v} + \rho(t)\right) + k_0 v = -\mu_1 T\left(\frac{\partial\Phi}{\partial t}(L,t)\right), \quad v = U_b - \rho_0 t,$$

where U_b denotes the bit axial position and $\rho(t)$ the rate of penetration.

Torsional and axial models are coupled through the frictional torque T, arising from the bit-rock interaction, which can be modeled as:

$$T\left(\frac{\partial\Phi}{\partial t}(L,t)\right) = c_b\frac{\partial\Phi}{\partial t}(L,t) + T_{nl}\left(\frac{\partial\Phi}{\partial t}(L,t)\right),$$

where the nonlinear term is defined as:

$$T_{nl}\left(\tfrac{\partial\Phi}{\partial t}(L,t)\right) = W_{ob}R_b\mu_b\left(\tfrac{\partial\Phi}{\partial t}(L,t)\right)\operatorname{sgn}\left(\tfrac{\partial\Phi}{\partial t}(L,t)\right),$$

$$\mu_b\left(\tfrac{\partial\Phi}{\partial t}(L,t)\right) = \mu_{cb} + (\mu_{sb} - \mu_{cb})e^{-\gamma_b\left|\frac{\partial\Phi}{\partial t}(L,t)\right|}.$$

Consider the normalized rod length $\zeta = s/L$. For the sake of notation simplicity denote $\frac{\partial\Phi}{\partial t}, \frac{\partial\Phi}{\partial\zeta}, \frac{\partial^2\Phi}{\partial t^2}, \frac{\partial^2\Phi}{\partial\zeta^2}$ by $\Phi_t, \Phi_\zeta, \Phi_{tt}, \Phi_{\zeta\zeta}$, respectively. The coupled wave-ODE model reads:

$$\Phi_{tt}(\zeta,t) = a\Phi_{\zeta\zeta}(\zeta,t) - d\Phi_t(\zeta,t), \quad \zeta \in (0,1), \tag{12.1a}$$

$$\Phi_\zeta(0,t) = g\left(\Phi_t(0,t) - u_1(t)\right), \tag{12.1b}$$

$$\Phi_\zeta(1,t) = -h\Phi_{tt}(1,t) - k\Phi_t(1,t) - qT_{nl}(\Phi_t(1,t)) + \omega(t), \tag{12.1c}$$

$$\dot{x}(t) = Ax(t) + Bu_2(t) + E_1\Phi_t(1,t) + E_2T_{nl}(\Phi_t(1,t)). \tag{12.1d}$$

The controllers u_1 and u_2 correspond to the angular velocity provided by the rotary table Ω and to the rate of penetration imposed at the top end ρ. The vector $x(t)$ is defined as $x(t) = [v(t)\ \dot{v}(t)]^T$. The model parameters are given by:

$$a = \frac{GJ}{IL^2}, \quad d = \frac{\gamma}{I}, \quad g = \frac{\beta L}{GJ}, \quad h = \frac{I_B L}{GJ}, \quad k = \frac{c_b L}{GJ}, \quad q = \frac{L}{GJ},$$

and the constant matrices are defined as:

$$A = \begin{bmatrix} 0 & 1 \\ -\frac{k_0}{m_0} & -\frac{c_0}{m_0} \end{bmatrix}, \quad B = \begin{bmatrix} 0 \\ -\frac{c_0}{m_0} \end{bmatrix}, \quad E_1 = \begin{bmatrix} 0 \\ -\frac{\mu_1 c_b}{m_0} \end{bmatrix}, \quad E_2 = \begin{bmatrix} 0 \\ -\frac{\mu_1}{m_0} \end{bmatrix}.$$

An additive variable $\omega(t)$ satisfying:

$$\omega(t)^2 \leqslant \bar{\varepsilon}, \tag{12.2}$$

is chosen to represent external disturbances and unmodeled dynamics.

12.1.1 Ultimate Boundedness Analysis

The following result, inspired by the results on input-to-state stability presented in [101], provides the conditions on a Lyapunov functional proposal to guarantee ultimate boundedness of the solutions of a boundary value problem in finite time, particularly, we consider a system of the form (12.1a–12.1d) subject to perturbations or disturbances satisfying (12.2).

Lemma 12.1.1 *Let a Lyapunov functional $V(\tilde{\Phi}(\cdot,t))$ satisfying*

$$\int_0^1 \tilde{\Phi}^T(\zeta,t)P\tilde{\Phi}(\zeta,t)d\zeta \leqslant V(\tilde{\Phi}(\cdot,t)) \leqslant \bar{V}(\tilde{\Phi}(\cdot,t)) \tag{12.3}$$

and

$$\frac{d}{dt}V(\tilde{\Phi}(\cdot,t)) + \sigma V(\tilde{\Phi}(\cdot,t)) \leqslant \bar{\beta}\bar{\varepsilon} \quad \forall t \geq 0, \; \sigma > 0, \; \bar{\beta} > 0, \tag{12.4}$$

where $\tilde{\Phi}(\zeta,t) = \left[\Phi_\zeta(\zeta,t) \;\; \Phi_t(\zeta,t) \;\; x_t\right]^T$ and P is a symmetric positive definite matrix. Then for any initial function $\tilde{\varphi}(\zeta,t) = \left[\varphi_\zeta(\zeta) \;\; \varphi_1(\zeta) \;\; x_0\right]^T$, the ultimate bound defined by

$$\int_0^1 \tilde{\Phi}^T(\zeta,t)P\tilde{\Phi}(\zeta,t)d\zeta \leqslant e^{-\sigma t}\bar{V}(\tilde{\varphi}(\zeta,t)) + \frac{\bar{\beta}\bar{\varepsilon}}{\sigma}, \tag{12.5}$$

is satisfied for $t \geq T_a(\tilde{\varphi},\varsigma) > 0$ where

$$T_a(\tilde{\varphi},\varsigma) = \frac{1}{\sigma}\ln\left(\frac{1}{\varsigma}\bar{V}(\tilde{\varphi}(\zeta,t)) - \frac{\bar{\beta}\bar{\varepsilon}}{\varsigma\sigma}\right).$$

Proof Multiplying (12.4) by $e^{\sigma\theta}$ gives

$$\frac{d}{d\theta}\left(e^{\sigma\theta}V(\tilde{\Phi}(\cdot,t))\right) \leqslant \bar{\beta}\bar{\varepsilon}e^{\sigma\theta},$$

an integration of the above expression yields:

$$V(\tilde{\Phi}(\cdot,t)) \leqslant e^{-\sigma t}V(\tilde{\varphi}(\zeta,t)) + \frac{\bar{\beta}\bar{\varepsilon}}{\sigma}\left(1-e^{-\sigma t}\right).$$

It follows from (12.3) that

$$\int_0^1 \tilde{\Phi}^T(\zeta,t)P\tilde{\Phi}(\zeta,t)d\zeta \leqslant e^{-\sigma t}\bar{V}(\tilde{\varphi}(\zeta,t)) + \frac{\bar{\beta}\bar{\varepsilon}}{\sigma}\left(1-e^{-\sigma t}\right),$$

then we have

$$\int_0^1 \tilde{\Phi}^T(\zeta,t)P\tilde{\Phi}(\zeta,t)d\zeta \leqslant e^{-\sigma t}\left(\bar{V}(\tilde{\varphi}(\zeta,t)) - \frac{\bar{\beta}\bar{\varepsilon}}{\sigma}\right) + \frac{\bar{\beta}\bar{\varepsilon}}{\sigma}. \tag{12.6}$$

For the initial conditions fulfilling $\bar{V}(\tilde{\varphi}(\zeta,t)) \leqslant \frac{\bar{\beta}\bar{\varepsilon}}{\sigma}$, the inequality

$$\int_0^1 \tilde{\Phi}^T(\zeta,t)P\tilde{\Phi}(\zeta,t)d\zeta \leqslant \frac{\bar{\beta}\bar{\varepsilon}}{\sigma}, \quad \forall t \geq 0,$$

holds; otherwise, if $\bar{V}(\tilde{\varphi}(\zeta,t)) > \frac{\bar{\beta}\bar{\varepsilon}}{\sigma}$, we look for the time instant $T_a(\tilde{\varphi},\varsigma)$ at which the ultimate bound (12.5) is satisfied. In view of (12.6), $T_a(\tilde{\varphi},\varsigma)$ should guarantee $\int_0^1 \tilde{\Phi}^T(\zeta,t)P\tilde{\Phi}(\zeta,t)d\zeta \leqslant \varsigma + \frac{\bar{\beta}\bar{\varepsilon}}{\sigma}$ for a small enough $\varsigma > 0$, i.e.,

$$\varsigma = e^{-\sigma T_a}\left(\bar{V}(\tilde{\varphi}(\zeta,t)) - \frac{\bar{\beta}\bar{\varepsilon}}{\sigma}\right).$$

From the above expression we have

$$e^{\sigma T_a} = \frac{1}{\varsigma}\left(\bar{V}(\tilde{\varphi}(\zeta,t)) - \frac{\bar{\beta}\bar{\varepsilon}}{\sigma}\right),$$

and the result follows.

12.1.2 Practical Stabilization Conditions

The main idea in designing the stabilizing controllers is to propose a suitable Lyapunov-Krasovskii functional such that conditions (12.3) and (12.4) of Lemma 12.1.1 are satisfied along the trajectories of the closed-loop system.

Consider the Lyapunov-Krasovskii functional

$$V(\tilde{\Phi}(\cdot,t)) = \int_0^1 \tilde{\Phi}^T(\zeta,t)P\tilde{\Phi}(\zeta,t)d\zeta,$$

with

$$\tilde{\Phi}(\zeta,t) = \left[\Phi_\zeta(\zeta,t)\ \Phi_t(\zeta,t)\ x(t)\right]^T,$$

and

$$P = \begin{bmatrix} ap & \chi(\zeta+1) & 0 \\ \chi(\zeta+1) & p & 0 \\ 0 & 0 & R \end{bmatrix} > 0, \quad p > 0, \quad a > 0, \quad R > 0.$$

Notice that condition (12.3) is satisfied with

$$\bar{V}(\tilde{\Phi}(\cdot,t)) = \lambda_{\max}(P)\int_0^1 \left\|\tilde{\Phi}(\zeta,t)\right\|^2 d\zeta$$

where $\lambda_{\max}(P)$ denotes the maximum eigenvalue of matrix P. Now, we establish the conditions under which the inequality (12.4) is satisfied. The proposed functional can be rewritten as:

$$V(\Phi_\zeta(\cdot,t),\Phi_t(\cdot,t),x_t) = V_1(\Phi_\zeta(\zeta,t))+V_2(\Phi_t(\zeta,t))+V_3(\Phi_\zeta(\zeta,t),\Phi_t(\zeta,t))+V_4(x_t)$$

where

$$\begin{aligned}
V_1(\Phi_\zeta(\zeta,t)) &= pa\int_0^1 \Phi_\zeta^2(\zeta,t)d\zeta,\\
V_2(\Phi_t(\zeta,t)) &= p\int_0^1 \Phi_t^2(\zeta,t)d\zeta,\\
V_3(\Phi_\zeta(\zeta,t),\Phi_t(\zeta,t)) &= 2\chi\int_0^1 (\zeta+1)\,\Phi_\zeta(\zeta,t)\Phi_t(\zeta,t)d\zeta,\\
V_4(x_t) &= x^T(t)Rx(t).
\end{aligned}$$

Taking time derivatives of each term of V, we obtain:

$$\dot{V}_1 = 2pa \int_0^1 \Phi_\zeta(\zeta,t)\Phi_{\zeta t}(\zeta,t)d\zeta, \tag{12.7}$$

$$\dot{V}_2 = 2p \int_0^1 \Phi_t(\zeta,t)\Phi_{tt}(\zeta,t)d\zeta. \tag{12.8}$$

Substituting the wave equation (12.1a) into (12.8) yields:

$$\dot{V}_2 = 2pa \int_0^1 \Phi_t(\zeta,t)\Phi_{\zeta\zeta}(\zeta,t)d\zeta - 2pd \int_0^1 \Phi_t^2(\zeta,t)d\zeta,$$

integration by parts of $\int_0^1 \Phi_t(\zeta,t)\Phi_{\zeta\zeta}(\zeta,t)d\zeta$ gives:

$$\int_0^1 \Phi_t(\zeta,t)\Phi_{\zeta\zeta}(\zeta,t)d\zeta = \Phi_t(\zeta,t)\Phi_\zeta(\zeta,t)\big|_0^1 - \int_0^1 \Phi_{t\zeta}(\zeta,t)\Phi_\zeta(\zeta,t)d\zeta.$$

Then, we obtain:

$$\dot{V}_2 = 2pa\ \Phi_t(\zeta,t)\Phi_\zeta(\zeta,t)\big|_0^1 - 2pa \int_0^1 \Phi_{t\zeta}(\zeta,t)\Phi_\zeta(\zeta,t)d\zeta \\ -2pd \int_0^1 \Phi_t^2(\zeta,t)d\zeta. \tag{12.9}$$

The derivative of V_3 is:

$$\dot{V}_3 = 2\chi \int_0^1 (\zeta+1)\,\Phi_\zeta(\zeta,t)\Phi_{tt}(\zeta,t)d\zeta + 2\chi \int_0^1 (\zeta+1)\,\Phi_{\zeta t}(\zeta,t)\Phi_t(\zeta,t)d\zeta,$$

substituting the wave equation (12.1a) yields:

$$\dot{V}_3 = 2a\chi \int_0^1 (\zeta+1)\,\Phi_\zeta(\zeta,t)\Phi_{\zeta\zeta}(\zeta,t)d\zeta - 2d\chi \int_0^1 (\zeta+1)\,\Phi_\zeta(\zeta,t)\Phi_t(\zeta,t)d\zeta \\ +2\chi \int_0^1 (\zeta+1)\,\Phi_{\zeta t}(\zeta,t)\Phi_t(\zeta,t)d\zeta. \tag{12.10}$$

Integration by parts of $\int_0^1 (\zeta+1)\,\Phi_\zeta(\zeta,t)\Phi_{\zeta\zeta}(\zeta,t)d\zeta$ gives:

$$\int_0^1 (\zeta+1)\,\Phi_\zeta(\zeta,t)\Phi_{\zeta\zeta}(\zeta,t)d\zeta = (\zeta+1)\,\Phi_\zeta^2(\zeta,t)\Big|_0^1 - \int_0^1 \Phi_\zeta^2(\zeta,t)d\zeta \\ - \int_0^1 (\zeta+1)\,\Phi_{\zeta\zeta}(\zeta,t)\Phi_\zeta(\zeta,t)d\zeta,$$

then, we obtain

$$2\int_0^1 (\zeta+1)\,\Phi_\zeta(\zeta,t)\Phi_{\zeta\zeta}(\zeta,t)d\zeta = (\zeta+1)\,\Phi_\zeta^2(\zeta,t)\Big|_0^1 - \int_0^1 \Phi_\zeta^2(\zeta,t)d\zeta. \tag{12.11}$$

Now, observe that

$$2\Phi_{\zeta t}(\zeta,t)\Phi_t(\zeta,t) = \frac{\partial}{\partial\zeta}\left(\Phi_t^2(\zeta,t)\right),$$

then,

$$2\int_0^1 (\zeta+1)\,\Phi_{\zeta t}(\zeta,t)\Phi_t(\zeta,t)d\zeta = \int_0^1 (\zeta+1)\,\frac{\partial}{\partial\zeta}\left(\Phi_t^2(\zeta,t)\right)d\zeta.$$

Integration by parts of $\int_0^1 (\zeta+1)\,\frac{\partial}{\partial\zeta}\left(\Phi_t^2(\zeta,t)\right)d\zeta$ gives

$$2\int_0^1 (\zeta+1)\,\Phi_{\zeta t}(\zeta,t)\Phi_t(\zeta,t)d\zeta = (\zeta+1)\,\Phi_t^2(\zeta,t)\Big|_0^1 - \int_0^1 \Phi_t^2(\zeta,t)d\zeta. \tag{12.12}$$

Substituting (12.11) and (12.12) into (12.10) yields:

$$\begin{aligned}\dot{V}_3 = a\chi\ (\zeta+1)\,\Phi_\zeta^2(\zeta,t)\Big|_0^1 - 2d\chi\int_0^1 (\zeta+1)\,\Phi_\zeta(\zeta,t)\Phi_t(\zeta,t)d\zeta \\ +\chi\ (\zeta+1)\,\Phi_t^2(\zeta,t)\Big|_0^1 - a\chi\int_0^1 \Phi_\zeta^2(\zeta,t)d\zeta - \chi\int_0^1 \Phi_t^2(\zeta,t)d\zeta.\end{aligned} \tag{12.13}$$

The derivative of V_4 along the trajectories of (12.1d) is given by:

$$\begin{aligned}\dot{V}_4(x_t) &- \dot{x}^T(t)Rx(t) + x^T(t)R\dot{x}(t) \\ &= \left[Ax(t) + Bu_2(t) + E_1\Phi_t(1,t) + E_2T_{nl}(\Phi_t(1,t))\right]^T Rx(t) \\ &\quad +x^T(t)R\left[Ax(t) + Bu_2(t) + E_1\Phi_t(1,t) + E_2T_{nl}(\Phi_t(1,t))\right].\end{aligned}$$

We seek the stabilization of axial drilling dynamics by means of a controller of the form:

$$u_2(t) = c_2x(t), \quad c_2 \in \mathbb{R}^{1\times 2}, \tag{12.14}$$

guaranteeing the quadratic form of $\dot{V}_4(x_t)$, i.e.,

$$\dot{V}_4(x_t) = x^T(t)\Upsilon_1x(t) + \Upsilon_2^T Rx(t) + x^T(t)R\Upsilon_2, \tag{12.15}$$

where

$$\Upsilon_1 = A^T R + RA + c_2^T B^T R + RBc_2,$$
$$\Upsilon_2 = E_1 \Phi_t(1,t) + E_2 T_{nl}(\Phi_t(1,t)).$$

In view of (12.7), (12.9), (12.13) and (12.15) the time derivative of V is given by:

$$\begin{aligned}
\dot{V} = {} & 2pa \int_0^1 \Phi_\zeta(\zeta,t)\Phi_{\zeta t}(\zeta,t)d\zeta - 2pa \int_0^1 \Phi_{t\zeta}(\zeta,t)\Phi_\zeta(\zeta,t)d\zeta \\
& -2d\chi \int_0^1 (\zeta+1)\,\Phi_\zeta(\zeta,t)\Phi_t(\zeta,t)d\zeta - \chi \int_0^1 \Phi_t^2(\zeta,t)d\zeta \\
& -2pd \int_0^1 \Phi_t^2(\zeta,t)d\zeta - a\chi \int_0^1 \Phi_\zeta^2(\zeta,t)d\zeta \\
& +2pa\ \Phi_t(\zeta,t)\Phi_\zeta(\zeta,t)\Big|_0^1 + a\chi\ (\zeta+1)\,\Phi_\zeta^2(\zeta,t)\Big|_0^1 + \chi\ (\zeta+1)\,\Phi_t^2(\zeta,t)\Big|_0^1 \\
& +x^T(t)\Upsilon_1 x(t) + \Upsilon_2^T Rx(t) + x^T(t)R\Upsilon_2.
\end{aligned}$$

In view of the second condition of Lemma 12.1.1, we derive the following expression:

$$\begin{aligned}
\dot{V} + \sigma V = {} & -2pd \int_0^1 \Phi_t^2(\zeta,t)d\zeta - 2d\chi \int_0^1 (\zeta+1)\,\Phi_\zeta(\zeta,t)\Phi_t(\zeta,t)d\zeta \\
& -a\chi \int_0^1 \Phi_\zeta^2(\zeta,t)d\zeta - \chi \int_0^1 \Phi_t^2(\zeta,t)d\zeta + x^T(t)\Upsilon_1 x(t) \qquad (12.16) \\
& +\Upsilon_2^T Rx(t) + x^T(t)R\Upsilon_2 + \sigma pa \int_0^1 \Phi_\zeta^2(\zeta,t)d\zeta \\
& +\sigma p \int_0^1 \Phi_t^2(\zeta,t)d\zeta + 2\sigma\chi \int_0^1 (\zeta+1)\,\Phi_\zeta(\zeta,t)\Phi_t(\zeta,t)d\zeta \\
& +\sigma x^T(t)Rx(t) + \Pi,
\end{aligned}$$

where:

$$\begin{aligned}
\Pi = {} & 2pa\ \Phi_t(\zeta,t)\Phi_\zeta(\zeta,t)\Big|_0^1 + a\chi\ (\zeta+1)\,\Phi_\zeta^2(\zeta,t)\Big|_0^1 + \chi\ (\zeta+1)\,\Phi_t^2(\zeta,t)\Big|_0^1 \\
= {} & 2\chi\Phi_t^2(1,t) + 2a\chi\Phi_\zeta^2(1,t) + 2pa\Phi_t(1,t)\Phi_\zeta(1,t) - \chi\Phi_t^2(0,t) - a\chi\Phi_\zeta^2(0,t) \\
& -2pa\Phi_t(0,t)\Phi_\zeta(0,t).
\end{aligned}$$

The introduction of the boundary conditions (12.1b) and (12.1c) yields:

$$\begin{aligned}\Pi &= 2\chi\Phi_t^2(1,t) + 2a\chi\left[-h\Phi_{tt}(1,t) - k\Phi_t(1,t) - qT_{nl}(\Phi_t(1,t)) + \omega(t)\right]^2 \\ &\quad + 2pa\Phi_t(1,t)\left[-h\Phi_{tt}(1,t) - k\Phi_t(1,t) - qT_{nl}(\Phi_t(1,t)) + \omega(t)\right] \\ &\quad - \chi\Phi_t^2(0,t) - a\chi\left(g\left(\Phi_t(0,t) - u_1(t)\right)\right)^2 \\ &\quad - 2pa\Phi_t(0,t)\left(g\left(\Phi_t(0,t) - u_1(t)\right)\right).\end{aligned}$$

Our purpose is to rearrange the terms of (12.16) into symmetric matrices. According to condition (12.4), the negative definiteness of these matrices guarantee ultimate boundedness of the system trajectories. The choice of the controller u_1 is based on the fact that a necessary condition for a symmetric matrix to be negative definite is that all the diagonal entries be negative.

For the stabilization of torsional trajectories we propose the following controller structure:

$$u_1(t) = c_{11}\Phi_{tt}(1,t) + c_{12}\Phi_t(1,t) + c_{13}T_{nl}(\Phi_t(1,t)) + c_{14}\Phi_t(0,t), \qquad (12.17)$$

with $c_{1i} \in \mathbb{R}^{1\times 1}$, $i = 1, \ldots, 4$. Then, Π is written as:

$$\begin{aligned}\Pi &= 2a\chi\left[h^2\Phi_{tt}^2(1,t) + 2hk\Phi_t(1,t)\Phi_{tt}(1,t) + 2hq\Phi_{tt}(1,t)T_{nl}(\Phi_t(1,t))\right. \\ &\quad -2h\Phi_{tt}(1,t)\omega(t) + k^2\Phi_t^2(1,t) + 2kq\Phi_t(1,t)T_{nl}(\Phi_t(1,t)) \\ &\quad \left.-2k\Phi_t(1,t)\omega(t) + q^2T_{nl}^2(\Phi_t(1,t)) - 2qT_{nl}(\Phi_t(1,t))\omega(t) + \omega(t)^2\right] \\ &\quad +2pa\Phi_t(1,t)\left[-h\Phi_{tt}(1,t) - k\Phi_t(1,t) - qT_{nl}(\Phi_t(1,t)) + \omega(t)\right] \\ &\quad -a\chi g^2\left[c_{11}^2\Phi_{tt}^2(1,t) + 2c_{11}c_{12}\Phi_t(1,t)\Phi_{tt}(1,t) + c_{12}^2\Phi_t^2(1,t)\right. \\ &\quad +2c_{11}c_{13}\Phi_{tt}(1,t)T_{nl}(\Phi_t(1,t)) - 2c_{11}\left(1-c_{14}\right)\Phi_{tt}(1,t)\Phi_t(0,t) \\ &\quad +2c_{12}c_{13}\Phi_t(1,t)T_{nl}(\Phi_t(1,t)) - 2c_{12}\left(1-c_{14}\right)\Phi_t(1,t)\Phi_t(0,t) \\ &\quad \left.+c_{13}^2T_{nl}^2(\Phi_t(1,t)) + \left(1-c_{14}\right)^2\Phi_t^2(0,t) - 2c_{13}\left(1-c_{14}\right)T_{nl}(\Phi_t(1,t))\Phi_t(0,t)\right] \\ &\quad -2pag\Phi_t(0,t)\left[\Phi_t(0,t) - c_{11}\Phi_{tt}(1,t) - c_{12}\Phi_t(1,t) - c_{13}T_{nl}(\Phi_t(1,t))\right] \\ &\quad +2pagc_{14}\Phi_t^2(0,t) + 2\chi\Phi_t^2(1,t) - \chi\Phi_t^2(0,t).\end{aligned}$$

The restriction (12.2) implies that $\omega(t)^2 - \bar{\varepsilon} \leqslant 0$, then, for any positive $\bar{\beta}$, the following inequality is satisfied:

$$-\bar{\beta}(\omega(t)^2 - \bar{\varepsilon}) \geq 0. \qquad (12.18)$$

In order to take into account the perturbation restriction, the term (12.18) is added to (12.16).

After symmetrization of the cross terms, we obtain:

$$\frac{d}{dt}V + \sigma V - \bar{\beta}\bar{\varepsilon} \leqslant \int_0^1 \eta_1^T\Psi_1\eta_1 d\zeta + \eta_2^T\Psi_2\eta_2,$$

where

$$\eta_1 = \left[\Phi_t(\zeta, t)\ \Phi_\zeta(\zeta, t)\right]^T,$$
$$\eta_2 = \left[\Phi_{tt}(1, t)\ \Phi_t(1, t)\ T_{nl}(\Phi_t(1, t))\ \Phi_t(0, t)\ x(t)\ \omega(t)\right]^T,$$

and

$$\Psi_1 = \begin{bmatrix} -2pd - \chi + \sigma p & -d\chi\,(\zeta + 1) + \sigma\chi\,(\zeta + 1) \\ * & -a\chi + \sigma pa \end{bmatrix}, \tag{12.19}$$

$$\Psi_2 = \begin{bmatrix} \Psi_{11} & \Psi_{12} & \Psi_{13} & a\chi g^2 c_{11}(1 - c_{14}) + pagc_{11} & 0 & -2a\chi h \\ * & \Psi_{22} & \Psi_{23} & \chi g^2 c_{12}(1 - c_{14}) + pagc_{12} & E_1{}^T R & a(p - 2\chi k) \\ * & * & \Psi_{33} & a\chi g^2 c_{13}(1 - c_{14}) + pagc_{13} & E_2^T R & -2a\chi q \\ * & * & * & -a\chi g^2(1 - c_{14})^2 - \chi - 2pag + pagc_{14} & 0 & 0 \\ * & * & * & * & \Upsilon_1 + \sigma R & 0 \\ * & * & * & * & * & 2a\chi - \bar{\beta} \end{bmatrix}, \tag{12.20}$$

with

$$\begin{aligned} \Psi_{11} &= 2a\chi h^2 - a\chi g^2 c_{11}^2 & \Psi_{23} &= 2a\chi kq - paq - a\chi g^2 c_{12} c_{13} \\ \Psi_{12} &= 2a\chi hk - pah - a\chi g^2 c_{11} c_{12} & \Psi_{33} &= 2a\chi q^2 - a\chi g^2 c_{13}^2 \\ \Psi_{13} &= 2a\chi hq - a\chi g^2 c_{11} c_{13} & \Psi_{34} &= a\chi g^2 c_{13}(1 - c_{14}) + pagc_{13} \\ \Psi_{22} &= 2a\chi k^2 + 2\chi - 2pak - a\chi g^2 c_{12}^2 & \Upsilon_1 &= A^T R + RA + c_2^T B^T R + RBc_2. \end{aligned}$$

The fulfillment of inequality (12.4) is conditioned to the negative definiteness of matrices Ψ_1 and Ψ_2 which entails the feasibility problem of the LMI $\Psi_1 < 0$ and of the BMI $\Psi_2 < 0$. Notice that the particular structure of the controller u_1 with appropriate values of c_{1i}, $i = 1, \ldots, 4$, ensures the negativity of the diagonal terms of matrix Ψ_2.

Summarizing the above ideas, the result on the practical stabilization of the wave–ODE system (12.1a–12.1d) is stated as follows:

Theorem 12.1.2 *The trajectories of the drilling system described by the coupled wave–ODE system* (12.1a–12.1d) *in closed loop with the controllers* (12.14) *and* (12.17) *admit the ultimate bound* (12.5) *if the matrix inequalities* $P > 0$, $\Psi_1 < 0$ *and* $\Psi_2 < 0$ *are satisfied for some* $p > 0$, $\chi > 0$, $\bar{\beta} > 0$, $R > 0$ *and any* $c_2 \in \mathbb{R}^{1\times 2}$, $c_{1i} \in \mathbb{R}^{1\times 1}$, $i = 1, \ldots, 4$.

12.1.3 Elimination of Coupled Vibrations

In what follows, the effectiveness of the proposed control approach is highlighted through simulations results. The numerical values of the system physical parameters are given in Table C.1 of Appendix C.

The additive variable $\omega(t)$ accounting for external disturbances and modeling errors is assumed to satisfy $\omega(t)^2 \leqslant \bar{\varepsilon}$, $\bar{\varepsilon} = 0.5$. The initial conditions are such that $\int_0^1 \|\tilde{\varphi}(\zeta, t)\|^2 \, d\zeta = 3.25$. Torsional and axial reference velocities are $10\,\text{rad s}^{-1}$ and $0.1\,\text{m s}^{-1}$, respectively.

Figures 12.1a and 12.2a show angular and axial velocities of the drill bit displaying stick-slip and bit-bounce behavior.

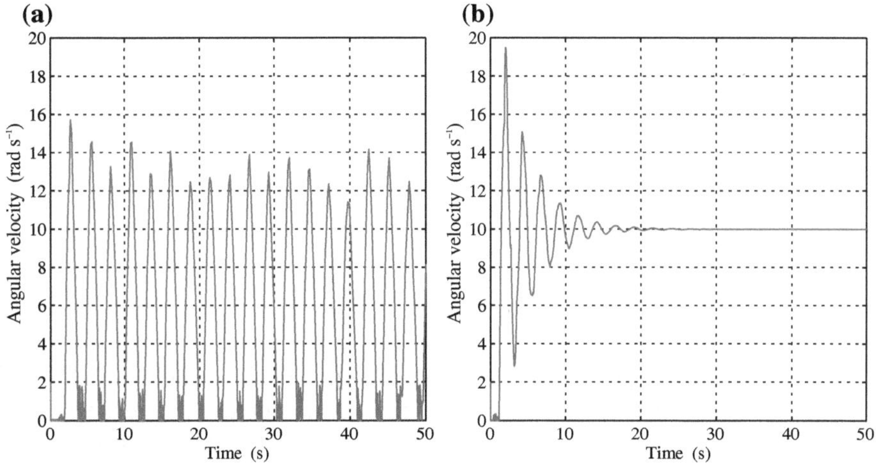

Fig. 12.1 Simulation of the coupled torsional–axial model (12.1a–12.1d). Angular velocity at the bottom extremity for a reference velocity of $10\,\text{rad s}^{-1}$. **a** Trajectory without control actions (stick-slip). **b** Elimination of stick-slip oscillations by means of the controllers (12.14) and (12.17)

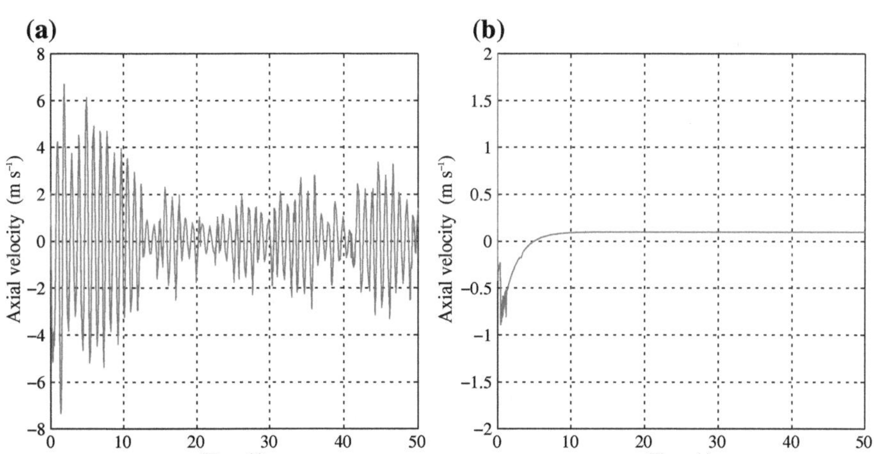

Fig. 12.2 Simulation of the coupled torsional–axial model (12.1a–12.1d). Axial velocity at the bottom extremity for a reference velocity of $0.1\,\text{m s}^{-1}$. **a** Trajectory without control actions (bit-bounce). **b** Elimination of bit-bounce oscillations by means of the controllers (12.14) and (12.17)

The proposed control approach ensures ultimate boundedness of a measure involving torsional–axial drilling system trajectories, the strategy succeeds in eliminating torsional and axial vibrations, see Figs. 12.1b and 12.2b.

Controllers (12.14) and (12.17) are synthesized by finding a feasible solution to the matrix inequalities stated in Theorem 12.1.2; the package "PENBMI" of MATLAB was used for this purpose.

The maximum exponential decay rate for which the conditions of Theorem 12.1.2 are satisfied is $\sigma = 0.8$. A feasible result of the LMI $\Psi_1 < 0$ with Ψ_1 given in (12.19) is:

$$p = 0.7406, \quad \chi = 0.9559.$$

Using the above values, a feasible result of the BMI $\Psi_2 < 0$ with Ψ_2 given in (12.20) is:

$$\begin{aligned} c_{11} &= -0.0067, & c_{12} &= -0.04699, & c_{13} &= -0.0548, \\ c_{14} &= 0.6642, & c_2 &= [0\ 2.0234], & \bar{\beta} &= 16.8149, \end{aligned} \tag{12.21}$$

$$R = \begin{bmatrix} 551.7373 & 11.7383 \\ 11.7383 & 13.4465 \end{bmatrix}.$$

The trajectories of the controllers $u_1(t)$ and $u_2(t)$ are shown in Fig. 12.3. Notice that the controllers' trajectories do not exhibit large amplitude variations which is a favorable feature regarding the saturation constraints on the control inputs.

An ultimate bound on the drilling system trajectories is given by:

$$\int_0^1 \tilde{\Phi}^T(\zeta,t)P\tilde{\Phi}(\zeta,t)d\zeta \leqslant e^{-\sigma t}\bar{V}(\tilde{\varphi}(\zeta,t)) + \frac{\bar{\beta}\bar{\varepsilon}}{\sigma} = 1.79 \times 10^3 e^{-0.8t} + 10.5093$$

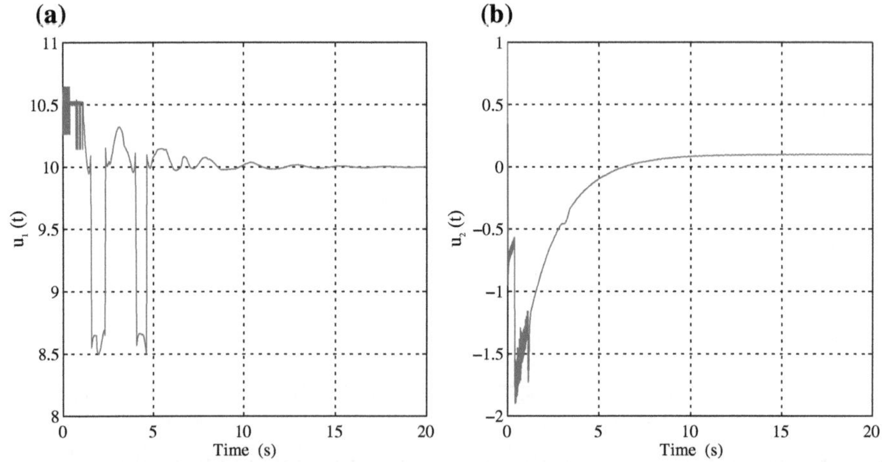

Fig. 12.3 Simulation of the coupled torsional–axial model (12.1a–12.1d). **a** Trajectory of the controller $u_1(t)$ defined in (12.17). **b** Trajectory of the controller $u_2(t)$ defined in (12.14)

where

$$\bar{V}(\tilde{\varphi}(\zeta,t)) = \lambda_{\max}(P)\int_0^1 \|\tilde{\varphi}(\zeta,t)\|^2\,d\zeta$$

for $t \geq T_a(\tilde{\varphi},\varsigma) > 0$, where

$$T_a(\tilde{\varphi},\varsigma) = \tfrac{1}{\sigma}\ln\left(\tfrac{1}{\varsigma}\lambda_{\max}(P)\int_0^1 \|\tilde{\varphi}(\zeta,t)\|^2\,d\zeta - \tfrac{\bar{\beta}\bar{\varepsilon}}{\varsigma\sigma}\right).$$

For $\varsigma = 0.1$, $T_a(\tilde{\varphi},\varsigma) = 12.2361$s.

12.2 Attractive Ellipsoid Method-Based Control

This section addresses the problem of control design for the suppression of drilling vibrations via the attractive ellipsoid method. Based on a combination of the Lyapounov method and of the attractive sets principle, we develop an effective methodology allowing the control synthesis for the drilling system stabilization through the solution of an optimization problem subject to bilinear matrix constraints. The proposed strategy, guaranteeing the suppression of axial–torsional coupled oscillations, allows to determine the minimum attractive ellipsoid for the trajectories of the closed-loop system.

Usually, the mathematical model of any physically motivated system is subject to perturbations due to uncertainties in the system parameters, measurement errors, or external disturbances. In [28], a control strategy allowing the system stability despite model perturbations is introduced; such a strategy guarantees that the system trajectories are retained within a given domain (invariant set) and ensures the achievement of some target set. The relation of invariant sets with the Lyapunov theory studied in [36] raises the possibility of using it for the control design of dynamic systems, robustness analysis, and suppression of disturbances.

Applying these ideas, it is possible to tackle the problem of drilling vibrations. Consider the following lemma for further developments.

Lemma 12.2.1 ([234]) *Let a functional $V(x_t)$ satisfying*

$$\frac{d}{dt}V(x_t) + \sigma V(x_t) \leqslant \bar{\beta},\ \forall t \geqslant 0,\ \sigma > 0,\ \bar{\beta} > 0,$$

it follows that,

$$\overline{\lim_{t\to\infty}} V(x_t) \leqslant \frac{\bar{\beta}}{\sigma}.$$

Notice that $V(x_t)$ defines an attractive set.

Consider the following time-delay system of neutral-type:

$$\dot{x}(t) + D\dot{x}(t-h) = A_0 x(t) + A_1 x(t-h) + n(t), \qquad (12.22)$$
$$x(\theta) = \varphi(\theta), \quad \theta \in [-h, 0],$$

where $h \geqslant 0$ is the time-delay, $D \in \mathbb{R}^{n\times n}$ is Schur stable, $A_0,\ A_1 \in \mathbb{R}^{n\times n}$, $n(t)$ satisfies $\|n(t)\| \leqslant \zeta, \zeta > 0, t \geqslant 0$.

Definition 12.2.2 ([164]) An ellipsoid centered at the origin is a set in $\mathbb{R}^n$ such that

$$\mathscr{E}_M = \left\{ x \in \mathbb{R}^n : x^T M x \leqslant 1 \right\},$$

where M is a symmetric positive definite matrix.

Definition 12.2.3 (*Invariant ellipsoid*, [215]) An ellipsoid $\mathscr{E}$ is positive invariant for the system (12.22) if $\varphi(\theta) \in \mathscr{E}, \theta \in [-h, 0]$ implies that $x(t, \varphi) \in \mathscr{E}, t \geqslant 0$ for every trajectory of the system.

Definition 12.2.4 (*Attractive ellipsoid*, [215]) An ellipsoid $\mathscr{E}$ is an attractive domain for the system (12.22) if

(1) $\varphi(\theta) \in \mathscr{E}, \theta \in [-h, 0]$ implies that $x(t, \varphi) \in \mathscr{E}, t \geqslant 0$,
(2) $\varphi(\theta) \in \mathbb{R}^n \backslash \mathscr{E}$, for some $\theta \in [-h, 0]$ implies that there exists T_a, $0 \leqslant T_a < \infty$, such that $x(t, \varphi) \in \mathscr{E}, t \geqslant 0$.

Lemma 12.2.5 *Let the functional $V(x_t)$ satisfying*

$$x^T(t) P x(t) \leqslant V(x_t) \leqslant \alpha \|x_t\|_h^2, \qquad (12.23)$$

where P is a symmetric positive definite matrix and

$$\frac{d}{dt} V(x_t) + \sigma V(x_t) \leqslant \bar{\beta},\ \forall t \geqslant 0,\ \sigma > 0,\ \bar{\beta} > 0. \qquad (12.24)$$

Then for any initial function $\varphi \in \mathscr{PC}([-h, 0], \mathbb{R}^n)$, the solution $x(t, \varphi)$ is retained within the ellipsoid $\mathscr{E}_{\tilde{P}} = \left\{ x \in \mathbb{R}^n : x^T \tilde{P} x \leqslant 1 \right\}$, $\tilde{P} = \left(\frac{\sigma}{\bar{\beta}} + \varsigma \right) P$, $\varsigma > 0$ for $t \geqslant T_a(\varphi, \varsigma) > 0$ where

$$T_a(\varphi, \varsigma) = \frac{1}{\sigma} \ln \left(\frac{\alpha \|\varphi\|_h^2}{\varsigma} - \frac{\bar{\beta}}{\varsigma \sigma} \right).$$

Proof Multiplying (12.24) by $e^{\sigma\theta}$ gives

$$\frac{d}{d\theta} \left(e^{\sigma\theta} V(x_t) \right) \leqslant \bar{\beta} e^{\sigma\theta}.$$

Integrating the above expression from 0 to "t" yields

$$V(x_t) \leqslant \mathrm{e}^{-\sigma t} V(0, x_0) + \frac{\bar{\beta}}{\sigma}\left(1 - \mathrm{e}^{-\sigma t}\right).$$

It follows from (12.23) that

$$x^T(t)Px(t) \leqslant \mathrm{e}^{-\sigma t} V(0, x_0) + \frac{\bar{\beta}}{\sigma}\left(1 - \mathrm{e}^{-\sigma t}\right) \leqslant \alpha \mathrm{e}^{-\sigma t} \left\|\varphi\right\|_h^2 + \frac{\bar{\beta}}{\sigma}\left(1 - \mathrm{e}^{-\sigma t}\right).$$

Then, we have

$$x^T(t)Px(t) \leqslant \mathrm{e}^{-\sigma t}\left(\alpha \left\|\varphi\right\|_h^2 - \frac{\bar{\beta}}{\sigma}\right) + \frac{\bar{\beta}}{\sigma}. \tag{12.25}$$

For the initial conditions fulfilling $\alpha \left\|\varphi\right\|_h^2 \leqslant \frac{\bar{\beta}}{\sigma}$, it yields

$$x^T(t)Px(t) \leqslant \frac{\bar{\beta}}{\sigma} \quad \forall t \geqslant 0,$$

equivalently $x^T(t)\frac{\sigma}{\bar{\beta}}Px(t) \leqslant 1$, for all $t \geqslant 0$. Otherwise, if

$$\alpha \left\|\varphi\right\|_h^2 > \frac{\bar{\beta}}{\sigma},$$

we look for the time instant $T_a(\varphi, \varsigma)$ at which the solution $x(t, \varphi)$ enters into the ellipsoid $\mathscr{E}_{\tilde{P}}$. In view of (12.25), $T_a(\varphi, \varsigma)$ should guarantee

$$x^T(t)Px(t) \leqslant \varsigma + \frac{\bar{\beta}}{\sigma},$$

for a small enough $\varsigma > 0$, i.e.,

$$e^{-\sigma T_a}\left(\alpha \left\|\varphi\right\|_h^2 - \frac{\bar{\beta}}{\sigma}\right) = \varsigma.$$

From the above expression we have

$$\frac{1}{\varsigma}\left(\alpha \left\|\varphi\right\|_h^2 - \frac{\bar{\beta}}{\sigma}\right) = e^{\sigma T_a},$$

and the result follows.

12.2.1 Stabilizing Feedback Controllers

Consider a neutral-type time-delay system of the form:

$$\begin{aligned}\dot{x}(t) + D\dot{x}(t-2\tau) = A_0 x(t) + A_1 x(t-2\tau) + B_1 u_1(t-\tau) \\ +B_2 u_2(t) + C_0 f(x(t)) + C_1 f(x(t-2\tau)) + \omega(t),\end{aligned} \tag{12.26}$$

where τ is a constant time-delay, $D \in \mathbb{R}^{n\times n}$ is Schur stable, $A_0,\ A_1 \in \mathbb{R}^{n\times n}$, $B \in \mathbb{R}^{n\times m}$, f satisfies $\|f(x(t))\| \leqslant \zeta, \zeta > 0, t \geqslant 0$, and $\omega(t)$ is such that:

$$\|\omega(t)\|_{K_\omega} = \omega(t)^T K_\omega \omega(t) \leqslant 1, \quad t \geqslant 0. \tag{12.27}$$

In view of the application under consideration, we propose the following structure for u_1:

$$u_1(t-\tau) = K_0 \dot{x}(t-2\tau) + K_1 x(t-2\tau), \tag{12.28}$$

and for $u_2(t)$ we consider:

$$u_2(t) = K_2 x(t). \tag{12.29}$$

Moreover, we assume that the control law u_2 is subject to the restriction:

$$|u_2(t)| \leqslant \bar{u}_2, \tag{12.30}$$

where $\bar{u}_2$ is the saturation level, i.e., the vector $u_2(t) = \text{col}\left\{u_2^1(t), \ldots, u_2^m(t)\right\}$ is subject to the following amplitude constraints:

$$\left|u_2^j(t)\right| \leqslant \bar{u}_2^j,\ \ 0 < \bar{u}_2^j,\ j = 1, \ldots, m.$$

Our aim is to find the conditions such that the ellipsoid

$$\mathscr{E}_{\tilde{P}} = \left\{x \in \mathbb{R}^n : x^T \tilde{P} x \leqslant 1\right\},$$

defines an attractive set for the trajectories of the system (12.26) in closed loop with (12.28) and (12.29), i.e.,

$$\overline{\lim_{t\to\infty}}\, x^T(t)\tilde{P}x(t) \leqslant 1.$$

Furthermore, the choice of the matrices $K_0, K_1, K_2 \in \mathbb{R}^{m\times n}$ and $\tilde{P}$ must guarantee the minimality of the ellipsoid $\mathscr{E}_{\tilde{P}}$. Given that the trace of the matrix $\tilde{P}$ is inversely related to the axes of the ellipsoid $\mathscr{E}_{\tilde{P}}$, the following optimization problem arises:

$$\min \operatorname{tr}(\tilde{P}^{-1}) \tag{12.31}$$

$$\text{subject to } \tilde{P} \in \sigma_1,\ K_0, K_1, K_2 \in \sigma_2,$$

where Σ_1, Σ_2, define the set of admissible matrices of dimension $n \times n$ and $m \times n$ respectively guaranteeing the invariance property of the ellipsoid $\mathscr{E}_{\tilde{P}}$.

Consider the following functional $V(x_t)$:

$$V(x_t) = x^T(t)Px(t) + \int_{t-2\tau}^{t} e^{\sigma(s-t)} x^T(s)Sx(s)ds \qquad (12.32)$$
$$+2\tau \int_{-2\tau}^{0} \int_{t+\theta}^{t} e^{\sigma(s-t)} \dot{x}^T(s)R\dot{x}(s)dsd\theta.$$

Notice that condition (12.23) of Lemma 12.2.5 is satisfied with

$$\alpha = \lambda_{\max}(P) + 2\tau\lambda_{\max}(S) + 4\tau^2\lambda_{\max}(R).$$

The time derivative of (12.32) satisfies

$$\begin{aligned}\frac{dV(x_t)}{dt} + \sigma V(x_t) \leqslant\ & 2x^T(t)P\dot{x}(t) + \sigma x^T(t)Px(t) \\ & + x^T(t)Sx(t) - e^{-\sigma 2\tau} x^T(t-2\tau)Sx(t-2\tau) \qquad (12.33) \\ & - e^{-\sigma 2\tau}(x(t) - x(t-2\tau))^T R(x(t) - x(t-2\tau)) \\ & + 4\tau^2 \dot{x}^T(t)R\dot{x}(t).\end{aligned}$$

According to the descriptor approach introduced in [97], we add to the right hand side of the inequality (12.33), the following null terms derived from the system dynamic and the controllers structure (12.28) and (12.29):

$$\begin{aligned}0 = 2[&P_2x(t) + P_3\dot{x}(t) + P_4\dot{x}(t-2\tau) + P_5u_1(t-\tau) \qquad (12.34) \\ &+ P_6u_2(t) + P_7f(x(t)) + P_8f(x(t-2\tau)) + P_9\omega(t)]^T \\ &[-\dot{x}(t) - D\dot{x}(t-2\tau) + A_0x(t) + A_1x(t-2\tau) + B_1u_1(t-\tau) \\ &+ B_2u_2(t) + C_0f(x(t)) + C_1f(x(t-2\tau)) + \omega(t)],\end{aligned}$$

$$\begin{aligned}0 = 2[&P_{10}u_1(t-\tau) + P_{11}\dot{x}(t-2\tau) + P_{12}x(t-2\tau)]^T \qquad (12.35) \\ &[-u_1(t-\tau) + K_0\dot{x}(t-2\tau) + K_1x(t-2\tau)],\end{aligned}$$

$$0 = 2[P_{13}u_2(t) + P_{14}x(t)]^T[-u_2(t) + K_2x(t)], \qquad (12.36)$$

where $P_2, P_3, P_4, P_9 \in \mathbb{R}^{n\times n}$, $P_5, P_6, P_7, P_8 \in \mathbb{R}^{n\times m}$, $P_{10}, P_{13} \in \mathbb{R}^{m\times m}$, $P_{11}, P_{12}, P_{14}, K_0, K_1, K_2 \in \mathbb{R}^{m\times n}$.

The condition (12.27) implies that $\omega(t)^T K_\omega \omega(t) - 1 \leqslant 0$, then for any $\bar{\beta} > 0$, we have that $-\bar{\beta}(\omega(t)^T K_\omega \omega(t) - 1) \geqslant 0$. In order to take into account the perturbation restriction we can add to (12.33) the term

$$-\bar{\beta}(\omega(t)^T K_\omega \omega(t) - 1). \qquad (12.37)$$

Notice that condition (12.24) is fulfilled.
Finally, we obtain after symmetrization of the cross terms:

$$\frac{dV(x_t)}{dt} + \sigma V(x_t) - \bar{\beta} \leqslant \eta^T \bar{\Psi} \eta,$$

where

$$\eta = (x(t)\; x(t-2\tau)\; \dot{x}(t)\; \dot{x}(t-2\tau)\; u_1(t-\tau)\; u_2(t)\; f(x(t))\; f(x(t-2\tau))\; \omega(t))^T,$$

and $\bar{\Psi}$ is a symmetric matrix with elements $\bar{\Psi}_{ij}$, $i = 1, \ldots, 9$, $j = 1, \ldots, 9$ defined as follows:

$$\begin{bmatrix} \bar{\Psi}_{11} & \bar{\Psi}_{12} & \bar{\Psi}_{13} & \bar{\Psi}_{14} & \bar{\Psi}_{15} & \bar{\Psi}_{16} & \bar{\Psi}_{17} & P_2^T C_1 + A_0^T P_8 & P_2^T + A_0^T P_9 \\ * & \bar{\Psi}_{22} & A_1^T P_3 & \bar{\Psi}_{24} & \bar{\Psi}_{25} & A_1^T P_6 & A_1^T P_7 & A_1^T P_8 & A_1^T P_9 \\ * & * & \bar{\Psi}_{33} & \bar{\Psi}_{34} & \bar{\Psi}_{35} & \bar{\Psi}_{36} & \bar{\Psi}_{37} & P_3^T C_1 - P_8 & P_3^T - P_9 \\ * & * & * & \bar{\Psi}_{44} & \bar{\Psi}_{45} & \bar{\Psi}_{46} & \bar{\Psi}_{47} & P_4^T C_1 - D^T P_8 & P_4^T - D^T P_9 \\ * & * & * & * & \bar{\Psi}_{55} & \bar{\Psi}_{56} & \bar{\Psi}_{57} & P_5^T C_1 + B_1^T P_8 & P_5^T + B_1^T P_9 \\ * & * & * & * & * & \bar{\Psi}_{66} & \bar{\Psi}_{67} & P_6^T C_1 + B_2^T P_8 & P_6^T + B_2^T P_9 \\ * & * & * & * & * & * & \bar{\Psi}_{77} & P_7^T C_1 + C_0^T P_8 & P_7^T + C_0^T P_9 \\ * & * & * & * & * & * & * & P_8^T C_1 + C_1^T P_8 & P_8^T + C_1^T P_9 \\ * & * & * & * & * & * & * & * & \bar{\Psi}_{99} \end{bmatrix},$$

where:

$$\begin{aligned}
\bar{\Psi}_{11} &= A_0^T P_2 + P_2^T A_0 + S + \sigma P - e^{-\sigma 2\tau} R + P_{14} K_2 + K_2^T P_{14} \\
\bar{\Psi}_{12} &= e^{-\sigma 2\tau} R + P_2^T A_1 \\
\bar{\Psi}_{13} &= P - P_2^T + A_0^T P_3 \\
\bar{\Psi}_{14} &= -P_2^T D + A_0^T P_4 \\
\bar{\Psi}_{15} &= P_2^T B_1 + A_0^T P_5 \\
\bar{\Psi}_{16} &= P_2^T B_2 + A_0^T P_6 + K_2^T P_{13} - P_{14}^T \\
\bar{\Psi}_{17} &= P_2^T C_0 + A_0^T P_7 \\
\bar{\Psi}_{22} &= -e^{-\sigma 2\tau}(S + R) + P_{12}^T K_1 + K_1^T P_{12} \\
\bar{\Psi}_{23} &= A_1^T P_3 \\
\bar{\Psi}_{24} &= A_1^T P_4 + K_1^T P_{11} + P_{12}^T K_0 \\
\bar{\Psi}_{25} &= A_1^T P_5 + K_1^T P_{10} - P_{12}^T \\
\bar{\Psi}_{33} &= 4\tau^2 R - P_3 - P_3^T \\
\bar{\Psi}_{34} &= -P_3^T D - P_4 \\
\bar{\Psi}_{35} &= P_3^T B_1 - P_5 \\
\bar{\Psi}_{36} &= P_3^T B_2 - P_6 \\
\bar{\Psi}_{37} &= P_3^T C_0 - P_7 \\
\bar{\Psi}_{44} &= -P_4^T D - D^T P_4 + P_{11}^T K_0 + K_0^T P_{11} \\
\bar{\Psi}_{45} &= P_4^T B - D^T P_5 + K_0^T P_{10} - P_{11}^T \\
\bar{\Psi}_{46} &= P_4^T B_2 - D^T P_6 \\
\bar{\Psi}_{47} &= P_4^T C_0 - D^T P_7 \\
\bar{\Psi}_{55} &= P_5^T B_1 + B_1^T P_5 - P_{10}^T - P_{10} \\
\bar{\Psi}_{56} &= P_5^T B_2 + B_1^T P_6 \\
\bar{\Psi}_{57} &= P_5^T C_0 + B_1^T P_7 \\
\bar{\Psi}_{66} &= P_6^T B_2 + B_2^T P_6 - P_{13}^T - P_{13} \\
\bar{\Psi}_{67} &= P_6^T C_0 + B_2^T P_7 \\
\bar{\Psi}_{77} &= P_7^T C_0 + C_0^T P_7 \\
\bar{\Psi}_{99} &= P_9^T + P_9 - \bar{\beta} K_\omega
\end{aligned}$$

Clearly, if $\bar{\Psi} < 0$, the condition (12.24) of Lemma 12.2.5 is satisfied.

The introduction of the additional inequality:

$$\mathscr{H} := \begin{bmatrix} H & I_n \\ I_n & P \end{bmatrix} > 0, \; H \geqslant P^{-1}$$

allows us to reduce the nonlinear optimization problem (12.31) to the linear problem min tr(H). By Schur complements, the trace minimization of H implies the trace minimization of P^{-1} [117].

To address the presence of physical constraints on the control inputs such as limits in force, torque, current, flow rate, etc., we combine the ideas introduced in [86, 98] with the attractive ellipsoid properties as follows. From (12.29) and (12.30) we get the inequality $k_2^j x \leqslant \bar{u}_2^j$, $j = 1, \ldots, m$, then for all $x \in \mathscr{E}_{\tilde{P}}$, $\mathscr{E}_{\tilde{P}} = \left\{ x \in \mathbb{R}^n : x^T \left(\frac{\sigma}{\bar{\beta}} + \varsigma \right) Px \leqslant 1 \right\}$ we have that:

$$2 \left| k_2^j x \right| \leqslant \bar{u}_2^j \left[1 + \left(\frac{\sigma}{\bar{\beta}} + \varsigma \right) x^T P x \right] \leqslant 2\bar{u}_2^j \quad j = 1, \ldots, m \tag{12.38}$$

we can write (12.38) as

$$\begin{pmatrix} 1 & \pm x^T \end{pmatrix} \begin{pmatrix} \bar{u}_2^j & k_2^j \\ * & \left(\frac{\sigma}{\bar{\beta}} + \varsigma \right) \bar{u}_2^j P \end{pmatrix} \begin{pmatrix} 1 \\ \pm x \end{pmatrix} \geqslant 0, \quad j = 1, \ldots, m,$$

by Schur complements, the latter inequalities are equivalent to the following ones:

$$\left(\frac{\sigma}{\bar{\beta}} + \varsigma \right) \bar{u}_2^j P - k_2^{jT} (\bar{u}_2^j)^{-1} k_2^j \geqslant 0,$$

$$\mathscr{I}_j := \begin{pmatrix} \left(\frac{\sigma}{\bar{\beta}} + \varsigma \right) \bar{u}_2^j P & k_2^{jT} \\ * & \bar{u}_2^j \end{pmatrix} \geqslant 0 \quad j = 1, \ldots, m,$$

where k_2^j is the jth row of the matrix K_2.

Summarizing the above ideas, the result on the minimum attractive ellipsoid for system (12.26) is stated as follows:

Theorem 12.2.6 *Let the optimization problem*

$$\min tr(H)$$

subject to

$$\begin{cases} \Lambda := \{H, P, S, R, K_0, K_1, K_2, P_k, \bar{\beta}\}, \ k = 1, \ldots, 14, \\ \bar{\Psi} < 0, \\ \mathscr{H} > 0, \\ \mathscr{I}_j \geqslant 0 \ \ j = 1, \ldots, m, \\ P > 0, \ S > 0, \ R > 0, \ \bar{\beta} > 0, \ \sigma > 0, \ \varsigma > 0, \end{cases}$$

with optimal solution $\hat{\Lambda} := \left\{\hat{H}, \hat{P}, \hat{S}, \hat{R}, \hat{K}_0, \hat{K}_1, \hat{K}_2, \hat{P}_k, \hat{\bar{\beta}}\right\}$, $k = 1, \ldots, 14$. *The ellipsoid* $\mathscr{E}(0, \tilde{P})$ *determined by the matrix* $\left(\frac{\sigma}{\hat{\bar{\beta}}} + \varsigma\right)\hat{P}$, *is a minimum attractive ellipsoid for system* (12.26) *in closed loop with the controllers* (12.28)–(12.30) *for* $t \geqslant T_a(\varphi, \varsigma) = \frac{1}{\sigma} \ln\left(\frac{\alpha\|\varphi\|_h^2}{\varsigma} - \frac{\bar{\beta}}{\varsigma\sigma}\right)$.

Dynamic systems are usually subject to uncertain time-varying delays. In [99] an overview on this topic is given. Here we consider the stability of a neutral-type time-delay system of the form:

$$\dot{x}(t) + D\dot{x}(t - g(t)) = A_0 x(t) + A_1 x(t - \tau(t)), \tag{12.39}$$

where $g(t)$ and $\tau(t)$ are uncertain time-varying delays such that

$$\tau(t) = h + \xi(t), \tag{12.40}$$

where $h > 0$ is a nominal constant value and ξ is a time-varying perturbation. Following [151], system (12.39) is represented as:

$$\dot{x}(t) + D\dot{x}(t - g(t)) = A_0 x(t) + A_1 x(t - h) + A_1 \left[x(t - h - \xi(t)) - x(t - h)\right],$$

which is equivalent to

$$\dot{x}(t) + D\dot{x}(t - g(t)) = A_0 x(t) + A_1 x(t - h) - A_1 \int_{t-h-\xi(t)}^{t-h} \dot{x}(s) ds. \tag{12.41}$$

In [99], two delay perturbation cases are considered:

- **Case 1**. $\xi(t)$ is a sign-varying piecewise-continuous function satisfying $|\xi(t)| \leqslant \mu < h$ with known upper bounds μ, i.e.,

$$\tau(t) \in [h - \mu, \ h + \mu].$$

- **Case 2.** $\xi(t)$ is a nonnegative piecewise-continuous function such that $\xi(t) \leqslant \mu$, and thus,

$$\tau(t) \in [h,\ h + \mu].$$

The main idea for the stability analysis of (12.41) is to consider a Lyapunov function of the form: $V = V_n + V_a$, where V_n is a nominal Lyapunov function from which the stability conditions of the nominal system ($\xi = 0$) are derived, and V_a consists of an additional term depending on μ.

Following these ideas, it is possible to determine additional conditions under which $\mathscr{E}(0, \tilde{P})$ is the minimum attractive ellipsoid for system (12.26)–(12.30) subject to uncertain time-varying state delay of the form (12.40). In this case we might choose V_n as in (12.32) with $2\tau = h$. For the Case 1, V_a may be defined as:

$$V_a = (h + \mu) \int_{-\mu}^{\mu} \int_{t+\theta-h}^{t} \mathrm{e}^{\sigma(s-t)} \dot{x}^T(s) R_a \dot{x}(s) \mathrm{d}s \mathrm{d}\theta, \ R_a > 0,$$

and for the Case 2, as follows:

$$V_a = (h + \mu) \int_{-\mu}^{0} \int_{t+\theta-h}^{t} \mathrm{e}^{\sigma(s-t)} \dot{x}^T(s) R_a \dot{x}(s) \mathrm{d}s \mathrm{d}\theta, \ R_a > 0.$$

Note that for $\mu \to 0$ we have $V_a \to 0$, and thus $V \to V_n$. The latter will guarantee that, if the conditions for the stability of the nominal system are feasible, then the stability conditions for the perturbed system will be feasible for small enough μ. The additional stability conditions will be derived according to the choice of V_a and to the system representation similar to (12.41).

12.2.2 Stick-slip and Bit-bounce Elimination

In view of the axial model (2.2), coupled to the torsional one (2.32), subject to the frictional torque given in (3.14) with the friction coefficient given in (3.15), the drillstring dynamics are described by the following equation:

$$\begin{aligned} \dot{x}(t) + D\dot{x}(t - 2\tau) &= A_0 x(t) + A_1 x(t - 2\tau) + B_1 u_1(t - \tau) + B_2 u_2(t) \\ &\quad + C_0 f(x_1(t) + \Omega_0) + C_1 f(x_1(t - 2\tau) + \Omega_0) + \omega(t), \end{aligned} \tag{12.42}$$

where $x(t) = \left[x_1(t)\ x_2(t)\ x_3(t)\right]^T$,

$$x_1(t) = \dot{\Phi}_b(t) - \Omega_0, \quad x_2(t) = v(t) = U_b(t) - \rho_0 t \quad x_3(t) = \dot{v}(t),$$

$$A_0 = \begin{bmatrix} -\Psi - \frac{c_b}{I_B} & 0 & 0 \\ 0 & 0 & 1 \\ -\frac{\mu_1 c_b}{m_0} & -\frac{k_0}{m_0} & -\frac{c_0}{m_0} \end{bmatrix}, \ A_1 = \begin{bmatrix} \frac{\Upsilon c_b}{I_B} - \Upsilon \Psi & 0 & 0 \\ 0 & 0 & 0 \\ \frac{\mu_1 c_b}{m_0} & 0 & 0 \end{bmatrix}, \ B_1 = \begin{bmatrix} \Pi \\ 0 \\ 0 \end{bmatrix},$$

$$B_2 = \begin{bmatrix} 0 \\ 0 \\ -\frac{c_0}{m_0} \end{bmatrix}, \ C_0 = \begin{bmatrix} -\frac{1}{I_B} \\ 0 \\ -\frac{\mu_1}{m_0} \end{bmatrix}, \ C_1 = \begin{bmatrix} \frac{\Upsilon}{I_B} \\ 0 \\ 0 \end{bmatrix}, \ D = \begin{bmatrix} -\Upsilon & 0 & 0 \\ 0 & 0 & 0 \\ 0 & 0 & 0 \end{bmatrix},$$

the time-delay is given by $\tau = \sqrt{\frac{I}{GJ}}L$, the control inputs u_1 and u_2 corresponds to the angular velocity at the surface $\Omega(t)$ and the rate of penetration $\rho(t)$, respectively. The nonlinear term f is given by:

$$f(x(t)) = W_{ob} R_b \left(\mu_{cb} + (\mu_{sb} - \mu_{cb}) e^{-\gamma_b |x(t)|} \right) \operatorname{sgn}(x(t)). \tag{12.43}$$

The parameters used in simulations are given in Table C.1 of Appendix C.

We assume that due to the physical constraints of the drilling system, the control input $u_2(t)$ is subject to the restriction $|u_2(t)| \leqslant \bar{u}_2$ where $\bar{u}_2 = 100$. We consider that the initial condition φ is such that $\|\varphi\|_h = 2$. Using Theorem 12.2.6 we obtain the synthesis of the controllers (12.28) and (12.29) by means of the computational package "PENBMI" of MATLAB.

The maximum exponential decay rate for which the conditions given in Theorem 12.2.6 are satisfied is $\sigma = 1.5$. The solution of the optimization problem is:

$$P = \begin{bmatrix} 6.1522 & 2.5914 & 0.2713 \\ 2.5914 & 13.0989 & 3.4865 \\ 0.2713 & 3.4865 & 1.4203 \end{bmatrix}, \quad \operatorname{eig}(P) = \begin{cases} 0.4238 \\ 5.4240 \\ 14.8236 \end{cases},$$

$$\lambda_{\max}(S) = 318.9226, \quad \lambda_{\max}(R) = 3.9407, \quad \bar{\beta} = 1.3620, \quad \alpha = 254.1923,$$

$$K_0 = \begin{bmatrix} -0.1272 & 0 & 0 \end{bmatrix}, \quad K_1 = \begin{bmatrix} 0.4429 & 0.0196 & 0.0088 \end{bmatrix}, \tag{12.44}$$

$$K_2 = \begin{bmatrix} 24.3378 & 305.7411 & 120.3589 \end{bmatrix}.$$

The trajectories of the drilling system (12.42), in closed loop with the controllers (12.28) and (12.29), with K_0, K_1, K_2 given in (12.44), are shown in Figs. 12.4b, 12.5b, and 12.6b.

Figure 12.7 shows the trajectories of the stabilizing controllers $u_1(t)$ and $u_2(t)$.

For $\varsigma = 0.1$, the trajectories of the drilling system are retained into the ellipsoid $\mathscr{E}_{\tilde{P}} = \left\{ x \in \mathbb{R}^n : x^T \tilde{P} x \leqslant 1 \right\}$ where

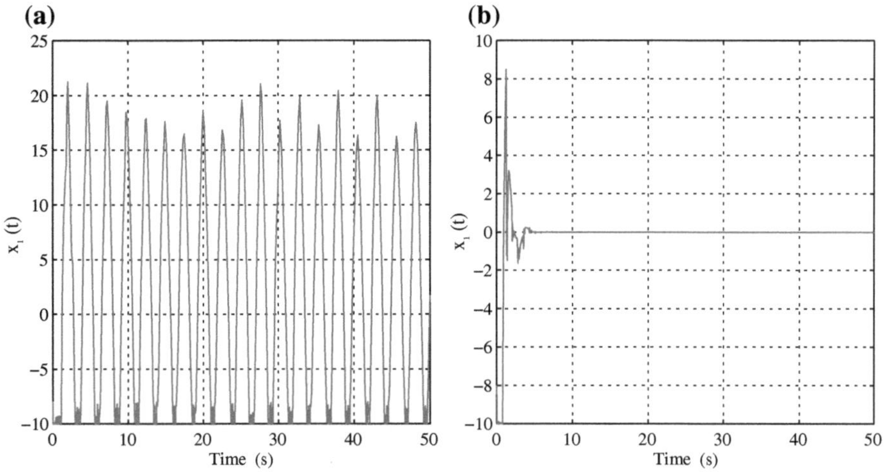

Fig. 12.4 Simulation of the coupled torsional–axial model (12.42). Angular variable $x_1(t) = \dot{\Phi}_b(t) - \Omega_0$ for a reference velocity of $10\,\text{rad s}^{-1}$. **a** Trajectory without control actions (stick-slip). **b** Elimination of stick-slip oscillations by means of the controllers (12.28) and (12.29)

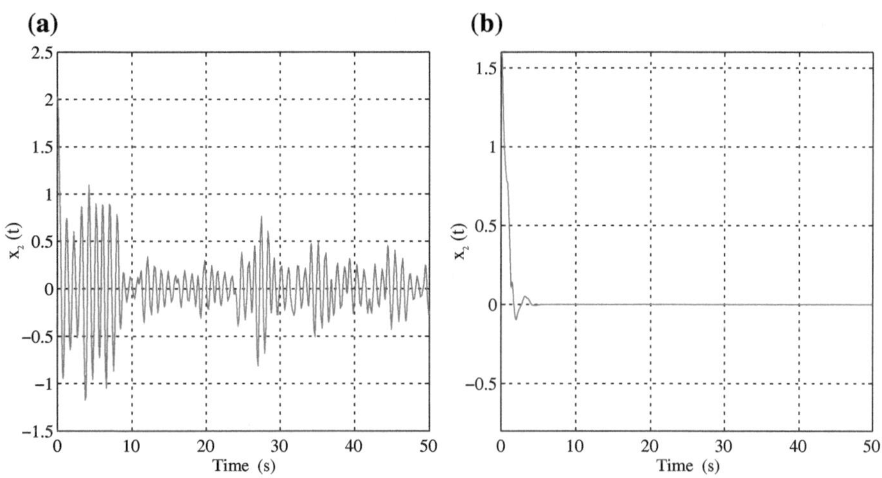

Fig. 12.5 Simulation of the coupled torsional–axial model (12.42). Axial variable $x_2(t) = \nu(t) = U_b(t) - \rho_0 t$ for a reference velocity of $0.1\,\text{m s}^{-1}$. **a** Trajectory without control actions (bit-bounce). **b** Elimination of bit-bounce oscillations by means of the controllers (12.28) and (12.29)

$$\tilde{P} = \begin{bmatrix} 7.3906 & 3.1130 & 0.3259 \\ 3.1130 & 15.7357 & 4.1883 \\ 0.3259 & 4.1883 & 1.7062 \end{bmatrix},$$

for $t \geqslant T_a(\varphi, \varsigma) = 0.6667 \ln \left(2541.9 \left\|\varphi\right\|_h^2 - 9.0800\right) = 6.1510\text{s}$. The minimum attractive ellipsoid obtained for the drilling system (12.42) is shown in Fig. 12.8.

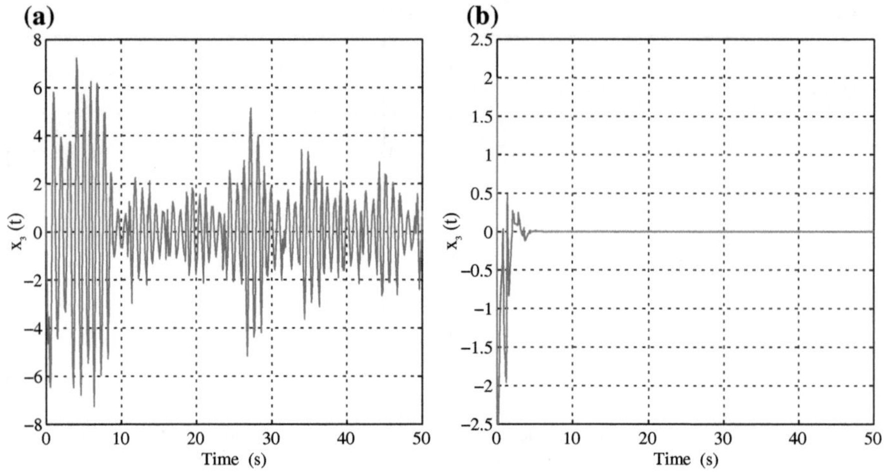

Fig. 12.6 Simulation of the coupled torsional–axial model (12.42). Axial variable $x_3(t) = \dot{v}(t)$ for a reference velocity of 0.1 m s^{-1}. **a** Trajectory without control actions (bit-bounce). **b** Elimination of bit-bounce oscillations by means of the controllers (12.28) and (12.29)

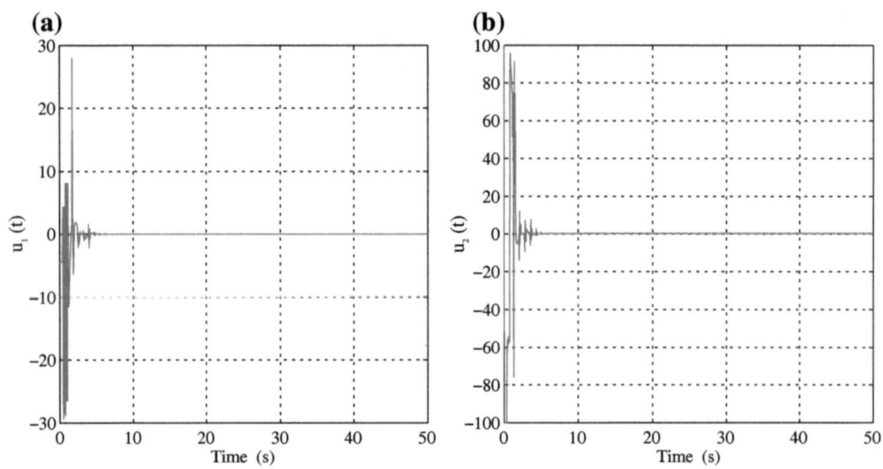

Fig. 12.7 Simulation of the coupled torsional–axial model (12.42). **a** Trajectory of the controller $u_1(t)$ defined in (12.28). **b** Trajectory of the controller $u_2(t)$ defined in (12.29)

Table 12.1 shows a comparative analysis of the results obtained with and without saturation restriction on the control input u_2 ($|u_2(t)| \leqslant 100$). Clearly, the absence of restrictions on u_2 provides better performance.

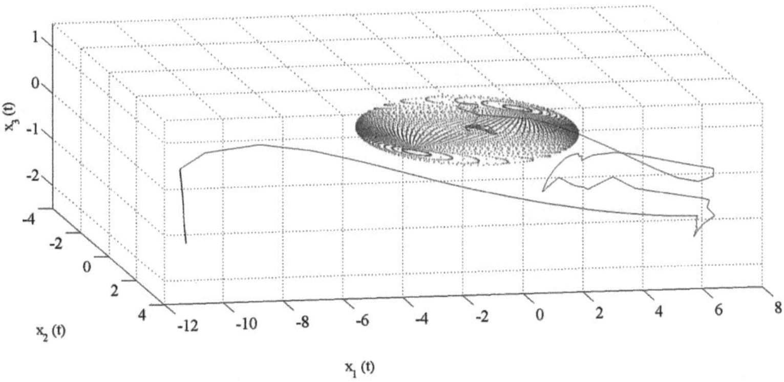

Fig. 12.8 Simulation of the coupled torsional–axial model (12.42). Phase portrait of the drillstring trajectories: $x_1(t)$ versus $x_2(t)$ versus $x_3(t)$ converging to the ellipsoid $\mathscr{E}(0, \tilde{P})$

Table 12.1 Numerical results determining the attractive ellipsoid method-based controller with and without saturated control input u_2

Control input	$u_{2_{\max}}(t)$	σ	α	$\bar{\beta}$	$T_a(\varphi, \varsigma)$
Restriction on $u_2(t)$	100	1.5	254.1923	1.3620	6.1510s
No restriction on $u_2(t)$	159	1.6	123.4176	1.6227	5.3140s

12.3 Notes and References

We have addressed the problem of control design for the practical stabilization of the drilling system based on two different modeling strategies: the coupled wave–ODE model and the coupled NDDE–ODE representation.

Through LMI–BMI techniques we have established feedback controllers guaranteeing ultimate boundedness of the system trajectories which inherently implies the system practical stabilization. The performance of the proposed control approaches is evidenced by system simulations showing an effective suppression of undesirable drilling vibrations.

The control approach for eliminating the stick-slip and bit-bounce based on the attractive ellipsoid method is presented in [257]. The idea of designing a controller that retains the system trajectories within a given target "tube" (invariant set) was introduced in [28]; a closed-loop control strategy for discrete-time systems in the presence of uncertainty is therein proposed. Furthermore, an algorithm for the construction of ellipsoidal approximations to the sets involved is therein given. The conditions under which the state of an uncertain system can be forced to stay in a specified region of the state space by using feedback control are investigated in [29]. The same problem is also analyzed in [116], where necessary and sufficient conditions and an algorithm which constructs the control are derived for open-loop and closed-loop control laws; additionally, ellipsoids that bound the state and control are also therein derived. An overview of the literature concerning positively invariant sets and their relation with the Lyapunov theory, allowing tackling problems concerning the analysis and control synthesis of dynamical systems is provided in [36].

Both control strategies presented in this chapter succeed in eliminating undesirable drilling vibrations; furthermore, the proposed methodologies can be easily extended to treat more general systems.

Different control methodologies to tackle oilwell drilling vibrations which lead to the system stabilization were presented in Chaps. 9–12. It was shown that all of them are effective in suppressing unwanted oscillations; however, in order to assess their benefits and drawbacks, a comparative analysis must be developed. For the sake of completeness, the following chapter presents a contrastive analysis of all proposed control schemes that allows comparing some controller properties such as their ability in eliminating vibrations in the presence of external disturbances and when drilling hard surfaces, their simplicity and feasibility of implementation, among others.

Chapter 13
Performance Analysis of the Controllers

Throughout the third part of the book, we have presented various control strategies aimed at eliminating harmful vibrations in the drilling system. It has been shown that the proposed control solutions stabilize the system, effectively suppressing the stick-slip and the bit-bounce, notwithstanding, it is important to provide a comparative analysis to highlight the advantages and vulnerabilities of each control method.

This chapter presents a contrastive study of the system response to the control laws, presented in Chaps. 9–12, which have been designed through diverse analysis approaches:

- **Low-order control scheme**. One of the most popular techniques used in practice to avoid drilling vibrations is the soft-torque control, discussed in Sect. 9.2.2. This control strategy is based on the idea that the torsional energy must be absorbed at the top extremity to avoid the reflection of waves back down to the rod. A different control approach involving a qualitative analysis of the system is proposed in Sect. 9.3. Through the center manifold theorem and normal forms theory, two low-order control schemes guaranteeing the asymptotic stability of the system were developed: time-delayed proportional and time-delayed PID controller. Simulation results show that the stick-slip and the bit-bounce phenomena are mitigated through these control approaches.
- **Flatness line based approach**. The differential flatness property of the system is exploited to design a pair of stabilizing controllers tackling the trajectory tracking problem. This approach gives rise to an exponential convergence of the error toward zero, allowing the suppression of undesirable drillstring vibrations.
- **Lyapunov techniques**. By using Lyapunov stability theory tools, diverse control laws to eliminate the stick-slip and the bit-bounce were designed. Asymptotic, exponential, and practical stability of the system is achieved via feedback linear and nonlinear controllers developed within the framework of different drilling system modeling approaches: switching system, polytopic approximation, coupled wave-ODE model, and NDDE–ODE representation.

M.B. Saldivar Márquez et al., *Analysis and Control of Oilwell Drilling Vibrations*, Advances in Industrial Control, DOI 10.1007/978-3-319-15747-4_13

13.1 Comparative Analysis Guidelines

In order to evaluate the performance of the proposed controllers, the response of the system will be analyzed through simulations. A reliable approximation of the phenomena taking place in a rotary oilwell drilling system is provided by undamped wave equations which capture the essential vibrational dynamics of the drilling rod. The infinite-dimensional model, chosen to develop the comparative analysis, is written as:

$$\begin{cases} \dfrac{\partial^2 \Phi}{\partial s^2}(s,t) = \tilde{c}^2 \dfrac{\partial^2 \Phi}{\partial t^2}(s,t), \quad \tilde{c} = \sqrt{\dfrac{\rho_a}{G}} \\ GJ\dfrac{\partial \Phi}{\partial s}(0,t) = \beta \dfrac{\partial \Phi}{\partial t}(0,t) - u_T(t) \\ GJ\dfrac{\partial \Phi}{\partial s}(L,t) = -I_B \dfrac{\partial^2 \Phi}{\partial t^2}(L,t) - T\left(\dfrac{\partial \Phi}{\partial t}(L,t)\right) \end{cases}$$

$$\begin{cases} \dfrac{\partial^2 U}{\partial s^2}(s,t) = c^2 \dfrac{\partial^2 U}{\partial t^2}(s,t), \quad c = \sqrt{\dfrac{\rho_a}{E}} \\ E\Gamma\dfrac{\partial U}{\partial s}(0,t) = \alpha \dfrac{\partial U}{\partial t}(0,t) - u_H(t) \\ E\Gamma\dfrac{\partial U}{\partial s}(L,t) = -M_B \dfrac{\partial^2 U}{\partial t^2}(L,t) - T\left(\dfrac{\partial \Phi}{\partial t}(L,t)\right) \end{cases}$$

where the rotary angle and the longitudinal position are denoted by $\Phi(s,t)$ and $U(s,t)$, respectively. The system is controlled from the surface ($s = 0$) through the torque provided by the rotor $u_T(t)$ and the upward hook force $u_H(t)$ which satisfy:

$$u_T(t) = \beta \Omega(t), \quad u_H(t) = \alpha \rho(t),$$

where Ω is the angular velocity delivered by the rotary table and ρ the rate of penetration imposed at the surface. The nonlinear aspect of the system giving rise to instabilities that trigger drillstring vibrations is given by the frictional torque T arising from the contact between the bit and the rock at the lower end of the rod ($s = L$).

A direct transformation through the d'Alembert method allows representing the PDE model by a pair of neutral-type time-delay equations:

$$\begin{cases} \ddot{\Phi}_b(t) - \Upsilon\ddot{\Phi}_b(t-2\tau) = -\Psi\dot{\Phi}_b(t) - \Upsilon\Psi\dot{\Phi}_b(t-2\tau) - \dfrac{1}{I_B}T\big(\dot{\Phi}_b(t)\big) \\ \qquad\qquad + \dfrac{1}{I_B}\Upsilon T\big(\dot{\Phi}_b(t-2\tau)\big) + \Pi\Omega(t-\tau) + \omega(t) \\ \ddot{U}_b(t) - \tilde{\Upsilon}\ddot{U}_b(t-2\tilde{\tau}) = -\tilde{\Psi}\dot{U}_b(t) - \tilde{\Upsilon}\tilde{\Psi}\dot{U}_b(t-2\tilde{\tau}) - \dfrac{1}{M_B}T\big(\dot{\Phi}_b(t)\big) \\ \qquad\qquad + \dfrac{1}{M_B}\tilde{\Upsilon}T\big(\dot{\Phi}_b(t-2\tilde{\tau})\big) + \tilde{\Pi}\rho(t-\tilde{\tau}) + \omega(t), \end{cases} \tag{13.1}$$

where $\dot{\Phi}_b(t)$ and $\dot{U}_b(t)$ are the angular and axial bit velocities and

$$\Pi = \frac{2\Psi\beta}{\beta + \tilde{c}GJ}, \quad \Upsilon = \frac{\beta - \tilde{c}GJ}{\beta + \tilde{c}GJ}, \quad \Psi = \frac{\tilde{c}GJ}{I_B}, \quad \tau = \tilde{c}L,$$

$$\tilde{\Pi} = \frac{2\tilde{\Psi}\alpha}{\alpha + cE\Gamma}, \quad \tilde{\Upsilon} = \frac{\alpha - cE\Gamma}{\alpha + cE\Gamma}, \quad \tilde{\Psi} = \frac{cE\Gamma}{M_B}, \quad \tilde{\tau} = cL.$$

In order to assess the robustness of the controllers, an additive variable $\omega(t)$, satisfying $|\omega(t)| \leqslant \bar{\varepsilon}$, has been introduced to account for external system perturbations and unmodeled dynamics.

An important aspect to be considered in the evaluation of the controllers' performance is their ability in suppressing drilling vibrations taking into account the rock hardness. The frictional torque model introduced in [243] allows considering certain properties related with the bit and the drilling surface characteristics. It assumes that the torque on bit T consist of a term T_c, accounting for the forces transmitted by the cutting process, and of a term T_f, to represent the forces arising from the frictional contact.

The frictional torque is modeled as follows:

$$\begin{cases} T = T_c + T_f, \\ T_f = \dfrac{R_b^2}{2}\varkappa\mu\varsigma l \dfrac{\alpha_0\|V_b(L,t)\|}{\sqrt{\|V_b(L,t)\|^2 + \varepsilon^2}}, \\ T_c = \dfrac{R_b^2}{2}\varepsilon d(t), \end{cases} \tag{13.2}$$

where $d(t)$ is the depth of cut (see Sect. 4.3), the vector $V_b(L,t)$ is composed by the helicoidal coordinates corresponding to the axial and angular velocities of the bit $(\dot{U}_b, \dot{\Phi}_b)$, R_b is the bit radius, l the length of the wearflat, ς the contact stress, $\varkappa$ is the bit geometry number, μ is a coefficient related to the internal friction angle of the rock and ε denotes the intrinsic specific energy.

The intrinsic specific energy ε is a significant measure of drilling performance, especially of the cutting efficiency of bits and rock hardness. It quantifies a complex process of rock destruction and generally depends on various factors, such as rock type, the rake angle of the cutter, the cutter material, and pressure on the rock surface [269]. The intrinsic specific energy is defined as the energy required to remove a unit volume of rock [282]. In 1982, Rabia concluded that the specific energy is not a fundamental intrinsic property of rocks [235], however, Reddish and Yasar [240], Detournay [73] and Ersoy [84] have found the correlation of the specific energy to the rock properties such as compressive strength and hardness of the rock. In a given drilling environment, a lower ε entails a more efficient drilling process.

In order to determine which control scheme provides better performance, several aspects should be considered; the main characteristics to be evaluated are:

1. The time required to eliminate the stick-slip and the bit-bounce.
2. The maximum value of torsional and axial velocities in the presence of drilling vibrations.
3. The ability of the controllers to eliminate drillstring vibrations in the presence of external disturbances.
4. The controllers performance when drilling hard surfaces.
5. The structural complexity of the control laws, their magnitude, and their feasibility of implementation.

Simulations of the model (13.1) with the frictional torque given in (13.2) under different control schemes are presented below. The numerical values of the system parameters are given in Table C.1 of Appendix C. The considered reference angular and axial velocities are $10\,\mathrm{rad\,s^{-1}}$ and $0.1\,\mathrm{m\,s^{-1}}$, respectively.

Note that the design of the flatness-based control laws was conducted using the model given in (2.13)–(2.17) which includes Newtonian type boundary conditions whereas model (13.1) with the frictional torque given in (13.2), on which all other control laws were developed, includes a simpler kinematic like top boundary condition. The model (2.13)–(2.17) leads to a non-formally stable neutral delay system, which most of the existing controllers in the literature are unable to stabilize. Thus the flatness-based controllers derived in Chap. 10 are not included in the comparisons of subsection 13.5.

13.2 Low-Order Controllers

Low-order controllers are usually desired since they are of highly reduced complexity which renders them fairly easy to implement and to understand. For this reason, we have reviewed some of the main reduced-order strategies to control drilling vibrations and we have proposed new simple methods. In this section we analyze the performance of certain low-order control schemes presented in Chap. 9. The control methods studied here are:

1. **Soft-torque control**. A common strategy to avoid the stick-slip vibration is the soft-torque controller, defined as follows:

$$\begin{cases} u_T(t) = \kappa_p \dot{\tilde{\xi}}(t) + \kappa_i\, \tilde{\xi}(t) \\ \tilde{\xi}(t) = \Omega_0\, t - \dfrac{h}{\omega_c} T_c(t) - \Phi_p + \xi_0, \end{cases} \tag{13.3}$$

where Φ_p denotes the angular position at the top extremity, κ_p and κ_i are the proportional and integral controller gains and T_c is the output of a low-pass filter applied to the contact torque (see Sect. 9.2.2).

2. **Delayed feedback proportional control**. It has been proven that a pair of controllers of the form:
$$\begin{cases} u_T(t) = p_\Omega \Phi_b(t-\tau) \\ u_H(t) = p_H U_b(t-\tilde{\tau}), \end{cases} \tag{13.4}$$
guarantees the global asymptotic stability of the system for some p_H and p_Ω meeting the requirements stated in Proposition 9.4.1.

3. **Delayed feedback PID control**. The controllers
$$\begin{cases} u_T(t) = \Omega_p \Phi_b(t-\tau) + \Omega_i \displaystyle\int_0^{t-\tau} \Phi_b(s)ds \\ u_H(t) = H_p U_b(t-\tilde{\tau}) + H_d \dot{U}_b(t-\tilde{\tau}), \end{cases} \tag{13.5}$$
guarantee the local asymptotic stability of the system for some H_d, H_p, Ω_i and Ω_p satisfying the conditions of Proposition 9.5.1.

Figure 13.1 shows the angular and axial trajectories of the system for $\omega(t) = 0$ in closed loop with the proposed low-order controllers. Notice that the delayed proportional feedback control approach suppresses the drilling vibrations faster than the soft-torque and than the delayed PID one.

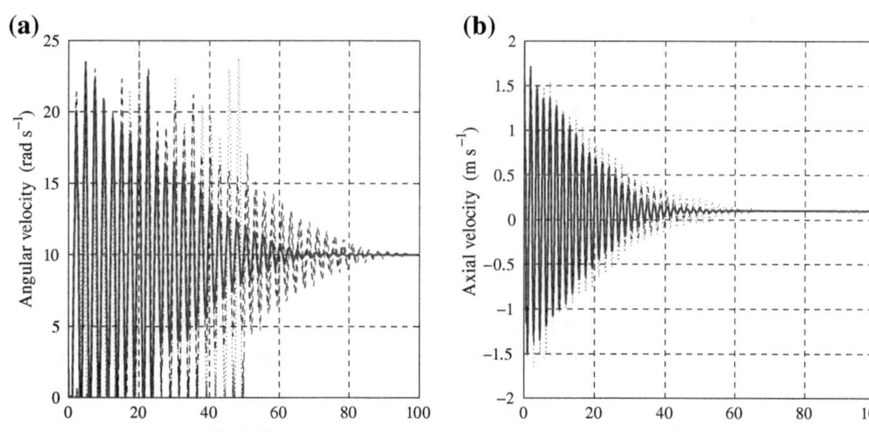

Fig. 13.1 Simulation of the model (13.1) with the frictional torque given in (13.2) with $\omega(t) = 0$. Reduction of stick-slip and bit-bounce oscillations by means of the low-order controllers proposed in Chap. 9: soft-torque (13.3) (*dotted line*), delayed proportional (13.4) (*solid line*) and delayed PID (13.5) (*dashed line*). **a** Bit angular velocity. **b** Bit axial velocity

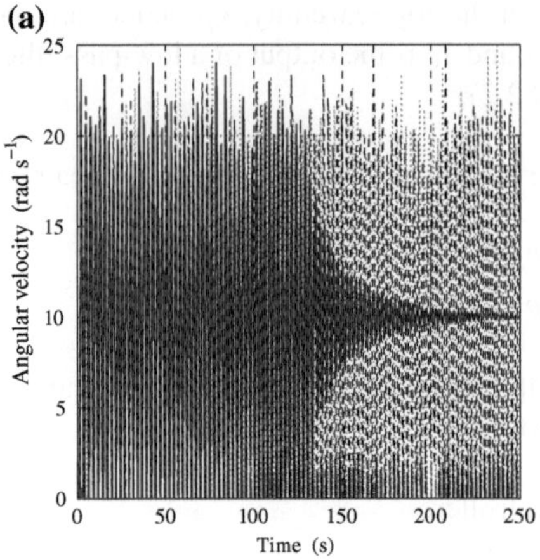

Fig. 13.2 Simulation of the model (13.1) with the frictional torque given in (13.2). Trajectories of the drilling system subject to bounded perturbations satisfying $|\omega(t)| \leqslant 0.1$, in closed loop with the low-order controllers proposed in Chap. 9: soft-torque (13.3) (*dotted line*), delayed proportional (13.4) (*solid line*) and delayed PID (13.5) (*dashed line*). **a** Bit angular velocity. **b** Bit axial velocity

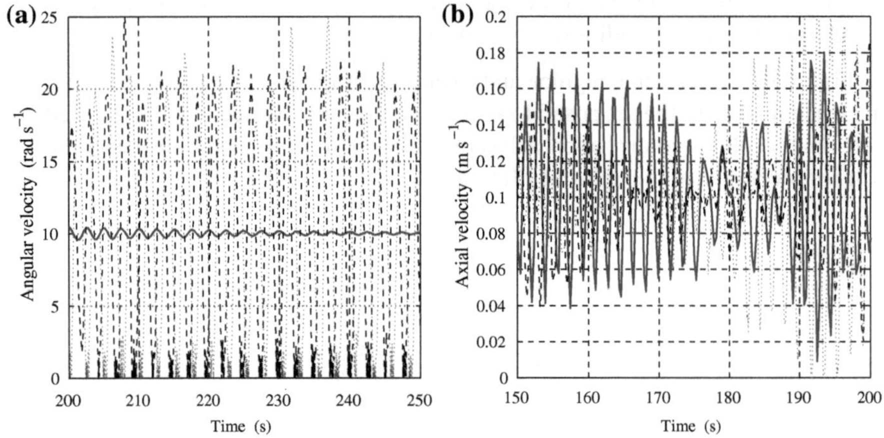

Fig. 13.3 Zoom-in of Fig. 13.2

Figure 13.2 shows the system trajectories under external perturbations such that $|\omega(t)| \leqslant 0.1$. When bounded perturbations are considered, the delayed feedback control scheme, takes longer, but is still able to eliminate torsional vibrations meanwhile the soft-torque and the delayed PID controllers fail to suppress the stick-slip. Figure 13.3 shows a zoom-in view of the system trajectories depicted in Fig. 13.2.

It is well known that low values of the intrinsic specific energy ε describe a scenario in which the drilling surface is not very hard and the bit characteristics lead to a satisfactory drilling operation. Under this scenario, the performance of the control laws must be improved. For the sake of model validation, simulations of the

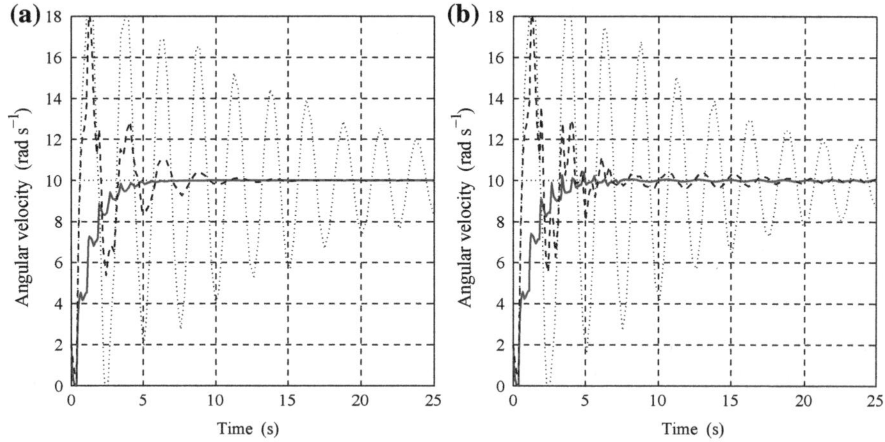

Fig. 13.4 Simulation of the model (13.1) with the frictional torque given in (13.2) in closed loop with the delayed proportional controllers (13.4). Bit angular velocity for different values of the intrinsic specific energy: $\varepsilon = 1 \times 10^3\,\mathrm{N\,m^{-2}}$ (*solid line*), $\varepsilon = 1 \times 10^4\,\mathrm{N\,m^{-2}}$ (*dashed line*), $\varepsilon = 1 \times 10^5\,\mathrm{N\,m^{-2}}$ (*dotted line*). **a** Trajectories of the unperturbed system, $\omega(t) = 0$. **b** Trajectories of the system subject to perturbations satisfying $|\omega(t)| \leqslant 1$

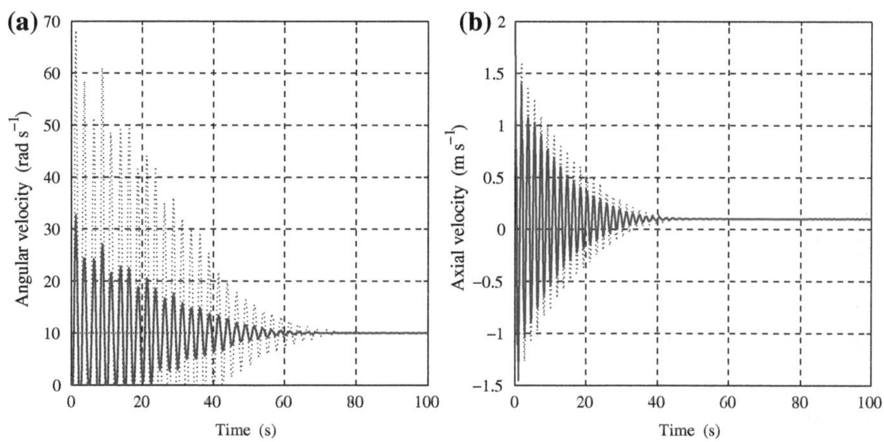

Fig. 13.5 Simulation of the model (13.1) with the frictional torque given in (13.2) with $\omega(t) = 0$ in closed loop with the delayed proportional controllers (13.4). System trajectories for large values of the intrinsic specific energy: $\varepsilon = 1 \times 10^6\,\mathrm{N\,m^{-2}}$ (*solid line*), $\varepsilon = 4 \times 10^6\,\mathrm{N\,m^{-2}}$ (*dotted line*). **a** Bit angular velocity. **b** Bit axial velocity

system for different values of the intrinsic specific energy ε are developed. Figure 13.4 shows the trajectories of the bit angular velocity under the delayed feedback control approach (13.4) for different low values of ε. Under this scenario, the stick-slip is effectively eliminated even in presence of external perturbations.

Figure 13.5 shows that when the intrinsic specific energy ε is higher, the angular velocity reaches higher values, meanwhile the axial behavior does not exhibit significant change.

13.3 Flatness-Based Control

As explained before, using the flatness-based approach, it is possible to solve trajectory tracking problems in a direct manner. For this reason, we have exploited the flatness property of the drilling system to design a pair of feedback controllers aimed at exponentially stabilizing the system trajectories (see Chap. 10). Note, as already pointed out in subsection 13.1, that the following controllers were developed based on a PDE system with inertial boundary conditions (2.13)–(2.17), whereas the other controllers of this chapter were designed using the simpler system (13.1) with the frictional torque given in (13.2). The proposed controllers guarantee that the errors between the reference trajectories $\dot{\Phi}_{br}$, $\dot{U}_{br}$ and the actual trajectories $\dot{\Phi}_b$, $\dot{U}_b$ exponentially converge to zero. The flatness-based control laws are defined as:

$$\begin{cases} u_T(t) = I_T\ddot{\Phi}_p(t) + a_1\dot{\Phi}_p(t) - a_1\left[\dot{\Phi}_b(t-\tau) + 0.5\gamma(t)\right] + I_B v(t) \\ \qquad + \tilde{p}F\left(\dot{\Phi}_b(t-\tau) + \gamma(t)\right) \\ v(t) = a_2\ddot{\Phi}_{br}(t+\tau) - a_2\lambda I + a_3\ddot{\Phi}_b(t-\tau) \\ \qquad - a_4\left[F\left(\dot{\Phi}_b(t-\tau) + \gamma(t)\right) - F\left(\dot{\Phi}_b(t-\tau)\right)\right] \\ I = 2\dot{\Phi}_p(t) - \dot{\Phi}_b(t-\tau) - \dot{\Phi}_{br}(t+\tau), \quad \gamma(t) = \int_{t-2\tau}^{t} v(\xi)d\xi, \end{cases} \tag{13.6}$$

$$\begin{cases} u_H(t) = M_T\ddot{U}_p(t) + b_1\dot{U}_p(t) - b_1\left[\dot{U}_b(t-\tilde{\tau}) + 0.5\bar{\gamma}(t)\right] + M_B\bar{v}(t)) \\ \qquad + pF\left(\dot{\Phi}_b(t-\tilde{\tau}) + \gamma(t)\right) \\ \bar{v}(t) = b_2\ddot{U}_{br}(t+\tilde{\tau}) - b_2\bar{\lambda}\bar{I} + b_3\ddot{U}_b(t-\tilde{\tau}) \\ \qquad - b_4\left[F\left(\dot{\Phi}_b(t-\tau) + \gamma(t)\right) - F\left(\dot{\Phi}_b(t-\tau)\right)\right] \\ \bar{I} = 2\dot{U}_p(t) - \dot{U}_b(t-\tilde{\tau}) - \dot{U}_{br}(t+\tilde{\tau}), \quad \bar{\gamma}(t) = \int_{t-2\tilde{\tau}}^{t} \bar{v}(\xi)d\xi, \end{cases} \tag{13.7}$$

where the constant parameters are defined below,

$$a_1 = \tau\bar{G}, \quad a_2 = \frac{\bar{G}\chi}{\lambda}, \quad a_3 = \frac{I_B\chi}{\tau}, \quad a_4 = \frac{\tilde{p}\chi}{\tau},$$

$$b_1 = \tilde{\tau}\bar{E}, \quad b_2 = \frac{\bar{E}\bar{\chi}}{\bar{\lambda}}, \quad b_3 = \frac{M_B\bar{\chi}}{\tilde{\tau}}, \quad b_4 = \frac{p\bar{\chi}}{\tilde{\tau}},$$

$$\bar{G} = \frac{GJ}{L}, \quad \chi = \frac{\lambda\tau}{\tau\bar{G} + \lambda I_B}, \quad \bar{E} = \frac{E\Gamma}{L}, \quad \bar{\chi} = \frac{\bar{\lambda}\tilde{\tau}}{\tilde{\tau}\bar{E} + \bar{\lambda}M_B}.$$

The variables $\dot{\Phi}_p$ and $\dot{U}_p$ denote the angular and axial velocities at the top extremity of the drillstring. It is worth mentioning that due to the flatness property of the

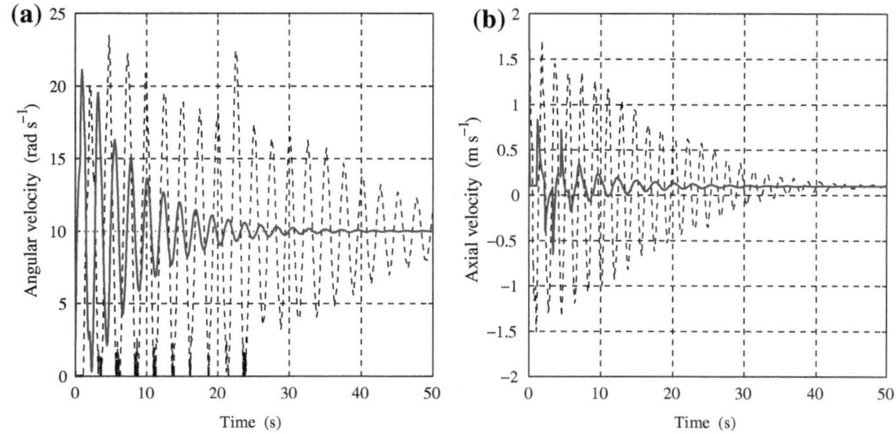

Fig. 13.6 Simulation of the model (13.1) with the frictional torque given in (13.2) with $\omega(t) = 0$. Trajectories of the drilling system in closed loop with the flatness-based controllers (13.6) and (13.7) (*solid line*) and with the delayed proportional controllers (13.4) (*dashed line*). **a** Bit angular velocity. **b** Bit axial velocity

system, these variables are explicitly characterized in terms of the flat outputs $\dot{\Phi}_b$ and $\dot{U}_b$. The exponential convergence rates are given by λ and $\bar{\lambda}$.

Figure 13.6 shows the system responses to the low-order controllers (13.4) and to the flatness-based control laws (13.6) and (13.7); improved results are obtained with the second approach. By means of the flatness-based controllers, the stick-slip and the bit-bounce are substantially reduced in a relatively short period of time. Clearly, the price paid for the control effectiveness is its structural complexity.

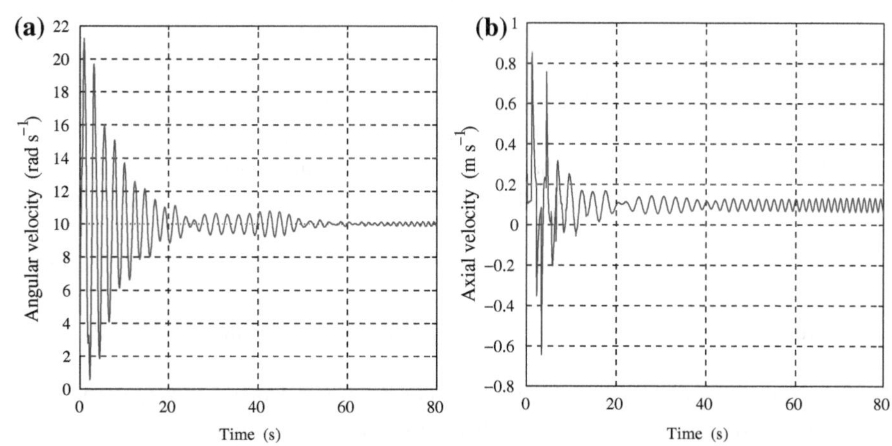

Fig. 13.7 Simulation of the model (13.1) with the frictional torque given in (13.2). Trajectories of the drilling system subject to external perturbations satisfying $|\omega(t)| \leqslant 1$ in closed loop with the flatness-based controllers (13.6) and (13.7). **a** Bit angular velocity. **b** Bit axial velocity

Figure 13.7 shows the system response to the flatness-based controllers in the presence of perturbations satisfying $|\omega(t)| \leqslant 1$; small variations of the angular and axial bit velocities can be observed.

13.4 Lyapunov-Based Controllers

Lyapunov's direct method constitutes an universally used tool for analyzing the stability of any dynamical system. Lyapunov's theory allows designing and synthesizing control laws that guarantee the system stabilization, and in our study case, the suppression of harmful drilling vibrations (see Chaps. 11 and 12).

Four different control schemes have been obtained through Lyapunov analyses; two of them are aimed at eliminating the stick-slip phenomenon leading to a exponentially stable closed loop system, the other two controllers take into account the axial drilling dynamics and suppress both the stick-slip and the bit-bounce, guaranteeing the system practical stability. These control schemes are derived from different modeling strategies:

1. **Switching system**. Since most of the proposed frictional torque models include the "sgn" function (see Chap. 11), the drilling system can be represented as an autonomous state-dependent switching system of the form (11.13) under the rule (11.15). Based on this system representation, the proposal of an energy functional allows to derive exponential stability conditions in terms of LMIs. The stabilizing controller leading to the suppression of stick-slip oscillations is given by:

$$\Omega(t) = -\lambda_0 \ddot{\Phi}_b(t-\tau) - \lambda_1 \dot{\Phi}_b(t-\tau) + \lambda_1 \Omega_0, \tag{13.8}$$

   where λ_0 and λ_1 must satisfy the conditions of Theorem 11.1.2.

2. **Multimodel representation**. The torsional nonlinear drilling model can be approximated by a set of linear models nonlinearly weighted of the form (11.33) and (11.34). The exponential stability analysis of the system is developed via Lyapunov techniques allowing the design of a nonlinear controller that ensures the stick-slip suppression. The feedback stabilizing control law is defined as follows:

$$\Omega(t) = -c_0 e^{-\gamma_b \dot{\Phi}_b(t-\tau)} + c_1 \dot{\Phi}_b(t-\tau) - c_1 \Omega_0, \tag{13.9}$$

   where c_0 and c_1 must fulfill the LMI-type conditions of Theorem 11.2.2.

3. **Coupled wave-ODE**. The dynamic behavior of a drillstring can be modeled by the wave equation, accounting for its torsional motion, coupled to an ODE to represent the rod axial movement (see the set of Eqs. 12.1). An ultimate boundedness analysis of the system trajectories, obtained by using an appropriate Lyapunov functional, gives rise to the design of a pair of feedback controllers able to practically stabilize the system and eliminate coupled drillstring vibrations. These controllers are of the form:

$$\begin{cases} \Omega(t) = c_{11}\ddot{\Phi}_b(t) + c_{12}\dot{\Phi}_b(t) + c_{13}T_{nl}(\dot{\Phi}_b(t)) + c_{14}\dot{\Phi}_p(t) \\ T_{nl}\left(\dot{\Phi}_b(t)\right) = W_{ob}R_b\left(\mu_{cb} + (\mu_{sb} - \mu_{cb})e^{-\gamma_b|\dot{\Phi}_b(t)|}\right)\text{sgn}\left(\dot{\Phi}_b(t)\right) \\ \rho(t) = c_{21}U_b(t) - c_{21}\rho_0 t + c_{22}\dot{U}_b(t) - c_{22}\rho_0 \end{cases} \tag{13.10}$$

where $\dot{\Phi}_p$ is the angular velocity at the top extremity and c_{1i}, c_{2j}, $i = 1, \ldots, 4$, $j = 1, 2$ are such that the conditions of Theorem 12.1.2 are satisfied.

4. **Coupled NDDE–ODE**. As discussed in Chap. 2, a neutral-type time-delay equation provides a reliable model of the torsional drilling dynamics and an ODE constitutes a simplified approximation of axial drillstring behavior. The coupling between these models is given by the frictional torque at the bottom extremity. The coupled NDDE–ODE is given in (12.42) and (12.43). The practical stabilization of this system is addressed via the attractive ellipsoid method; the aim of this strategy is the design of feedback controllers that drive the system trajectories into an invariant set. It has been shown that this strategy leads to the elimination of coupled drilling vibrations. The proposed controllers are defined as follows:

$$\begin{cases} \Omega(t) = k_{01}\ddot{\Phi}_b(t-\tau) + k_{11}\dot{\Phi}_b(t-\tau) - k_{11}\Omega_0 + k_{12}U_b(t-\tau) \\ \qquad - k_{12}\rho_0 t + k_{13}\dot{U}_b(t-\tau) - k_{13}\rho_0 \\ \rho(t) = k_{21}\dot{\Phi}_b(t) - k_{21}\Omega_0 + k_{22}U_b(t) - k_{22}\rho_0 t + k_{23}\dot{U}_b(t) - k_{23}\rho_0 \end{cases} \tag{13.11}$$

where k_{01}, k_{1i}, k_{2i}, $i = 1, 2, 3$ are such that the LMI–BMI type conditions of Theorem 12.2.6 are satisfied.

Figure 13.8 shows the system trajectories under two different control schemes: the flatness-based controllers (13.6) and (13.7) and the controller (13.8) obtained with the switching model approach. Notice that the second approach leads to the suppression of the stick-slip faster than the flatness-based control approach. Also note that although the control (13.8) is designed to only tackle torsional vibrations, it indirectly reduces the bit-bounce due to the intimate coupling between torsional and axial dynamics.

Observe in Fig. 13.9 the response of the system to the switching model-based control (13.8) in contrast with the response to the multimodel approximation-based controller (13.9). Note that the time required to eliminate the stick-slip is about the same in both cases, however, with the first approach, the maximum angular velocity that the bit reaches is higher. Both control strategies lead to the reduction of axial oscillations; the trajectories of the bit axial velocity are quite similar.

The trajectories of the control inputs corresponding to the switching and multimodel control approaches are shown in Fig. 13.10. Notice that the trajectories of the torsional control inputs do not differ greatly from the reference trajectory.

Figure 13.11 shows the system response to the control laws (13.8) and (13.9) in the presence of external perturbations satisfying $|\omega(t)| \leqslant 1$. The stick-slip and the

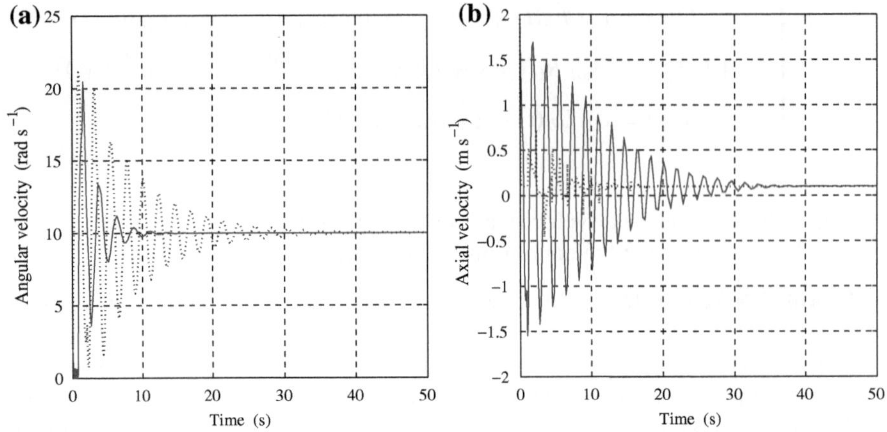

Fig. 13.8 Simulation of the model (13.1) with the frictional torque given in (13.2) with $\omega(t) = 0$. Trajectories of the system in closed loop with the switching system-based control (13.8) (*solid line*) and with the flatness-based controllers (13.6) and (13.7) (*dotted line*). **a** Bit angular velocity. **b** Bit axial velocity

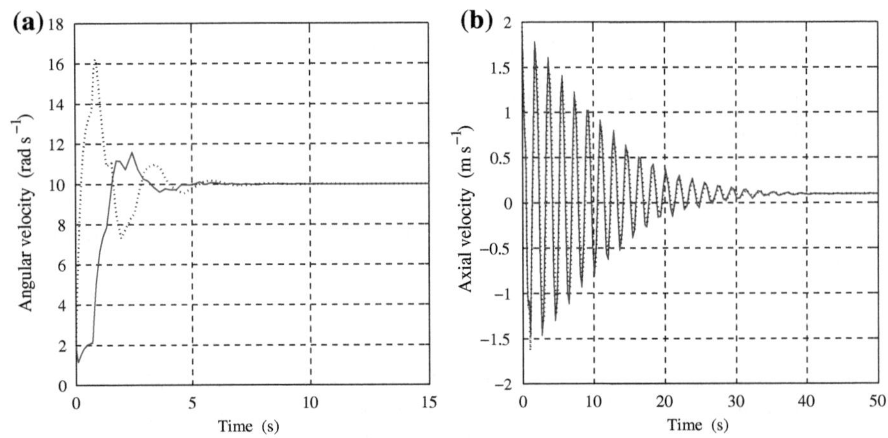

Fig. 13.9 Simulation of the model (13.1) with the frictional torque given in (13.2) with $\omega(t) = 0$. Trajectories of the system for $\varepsilon = 1 \times 10^5\,\text{N m}^{-2}$ in closed loop with the multimodel approximation-based controller (13.9) (*solid line*) and with the switching system-based control (13.8) (*dotted line*). **a** Bit angular velocity. **b** Bit axial velocity

bit-bounce are reduced, however, the angular and axial trajectories slightly fluctuate around the reference values.

Figure 13.12 shows the trajectories of the bit angular and axial velocities under the controllers (13.8) and (13.9) for $\varepsilon = 4 \times 10^6\,\text{N m}^{-2}$. Notice that in this case the switching system-based control performs better than the multimodel approximation-based controller; the torsional and angular reference trajectories are reached in less time by means of the control law (13.8).

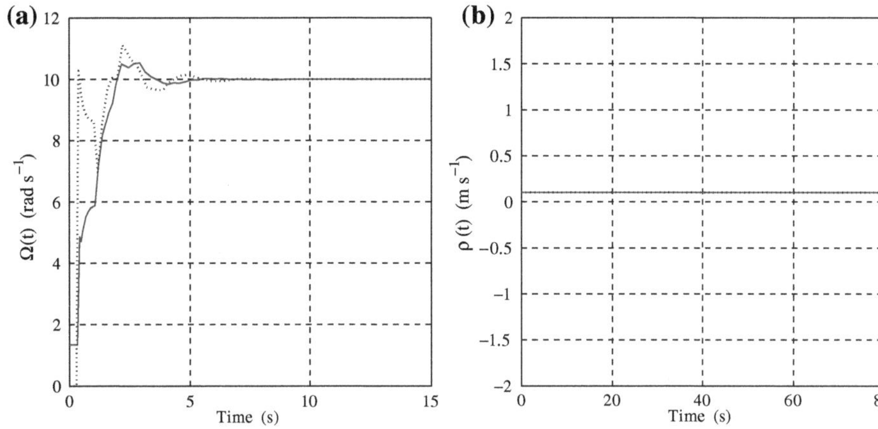

Fig. 13.10 Simulation of the model (13.1) with the frictional torque given in (13.2). Trajectories of the control inputs. Multimodel approximation-based controller (13.9) (*solid line*), switching model-based control (13.8) (*dotted line*). **a** Angular velocity provided by the rotary table. **b** Rate of penetration

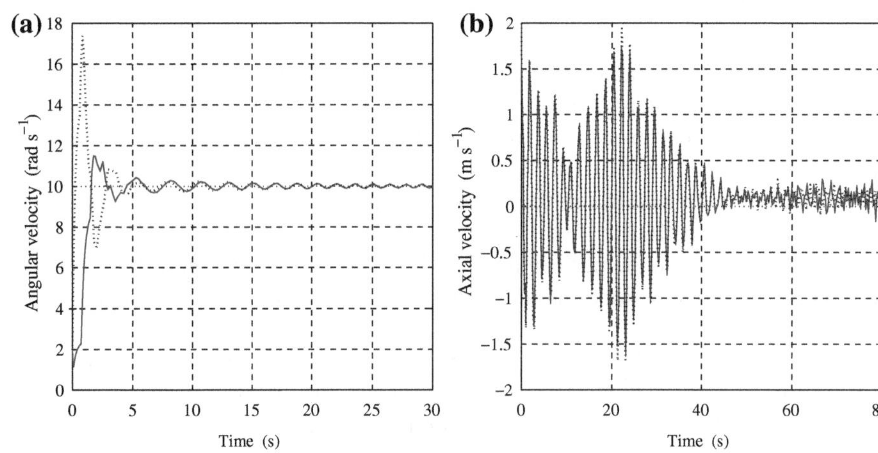

Fig. 13.11 Simulation of the model (13.1) with the frictional torque given in (13.2). Trajectories of the drilling system subject to external perturbations satisfying $|\omega(t)| \leqslant 1$ for $\varepsilon = 1 \times 10^5\,\mathrm{N\,m^{-2}}$ in closed loop with the multimodel approximation-based controller (13.9) (*solid line*) and with the switching system-based control (13.8) (*dotted line*). **a** Bit angular velocity. **b** Bit axial velocity

Notice that the structure of the Lyapunov-based controllers (13.8) and (13.9) is simple. The controller (13.8), obtained with the switching model, is linear and only requires the measurement of the angular velocity and acceleration of the bit. The controller (13.9), obtained with the polytopic approximation, includes a nonlinear term and only involves the angular velocity at the bottom end. These control schemes do not handle axial drilling dynamics (a constant rate of penetration ρ is considered), nevertheless, the bit-bounce is reduced, see Figs. 13.9b, 13.11b, and 13.12b.

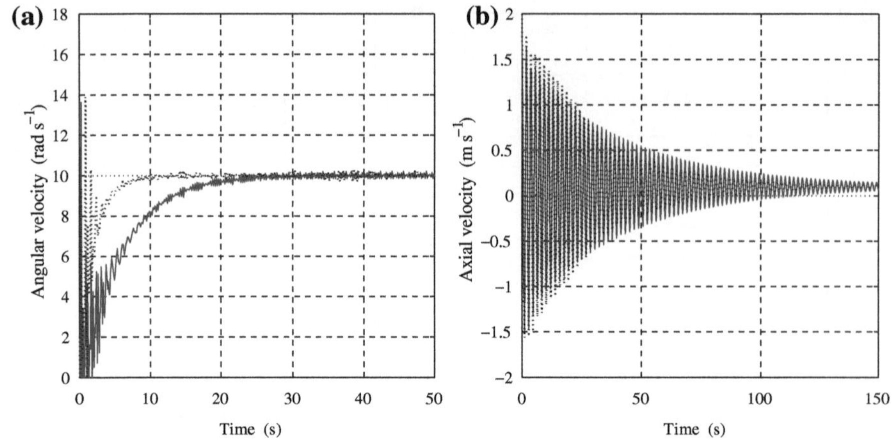

Fig. 13.12 Simulation of the model (13.1) with the frictional torque given in (13.2) with $\omega(t) = 0$. Trajectories of the system for $\varepsilon = 4 \times 10^6\,\mathrm{N\,m^{-2}}$ in closed loop with the multimodel approximation-based controller (13.9) (*solid line*) and with the switching system-based control (13.8) (*dotted line*). **a** Bit angular velocity. **b** Bit axial velocity

In contrast with the flatness-based controllers, both the multimodel-based control and the switching model-based control lead to a faster convergence of the torsional trajectories to the reference paths; meanwhile the axial trajectories are faster driven to the reference trajectories via the flatness control approach. Note, however, that the model being considered in the flatness-based design is notoriously difficult to control (see Remark 10.1).

Now, let us analyze the performance of the Lyapunov-based controllers aimed at eliminating coupled torsional-axial drilling vibrations: the wave-ODE model-based controllers (13.10) and the attractive ellipsoid method-based controllers (13.11).

Figure 13.13 shows the response of the system under the control laws (13.10) and (13.11) for $\varepsilon = 1 \times 10^5\,\mathrm{N\,m^{-2}}$, i.e., when a satisfactory drilling performance is expected due to the properties of the rock strength and the bit geometry. Observe that the reference angular velocity is achieved faster by means of the controllers (13.10) based on the wave-ODE system representation and the axial trajectory converges faster to the reference value via the controllers (13.11) derived from the attractive ellipsoid method.

A higher value of ε is considered in the simulation of Fig. 13.14. For $\varepsilon = 4 \times 10^6\,\mathrm{N\,m^{-2}}$, the controllers (13.11), obtained through the attractive ellipsoid method, provide significantly improved results. Torsional and axial dynamics are driven to the reference values in the few first seconds of the drilling process in the simulation. Notice that the axial velocity trajectory of the system in closed loop with the wave-ODE model-based controllers (13.10) is not significantly modified with the increase of ε.

A different case scenario is shown in Fig. 13.15; the system is assumed to be subject to bounded perturbations satisfying $|\omega(t)| \leqslant 1$ for $\varepsilon = 1 \times 10^5\,\mathrm{N\,m^{-2}}$.

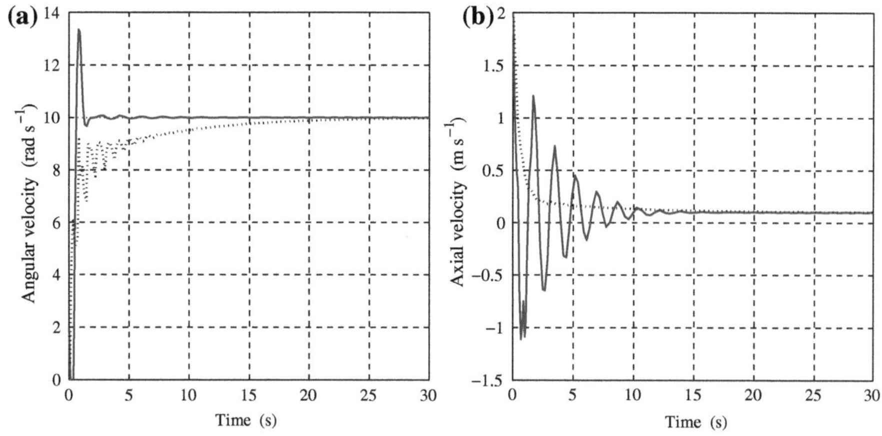

Fig. 13.13 Simulation of the model (13.1) with the frictional torque given in (13.2) with $\omega(t) = 0$. Trajectories of the system for $\varepsilon = 1 \times 10^5$ N m^{-2} in closed loop with the wave-ODE model-based controllers (13.10) (*solid line*) and with the attractive ellipsoid method-based controllers (13.11) (*dotted line*). **a** Bit angular velocity. **b** Bit axial velocity

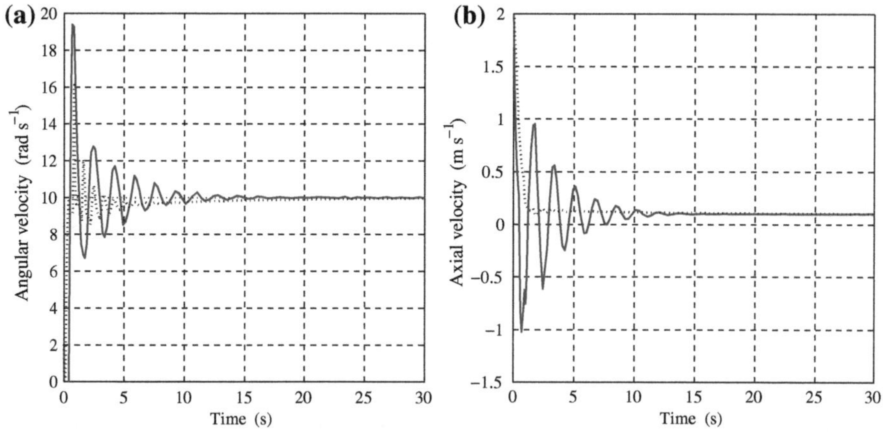

Fig. 13.14 Simulation of the model (13.1) with the frictional torque given in (13.2) with $\omega(t) = 0$. Trajectories of the system for $\varepsilon = 4 \times 10^6$ N m^{-2} in closed loop with the wave-ODE model-based controllers (13.10) (*solid line*) and with the attractive ellipsoid method-based controllers (13.11) (*dotted line*). **a** Bit angular velocity. **b** Bit axial velocity

The control approach based on the attractive ellipsoid method (13.11) performs better than the controllers (13.10); with the first approach the velocity trajectories accurately converge to the reference values.

An efficient performance is obtained with the attractive ellipsoid method-based controllers (13.11), however, a major drawback of this approach is the magnitude of the controller $\rho(t)$. Observe in Fig. 13.16, the trajectories of the controllers

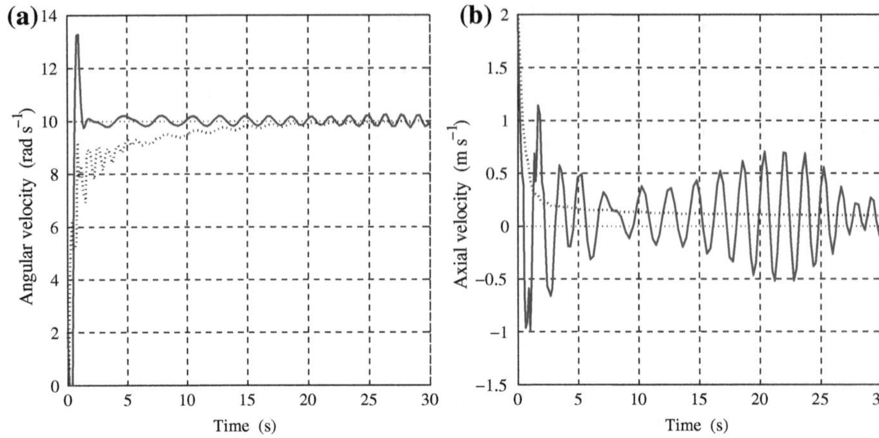

Fig. 13.15 Simulation of the model (13.1) with the frictional torque given in (13.2). Trajectories of the system subject to external perturbations satisfying $|\omega(t)| \leqslant 1$ for $\varepsilon = 1 \times 10^5\,\mathrm{N\,m^{-2}}$ in closed loop with the wave-ODE model-based controllers (13.10) (*solid line*) and with the attractive ellipsoid method-based controllers (13.11) (*dotted line*). **a** Bit angular velocity. **b** Bit axial velocity

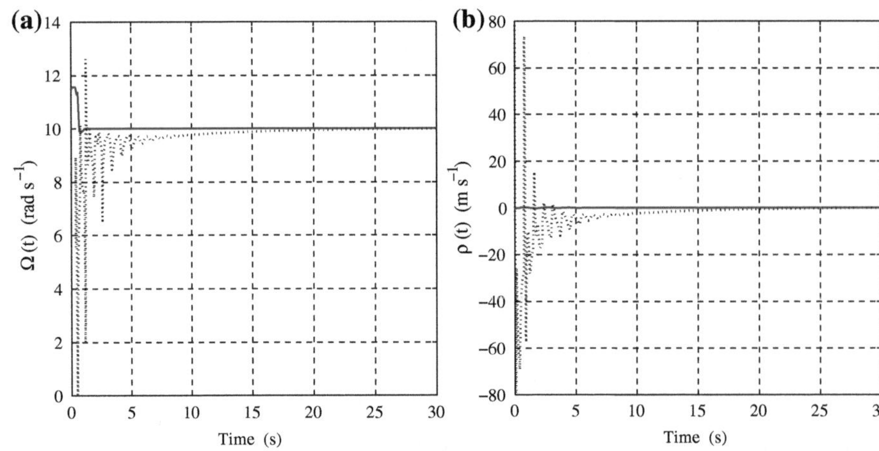

Fig. 13.16 Simulation of the model (13.1) with the frictional torque given in (13.2). Trajectories of the control inputs. Wave-ODE model-based controllers (13.10) (*solid line*), attractive ellipsoid method-based controllers (13.11) (*dotted line*). **a** Angular velocity provided by the rotary table. **b** Rate of penetration

(13.10) and (13.11). The trajectory corresponding to the attractive ellipsoid method-based axial controller greatly diverge from the reference path, then, considering that the actuator is restricted by physical limitations, this control approach may be impractical.

Figure 13.17 shows a zoom-in view of the controllers trajectories depicted in Fig. 13.16.

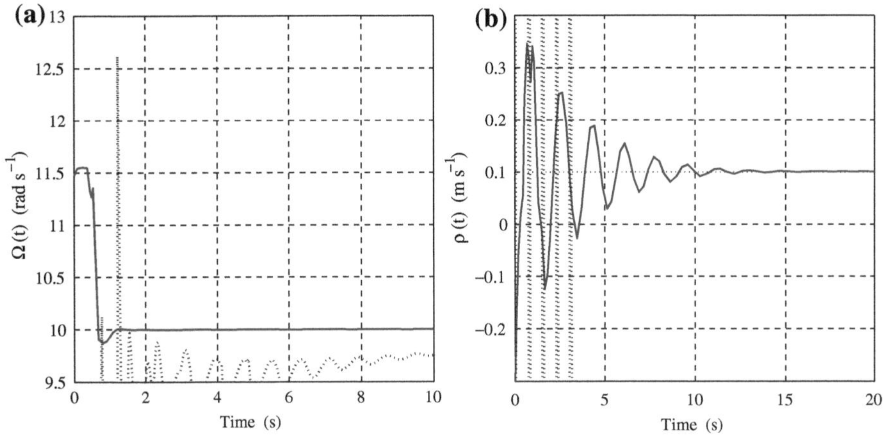

Fig. 13.17 Zoom-in of Fig. 13.16

13.5 Discussion and Graphical Comparisons

This section offers graphical comparisons of the performance of the different methods to suppress drilling vibrations addressed in this contribution. Data tables and radar charts allow examining, in a simple manner, the results obtained through seven control approaches which can be classified into two categories: low-order control and Lyapunov-based control.

The low-order controllers studied in this chapter (soft-torque, delayed proportional, and delayed PID) have similar performance when considering the unperturbed system under different values of the intrinsic specific energy; all of them are effective in reducing axial and torsional drilling vibrations. However, when considering external perturbations, the soft-torque and the delayed PID control laws are no longer able to deal with the stick-slip.

Four different control strategies derived from Lyapunov analyses were evaluated: switching system-based control, multimodel approximation-based control, wave-ODE model-based control and attractive ellipsoid method-based control. The Lyapunov-based approach provides improved results; the stick-slip is eliminated in considerably less time and the maximum bit angular velocity is reduced. According to the system simulations, when external perturbations and high values of the intrinsic specific energy are considered, the attractive ellipsoid method-based controller is the best control solution; however, its magnitude is large, therefore, considering the physical limitations of the actuators in real plants, the implementation of this control method could be impractical.

Tables 13.1 and 13.2 summarize some important results of the comparative analysis presented in this chapter; the first one provides data related to the torsional drilling dynamics and the second one gives information about the axial performance.

Table 13.1 Control strategies to suppress drilling vibrations: characteristics and numerical results

Control strategy $u_T(t)$, $\Omega(t)$	Controller form	Involved variables	Convergence time (s)	Maximum angular velocity (rad s^{-1})
Soft-torque control	Linear	$\Phi_p(t)$, $\dot{\Phi}_p(t)$ $T_c(t)$, $\dot{T}_c(t)$	90	23.6
Delayed feedback proportional control	Linear	$\Phi_b(t-\tau)$	69	24
Delayed feedback PID control	Linear	$\Phi_b(t-\tau)$ $\int_0^{t-\tau} \Phi_b(s)ds$	89	23.5
Switching system-based control	Linear	$\ddot{\Phi}_b(t-\tau)$ $\dot{\Phi}_b(t-\tau)$	7	16.1
Multimodel approximation-based control	Nonlinear	$\dot{\Phi}_b(t-\tau)$	6	11.5
Wave-ODE model-based control	Nonlinear	$\ddot{\Phi}_b(t)$, $\dot{\Phi}_b(t)$, $\dot{\Phi}_p(t)$	5	13.8
Attractive ellipsoid method-based control	Linear	$\ddot{\Phi}_b(t-\tau)$ $\dot{\Phi}_b(t-\tau)$ $\dot{U}_b(t-\tau)$ $U_b(t-\tau)$	18	10

Torsional dynamics data

Table 13.2 Control strategies to suppress drilling vibrations: characteristics and numerical results

Control strategy $u_H(t)$, $\rho(t)$	Controller form	Involved variables	Convergence time (s)	Maximum axial velocity (m s^{-1})
Soft-torque	–	–	61	1.6
Delayed feedback proportional control	Linear	$U_b(t-\tau)$	46	1.7
Delayed feedback PID control	Linear	$\dot{U}_b(t-\tilde{\tau})$ $U_b(t-\tilde{\tau})$	57	1.8
Switching system-based control	–	–	39	1.8
Multimodel approximation-based control	–	–	39	1.8
Wave-ODE model-based control	Linear	$\dot{U}_b(t)$, $U_b(t)$,	13	2
Attractive ellipsoid method-based control	Linear	$\dot{U}_b(t)$, $U_b(t)$ $\dot{\Phi}_b(t)$	8	2

Axial dynamics data

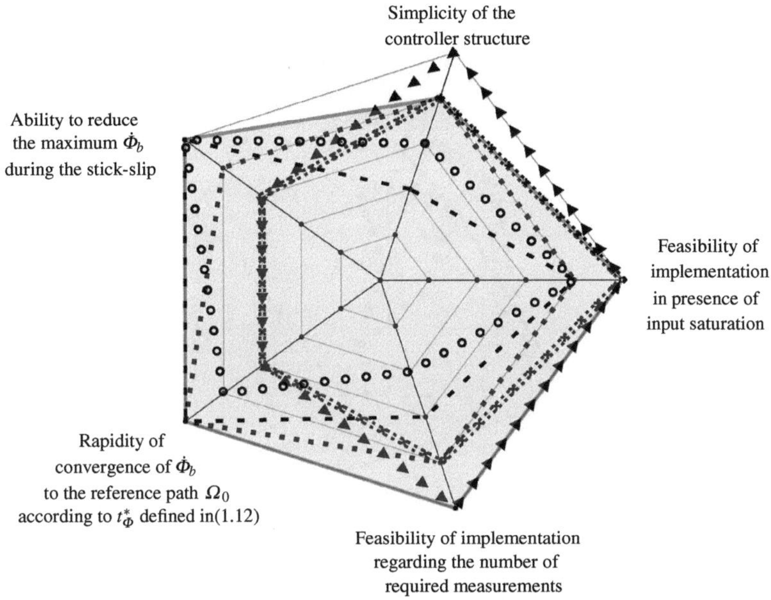

Fig. 13.18 Radar diagram illustrating the performance of the proposed controllers on various assessment aspects related to the drilling torsional dynamics. There are seven polygons with different contours; each one of them corresponds to a different control strategy: soft-torque control (13.3) (*double dotted contour*), delayed proportional controller (13.4) (*contour of triangles*), delayed PID control law (13.5) (*contour of crosses*), switching system-based control (13.8) (*dotted contour*), multimodel approximation-based controller (13.9) (*solid contour*), wave-ODE model-based controllers (13.10) (*dashed contour*), attractive ellipsoid method-based controllers (13.11) (*contour of circles*)

Data provided by Tables 13.1 and 13.2 include the measurements required to implement the control laws, the time at which the bit velocity converges to the reference value and the maximum bit speed after applying the stabilizing controller. Simulations were developed under the assumption that the system is not subject to perturbations ($\omega(t) = 0$) and that the intrinsic specific energy is $\varepsilon = 1 \times 10^5\,\mathrm{N\,m^{-2}}$.

The radar diagrams illustrated in Figs. 13.18, 13.19, and 13.20 allow to easily compare the controllers' performance according to their different abilities. The radar charts show five evaluation skills indicated at the vertices of regular pentagons. Each control proposal entails a different polygon which represents their performance; the better the control performance, the greater polygon area.

Figures 13.18, 13.19 and 13.20 depict seven polygons with different contours corresponding to each control law: soft-torque control (13.3) (double dotted contour), delayed proportional controller (13.4) (contour of triangles), delayed PID control law (13.5) (contour of crosses), switching system-based control (13.8) (dotted contour), multimodel approximation-based controller (13.9) (solid contour), wave-ODE model-based controllers (13.10) (dashed contour), attractive ellipsoid method-based controllers (13.11) (contour of circles). The shaded polygons correspond to the control strategies which perform better.

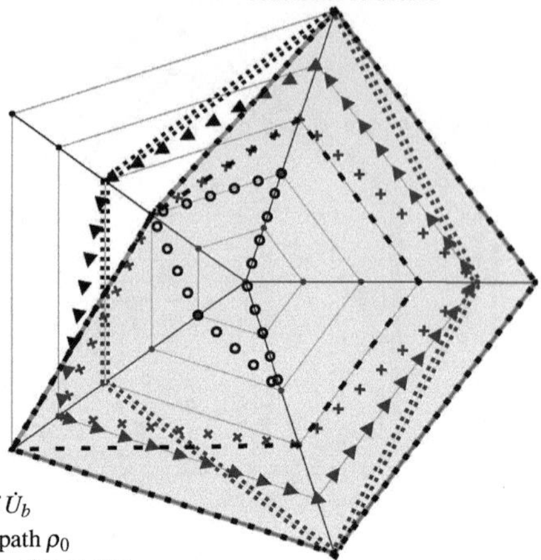

Fig. 13.19 Radar diagram illustrating the performance of the proposed controllers on various assessment aspects related to the drilling axial dynamics. There are seven polygons with different contours; each one of them corresponds to a different control strategy: soft-torque control (13.3) (*double dotted contour*), delayed proportional controller (13.4) (*contour of triangles*), delayed PID control law (13.5) (*contour of crosses*), switching system-based control (13.8) (*dotted contour*), multimodel approximation-based controller (13.9) (*solid contour*), wave-ODE model-based controllers (13.10) (*dashed contour*), attractive ellipsoid method-based controllers (13.11) (*contour of circles*)

The aspects compared in the radar diagrams of Figs. 13.18 and 13.19 include the simplicity of the control law, the ability to reduce the maximum angular and axial velocities after application of the vibration controller, the convergence time of the system trajectories to the reference values, the number of variables that the control laws involve and the implementation feasibility in the presence of constrained control inputs, i.e., the controllers' size. Figure 13.18 concerns the torsional dynamics, and Fig. 13.19 the axial behavior.

In order to properly evaluate the rapidity of convergence of the system trajectories toward the reference paths, the controllers energy must be taken into account. To this end, the following time measures are considered

$$t_\Phi^* = t_\Phi^c \int \|u_T(t)\| \,, \quad t_U^* = t_U^c \int \|u_H(t)\| \,, \tag{13.12}$$

where t_Φ^c and t_U^c denote the convergence time of the trajectories $\dot{\Phi}_b(t)$, $\dot{U}_b$ toward the prescribed values of Ω_0, ρ_0.

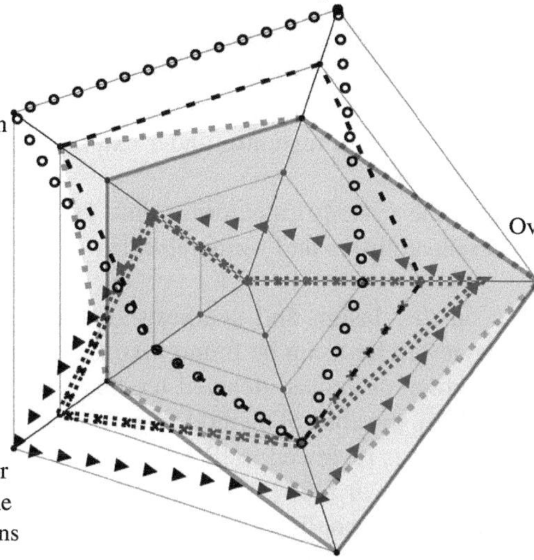

Fig. 13.20 Radar diagram illustrating the performance of the proposed controllers on various general assessment aspects. There are seven polygons with different contours; each one of them corresponds to a different control strategy: soft-torque control (13.3) (*double dotted contour*), delayed proportional controller (13.4) (*contour of triangles*), delayed PID control law (13.5) (*contour of crosses*), switching system-based control (13.8) (*dotted contour*), multimodel approximation-based controller (13.9) (*solid contour*), wave-ODE model-based controllers (13.10) (*dashed contour*), attractive ellipsoid method-based controllers (13.11) (*contour of circles*)

The diagram chart of Fig. 13.20 allows comparing certain general aspects of each control method such as the robustness against external perturbations, the ability to eliminate drilling vibrations when the intrinsic specific energy takes large values, the simplicity for calculating the controller gains and the performance of the controllers dealing with torsional and axial trajectories, i.e., the controllers' effectiveness according to the radar diagrams of Figs. 13.18 and 13.19.

The graphical diagrams show the strengths and vulnerabilities of each control method in the different evaluation aspects.

13.6 Notes and References

A comparative/contrastive performance analysis of the controllers arising in the third part of the book has been presented. Seven different control techniques aimed at suppressing drilling vibrations were tested via simulations of a pair of coupled neutral-type time-delay equations describing the torsional and axial dynamics of a drillstring in the perforation process. A frictional torque model that considers certain drilling bit and surface characteristics was chosen to represent the interaction between the cutting device and the rock. The controllers were evaluated by considering different values of the intrinsic specific energy, which is a parameter related to the rock strength. The system response under external perturbations bounded in magnitude was also analyzed.

Figures illustrating the system trajectories, data tables, and radar charts allow comparing the different abilities and vulnerabilities of the control proposals in a simple manner.

Deeper insights on the design and synthesis of the control laws to suppress drilling vibrations reviewed here can be found in different contributions of our own authorship. The low-order strategies: delayed feedback proportional and PID controllers are studied in [39], the control method based on the differential flatness property of the drilling system is reviewed in [260], the control law based on switching systems theory is presented in [254], the control solution derived from the multimodel approximation of the nonlinear drilling model is investigated in [255], the practical stabilization of the drilling system described by a coupled wave-ODE model is studied in [256] and the attractive ellipsoid method is addressed in [257].

These control strategies have been designed considering models that reflect the vibrational phenomena along the drillstring and they were evaluated using parameter values that approximate the physical conditions in a "real drilling operation," however, it is worthy of mention that the actual behavior of a realistic drilling system cannot be described in all its practical phases of operation by models inherently restricted to vibrations of coupled small amplitude axial and torsional motions.

The model limitations arising from physical simplifications hinder the study of certain topics of practical interest such as:

- The accuracy assessment of axial–torsional models against field logging data.
- The coupling of torsional and axial modes when one goes beyond the idealized small amplitude approximations.
- The effects of variation of mass-density, rotary inertia, and elastic modulus of the drillstring with length.
- The effects of wave reflection and transmission across junctions between the BHA and the drillstring.
- The relevance of nonplanar flexural vibrational modes (i.e., lateral but nonwhirling motions) of the drillstring to dynamic stability.
- The stability in the presence of bifurcations induced by torsional, axial, and flexural motions whereas parameters such as weight-on-bit and whirling speeds are varied.
- The effects on dynamical drillstring evolution of viscoelastic, anelastic, and hysteric damping.
- The effect of realistic hydrodynamic interactions of the drillstring with the bore hole casing via drilling fluid.

All these issues are out of the scope of the present monograph and remain open to further research.

Appendix A
Short Summary of Classical Local Bifurcations

The aim of this appendix is to define and provide the reader with the basic definitions we used in the manuscript in the field of dynamical systems. More precisely, we briefly present the elementary tools of bifurcations largely inspired from the presentations [52, 165, 121]. We refer the interested reader to the excellent book [165], where the bifurcation problem is treated in depth, and among other, a complete classification of codimension one and two local bifurcations are given.

For the sake of simplicity, consider the parameter-dependent finite-dimensional dynamical system:

$$\dot{x} = f(x, \alpha), \tag{A.1}$$

where $x \in \mathbb{R}^n$ is the phase variable and $\alpha \in \mathbb{R}^m$ is the parameter.

The system (A.1) is said to be topologically equivalent to

$$\dot{x} = g(x, \beta), \tag{A.2}$$

if there exists an appropriate homeomorphism of the parameter space $\sigma : \mathbb{R}^m \to \mathbb{R}^m$ such that $\beta = \sigma(\alpha)$ and there exists a parameter-dependent homeomorphism of the phase space $h_\alpha : \mathbb{R}^n \to \mathbb{R}^n$, mapping the orbits of the system (A.1) at a parameter value α onto the orbits of (A.2) at parameter value $\sigma(\alpha)$ preserving the direction of time.

Under the parameter variation, a change in the system topology can occur. The appearance of a topologically nonequivalent phase portrait under variation of parameters is called a *bifurcation*. Thus, a bifurcation is a change of the topological type of the system as its parameters pass through a bifurcation (critical) value.

A *bifurcation diagram* of the dynamical system is a partition of the parameter space induced by the topological equivalence relation. The codimension of a bifurcation in system is the difference between the dimension of the parameter space and the dimension of the corresponding bifurcation boundary.

By definition, topologically equivalent parameter-dependent systems have (topologically) equivalent bifurcation diagrams. For local bifurcations of equilibria and

M.B. Saldivar Márquez et al., *Analysis and Control of Oilwell Drilling Vibrations*,
Advances in Industrial Control, DOI 10.1007/978-3-319-15747-4

fixed points, universal bifurcation diagrams are provided by topological normal forms, which is a central notion in bifurcation theory.

A.1 Pitchfork Bifurcation

Let R be a real $n \times n$ matrix such that $R^2 = I$. The space $\mathbb{R}^n$ can be decomposed as $\mathbb{R}^n = X^+ \oplus X^-$, where $Rx = x$ for $x \in X^+$, and $Rx = -x$ for $x \in X^-$. Consider now the scalar system

$$\dot{x} = \alpha x - x^3, \quad x \in \mathbb{R}, \quad \alpha \in \mathbb{R}. \tag{A.3}$$

The system is obviously $\mathbb{Z}_2$-equivariant. Indeed, in this case, $Rx = -x$ (reflection). At $\alpha = 0$, system (A.3) has the fixed equilibrium $x^0 = 0$ with eigenvalue zero. The bifurcation diagram of (A.3) is given in Fig. A.1. A trivial equilibrium $x_0 = 0$ always exists. Furthermore, it is linearly stable for $\alpha < 0$ and unstable for $\alpha > 0$. There are also two stable nontrivial equilibria, $x^{1,2}(\alpha) = \pm\sqrt{\alpha}$, existing for $\alpha > 0$ and R-conjugate, $Rx^1(\alpha) = x^2(\alpha)$.

Any $\mathbb{Z}_2$-equivariant system

$$\dot{x} = \alpha x - x^3 + O(x^5), \tag{A.4}$$

is locally topologically equivalent near the origin to (A.3). The number and stability of the equilibria in (A.3) and (A.4) are the same for corresponding small parameter values with small $|\alpha|$.

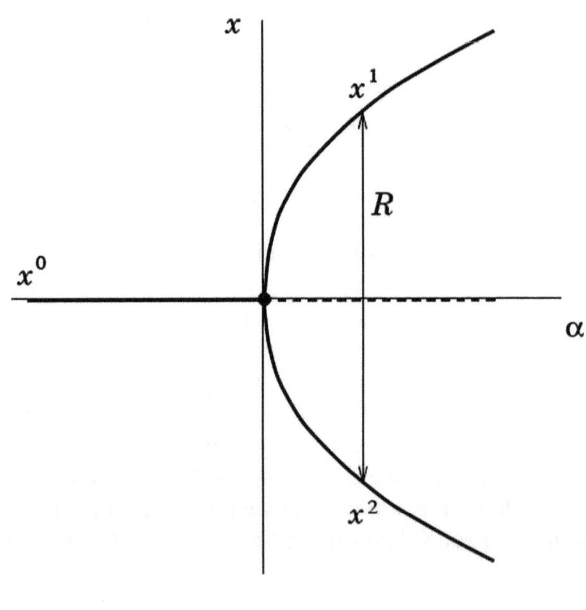

Fig. A.1 Pitchfork Bifurcation

Theorem A.1.1 (Bifurcations at a zero eigenvalue, [165]) *Suppose that a* $\mathbb{Z}_2$*-equivariant system*

$$\dot{x} = f(x, \alpha), \quad x \in \mathbb{R}^n, \quad \alpha \in \mathbb{R}$$

with smooth f*, such that* $Rf(x, \alpha) = f(Rx, \alpha)$*,* $R^2 = I$*, has at* $\alpha = 0$ *the fixed equilibrium* $x^0 = 0$ *with simple zero eigenvalue* $\lambda_1 = 0$*, and let* $v \in \mathbb{R}^n$ *be the corresponding eigenvector. Then the system has a one-dimensional R-invariant center manifold* W^c_α*, and one of the following alternatives generically takes place:*

(i) *(fold) If* $v \in X^+$*, then* $W^c_\alpha \subset X^+$ *for all sufficiently small* $|\alpha|$*, and the restriction of the system to* W^c_α *is locally topologically equivalent near the origin to the following normal form:*

$$\dot{\zeta} = \beta \pm \zeta^2.$$

(ii) *(Pitchfork)* $v \in X^-$*, then* $W^c_\alpha \subset X^-$ *for all sufficiently small* $|\alpha|$*, and the restriction of the system to* W^c_α *is locally topologically equivalent near the origin to the following normal form:*

$$\dot{\zeta} = \beta\zeta \pm \zeta^3.$$

In case (i), the standard fold bifurcation happens within the invariant subspace X^+, giving rise to two fixed-type equilibria. In case (ii), the pitchfork bifurcation resulting in the appearance of two R-conjugate equilibria, while the fixed equilibrium changes its stability. The genericity conditions include nonvanishing of the cubic term of the restriction of the system to the center manifold at $\alpha = 0$.

A.2 Andronov-Hopf Bifurcation

The case where two purely imaginary eigenvalues appear when varying the parameters is called the Andronov-Hopf bifurcation.

Consider a system

$$\dot{x} = f(x, \alpha), \quad x \in \mathbb{R}^2, \quad \alpha \in \mathbb{R} \tag{A.5}$$

with a smooth function f, which has at $\alpha = 0$ the equilibrium $x = 0$ with eigenvalues $\lambda_{1,2} = \pm i\omega_0$, $\omega_0 > 0$. By the Implicit Function Theorem, the system has a unique equilibrium $x_0(\alpha)$ in some neighborhood of the origin for all sufficiently small $|\alpha|$, since $\lambda = 0$ is not an eigenvalue of the Jacobian matrix. It is always possible to perform a coordinate shift, placing this equilibrium at the origin.

Theorem A.2.1 ([165]) *Consider a planar system* (A.5) *having at* $\alpha = 0$ *a non-hyperbolic equilibrium with eigenvalues* $i\omega_0$ *with* $\omega_0 > 0$*. For sufficiently small* $|\alpha|$ *the equilibrium* $x = 0$ *has eigenvalues* $\lambda_{1,2} = \mu(\alpha) \pm i\omega(\alpha)$ *with* $\mu(0) = 0$ *and* $\omega(0) = \omega_0 > 0$.

Let the following conditions be satisfied:

H1 $l_1(0) \neq 0$ *where* l_1 *is the first Lyapunov coefficient.*
H2 $\mu'(0)$.

Then there exists an invertible change of parameter, coordinates and time transforming the system into

$$\frac{d}{d\tilde{t}}\begin{pmatrix} y_1 \\ y_2 \end{pmatrix} = \begin{pmatrix} \beta & -1 \\ 1 & \beta \end{pmatrix}\begin{pmatrix} y_1 \\ y_2 \end{pmatrix} + \sigma(y_1^2 + y_2^2)\begin{pmatrix} y_1 \\ y_2 \end{pmatrix} + O(||y||^4), \tag{A.6}$$

where $\sigma = \text{sign}(l_1(0))$.

Furthermore, any such system satisfying these genericity conditions is locally topologically equivalent near the origin to one of the following normal forms:

$$\frac{d}{d\tilde{t}}\begin{pmatrix} y_1 \\ y_2 \end{pmatrix} = \begin{pmatrix} \beta & -1 \\ 1 & \beta \end{pmatrix}\begin{pmatrix} y_1 \\ y_2 \end{pmatrix} + \sigma(y_1^2 + y_2^2)\begin{pmatrix} y_1 \\ y_2 \end{pmatrix}, \tag{A.7}$$

The normal form can be studied using a polar representation $z = re^{i\theta}$. In this system of coordinates two types of behaviors occur depending on the sign of the first Lyapunov exponent.

- *Supercritical Andronov-Hopf bifurcation*: when $l_1(0) < 0$ the normal form has an equilibrium at the origin, which is asymptotically stable for $\beta \leqslant 0$ and unstable for $\beta > 0$. Moreover, there is a unique and stable limit cycle appearing for $\beta > 0$ and has radius $\sqrt{\beta}$.
- *Subcritical Andronov-Hopf bifurcation*: when $l_1(0) > 0$ the normal form has an equilibrium at the origin, which is asymptotically stable for $\beta \geqslant 0$. A unique and unstable limit cycle appears for $\beta < 0$.

A.3 Bogdanov-Takens Bifurcation

Bogdanov-Takens Singularity is known also as double zero singularity is characterized by a double zero eigenvalue with geometric multiplicity one.

The Bogdanov-Takens bifurcation is characterized by the normal form:

$$\begin{cases} \dot{x}_1 = x_2, \\ \dot{x}_2 = \beta_1 + \beta_2 x_1 + x_1^2 + s\, x_1 x_2. \end{cases} \tag{A.8}$$

Any equilibria of the system are located on the horizontal axis $x_2 = 0$, and satisfy the equation $\beta_1 + \beta_2 x_1 + x_1^2 = 0$.

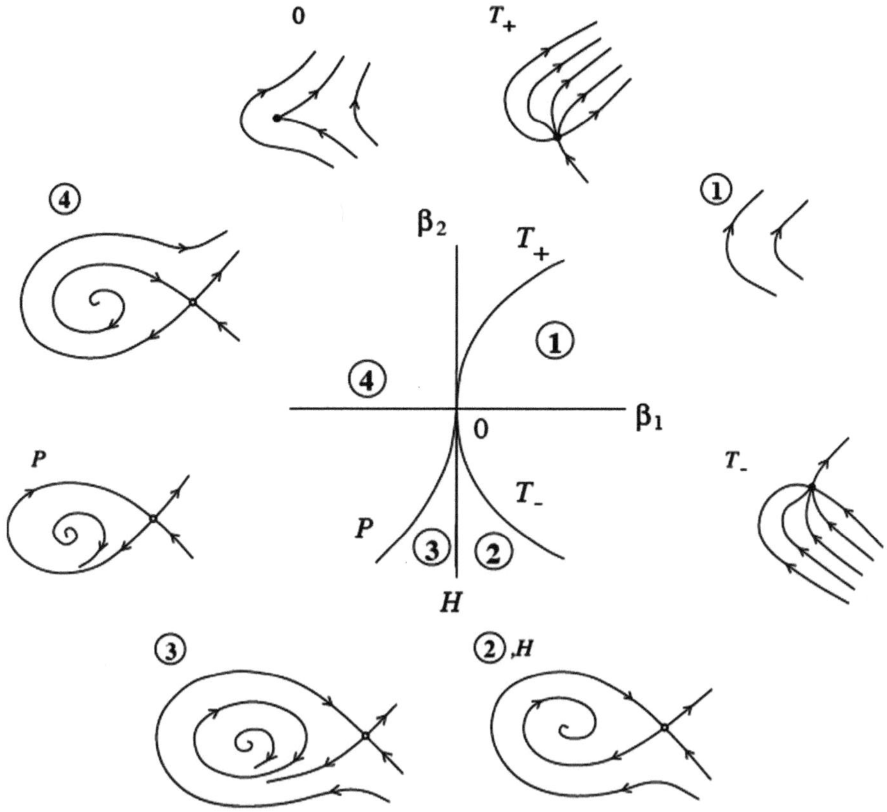

Fig. A.2 Bogdanov-Takens Bifurcation inspired by the presentation of [121, 165]

A complete description of the possible phase portraits is given in Fig. A.2 where the manifolds $T_\pm$ and P separating the regions of the parameter space are defined as follow: The discriminant parabola:

$$T = \left\{(\beta_1, \beta_2) \ : \ 4\beta_1 - \beta_2^2 = 0\right\} \tag{A.9}$$

corresponds to a fold bifurcation. The point $\beta = 0$ separate it into two branches T_+ and T_-. The separatrix P is the unique smooth curve corresponding to a saddle homoclinic bifurcation:

$$P = \left\{(\beta_1, \beta_2) \ : \ \beta_1 = -\frac{6}{25}\beta_2^2 + o(\beta_2^2), \ \ \beta_2 < 0\right\}. \tag{A.10}$$

Appendix B
Lyapunov Stability Theory

This appendix provides basic definitions of Lyapunov stability theory that are discussed in detail in classic literature on control systems theory such as [148] and [295].

B.1 Basic Definitions

Consider a dynamical system satisfying

$$\begin{cases} \dot{x} = f(x,t), & x \in \mathbb{R}^n \\ x(t_0) = x_0, \end{cases} \tag{B.1}$$

where $f(x,t)$ satisfies the standard conditions of existence and uniqueness of the solutions, i.e., $f(x,t)$ is Lipschitz continuous with respect to x and piecewise continuous in "t". The point $x^* \in \mathbb{R}^n$ is said to be an equilibrium point of (B.1) if $f(x,t) = 0$. Roughly speaking, an equilibrium point is locally stable if all solutions starting at nearby points stay nearby; otherwise, it is unstable. It is locally asymptotically stable if all solutions starting at nearby points not only stay nearby, but also tend to the equilibrium point as time approaches infinity. These concepts are formally defined below.

Definition B.1.1 (*Stability in the sense of Lyapunov*) An equilibrium point $x^* = 0$ of system (B.1) is stable (in the sense of Lyapunov) at $t = t_0$ if for any $\varepsilon > 0$ there exists a $\delta(t_0, \varepsilon) > 0$ such that

$$\|x(t_0)\| < \delta \implies \|x(t)\| < \varepsilon, \quad \forall t \geqslant t_0. \tag{B.2}$$

If condition (B.2) is satisfied for all t_0, i.e., $\delta(t_0, \varepsilon) = \delta(\varepsilon)$, then $x^* = 0$ is said to be uniformly Lyapunov stable which means that the equilibrium point is not losing stability.

M.B. Saldivar Márquez et al., *Analysis and Control of Oilwell Drilling Vibrations*,
Advances in Industrial Control, DOI 10.1007/978-3-319-15747-4

The concept of asymptotic stability defined below, entails stronger requirements on the system trajectories.

Definition B.1.2 (*Asymptotic stability*) An equilibrium point $x^* = 0$ of system (B.1) is asymptotically stable at $t = t_0$ if:

1. $x^* = 0$ is stable, and
2. $x^* = 0$ is locally attractive; i.e., there exists $\delta(t_0)$ such that

$$\|x(t_0)\| < \delta \implies \lim_{t\to\infty} x(t) = 0. \tag{B.3}$$

Uniform asymptotic stability is achieved if $x^* = 0$ is uniformly stable, and $x^* = 0$ is uniformly locally attractive; i.e., there exists $\delta(\varepsilon)$ for which condition (B.3) holds.

Definitions B.1.1 and B.1.2 provide local stability conditions, i.e., they describe the behavior of a system near an equilibrium point. An equilibrium point x^* is said to be globally stable if it is stable for all initial conditions $x_0 \in \mathbb{R}^n$.

Notice that Definition B.1.2 do not specify the convergence rate of the solution; exponential stability concept defines an exponential decaying rate on the system trajectories.

Definition B.1.3 (*Exponential stability*) An equilibrium point $x^* = 0$ of system (B.1) is exponentially stable if there exist constants $m, \alpha > 0$ and $\varepsilon > 0$ such that

$$\|x(t)\| < me^{-\alpha(t-t_0)} \|x(t_0)\| \tag{B.4}$$

for all $\|x(t_0)\| \leqslant \varepsilon$ and $t \geqslant t_0$. The largest constant α satisfying (B.4) is called the convergence rate.

Exponential stability implies uniform asymptotic stability.

B.2 Direct Method of Lyapunov

Lyapunov's direct method, also known as the second method of Lyapunov allows establishing the stability of a system without explicitly integrating the differential equation (B.1). The basic idea is determining stability through the analysis of the change rate of a "measure of energy" that the system involves. The following definitions allow characterizing this "measure of energy".

Let B_ε be a ball of size ε around the origin, i.e., $B_\varepsilon = \{x \in \mathbb{R}^n : \|x\| < \varepsilon\}$.

Definition B.2.1 (*Locally positive definite functions*) A continuous function $V : \mathbb{R}^n \times \mathbb{R}_+ \mapsto \mathbb{R}$ is a locally positive definite function if for some $\varepsilon > 0$ and some continuous, strictly increasing function $\alpha : \mathbb{R}_+ \mapsto \mathbb{R}$,

$$V(0, t) = 0 \quad \text{and} \quad V(x, t) \geqslant \alpha(\|x\|) \quad \forall x \in B_\varepsilon, \quad \forall t \geqslant 0. \tag{B.5}$$

A locally positive definite function is locally like an energy function. Functions which are globally like energy functions are called positive definite functions [204].

Definition B.2.2 (*Positive definite functions*) A continuous function $V : \mathbb{R}^n \times \mathbb{R}_+ \mapsto \mathbb{R}$ is a positive definite function if it satisfies the conditions (B.5) and, additionally, $\alpha(p) \to \infty$ as $p \to \infty$.

The following definition states an upper bound on the energy function.

Definition B.2.3 (*Decrescent functions*) A continuous function $V : \mathbb{R}^n \times \mathbb{R}_+ \mapsto \mathbb{R}$ is decrescent if for some $\varepsilon > 0$ and some continuous, strictly increasing function $\beta : \mathbb{R}_+ \mapsto \mathbb{R}$,

$$V(x,t) \leqslant \beta(\|x\|) \quad \forall x \in B_\varepsilon, \quad \forall t \geqslant 0.$$

Based on these definitions, the following theorem provides sufficient conditions for the system stability through the analysis of a suitable energy function.

Theorem B.2.4 (Basic theorem of Lyapunov) *Let $V(x,t)$ be a non-negative function with derivative $\dot{V}$ along the trajectories of the system.*

1. *If $V(x,t)$ is locally positive definite and $\dot{V}(x,t) \leqslant 0$ locally in x and for all "t", then the origin of the system is locally stable (in the sense of Lyapunov).*
2. *If $V(x,t)$ is locally positive definite and decrescent, and $\dot{V}(x,t) \leqslant 0$ locally in x and for all "t", then the origin of the system is uniformly locally stable (in the sense of Lyapunov).*
3. *If $V(x,t)$ is locally positive definite and decrescent, and $-\dot{V}(x,t)$ is locally positive definite, then the origin of the system is uniformly locally asymptotically stable.*
4. *If $V(x,t)$ is positive definite and decrescent, and $-\dot{V}(x,t)$ is positive definite, then the origin of the system is globally uniformly asymptotically stable.*

A Lyapunov exponential stability result is stated in the following theorem.

Theorem B.2.5 *An equilibrium point $x^* = 0$ of $\dot{x} = f(x,t)$ is exponentially stable if and only if there exists an $\varepsilon > 0$ and a function $V(x,t)$ which satisfies*

$$\alpha_1 \|x\|^2 \leqslant V(x,t) \leqslant \alpha_2 \|x\|^2$$

$$\dot{V}\big|_{\dot{x}=f(x,t)} \leqslant -\alpha_3 \|x\|^2$$

$$\left\| \frac{\partial V}{\partial x}(x,t) \right\| \leqslant \alpha_4 \|x\|$$

for some positive constants α_1, α_2, α_3, α_4 and $\|x\| \leqslant \varepsilon$.

The exponential convergence rate can be determined from the proof of Theorem B.2.5 (see [263]). It can be proven that (B.4) is satisfied with

$$m \leqslant \left(\frac{\alpha_2}{\alpha_1}\right)^{1/2} \quad \alpha \geqslant \frac{\alpha_3}{2\alpha_2}.$$

The equilibrium point $x^* = 0$ is globally exponentially stable if the bounds in Theorem B.2.5 hold for all x.

Appendix C
Drilling Model Parameters

This appendix presents the numerical values of the drilling system parameters reflecting typical operating conditions in real oilwell drilling platforms.

The numerical values of Table C.1, corresponding to the torsional dynamics, were taken from [53]; the ones corresponding to the axial drilling description, and to the simplified torque model (3.10), were taken from [39]; the ones corresponding to the lumped parameter model (9.1), were taken from [51]; the parameter values of the frictional torque models given in (3.14) with the friction coefficient given in (3.15), and in (13.2), were taken from [209] and [243], respectively.

For the sake of clarity, Table C.2 presents the notation used in the comprehensive model of the drilling system presented in Chap. 4.

M.B. Saldivar Márquez et al., *Analysis and Control of Oilwell Drilling Vibrations*,
Advances in Industrial Control, DOI 10.1007/978-3-319-15747-4

Table C.1 Numerical values of the drilling system parameters

Symbol	Parameter	Numerical value
L	String length	1172 m
G	Shear modulus	79.3×10^9 N m^{-2}
E	Young modulus	200×10^9 N m^{-2}
Γ	Drillstring's cross-section	35×10^{-4} m^2
J	Second moment of area	1.19×10^{-5} m^4
I	Inertia	0.095 Kg m
I_B	Lumped inertia at the BHA	89 Kg m^2
I_T	Lumped inertia at the top extremity	79 Kg m^2
M_B	Mass at the BHA	40000 Kg
ρ_a	Density	8000 Kg m^{-3}
β	Angular momentum	2000 N m s
α	Viscous friction coefficient	200.025 Kg s^{-1}
m_0	Mass	37278 Kg
c_0	Damping	16100 Kg s^{-1}
k_0	Spring constant	1.55×10^6 Kg s^{-2}
μ_1	Constant related to the bit geometry	257 m^{-1}
I_p	Top drive inertia	2122 Kg m^2
I_b	Bit inertia	374 Kg m^2
c	Torsional damping	23.2 N m s rad^{-1}
k	Torsional stiffness	473 N m rad^{-1}
d_p	Top drive local damping	425 N m s rad^{-1}
d_b	Bit local damping	50 N m s rad^{-1}
c_b	Damping constant	0.03 N m s rad^{-1}
R_b	Bit radius	0.155575 m
W_{ob}	Weight on the bit	97347 N
p	Friction force amplitude	35
$\bar{k}$	Constant of the friction top angle	0.3
ζ	Constant of the friction top angle	0.01
μ_{cb}	Coulomb friction coefficient	0.5
μ_{sb}	Static friction coefficient	0.8
γ_b	Velocity decrease rate	0.9
l	Length of the wearflat	1.2×10^{-3} m
μ	Ratio between horizontal and vertical friction force components	0.6
ς	Contact stress	60×10^6 N m^{-2}
$\varkappa$	Bit geometry number	1.1
α_0	Friction function amplitude	1
ε	Intrinsic specific energy	5.5×10^5 N m^{-2}

Table C.2 Parameters of the comprehensive drilling model described in Chap. 4

Symbol	Parameter
L_p	Drill pipes length
L_b	BHA length
L	Total drillstring length ($L = L_p + L_b$)
U_p, U_b	Drill pipes, BHA axial position
Φ_p, Φ_b	Drill pipes, BHA torsional position
$\varepsilon^i_{U_p}, \varepsilon^i_{\Phi_p}$	Internal damping coefficients
$\gamma^v_{U_p}, \gamma^v_{\Phi_p}$	Viscous damping coefficients
Γ_p, J_p	Cross-section and polar inertia moment of one drill pipe section
Γ_b, J_b	Cross-section and polar inertia moment of one drill collar section
r_{po}, r_{pi}	Outer, inner pipe radius
r_{bo}, r_{bi}	Outer, inner drill collar radius
$\Psi_{\Phi d}, \Psi_{Ud}$	D component of the induction motor flux
$L_{\Phi m}, L_{Um}$	Motor mutual inductance
$I_{\Phi d}, I_{\Phi q}$	d, q angular component of the induction motor stator current
I_{Ud}, I_{Uq}	d, q axial component of the induction motor stator current
I_T	Top drive inertia
u_T	Rotary table motor torque
u_H	Hook force acting at the top extremity
ζ_{rg_2}	Deformation of cables, crown and travelling blocks
ζ_{rg_1}	Deformation of other drilling rig elements
$k_{rg_{01}} \zeta_{rg_{ini}}$	Ground reaction force
$u_F(t)$	Tension force in the cable at the drawworks level
$M_{rg_i}, \gamma_{rg_1}, k_{rg_{ij}}$	Equivalent masses, damping and
	stiffness coefficients
M_T	Top drive mass
$T(t)$	Bit reaction torque
$W(t)$	Reaction force at the bit
T_c, W_c	Bottom hole cutting torque and force
T_f, W_f	Bottom hole friction torque and force
V_b	Helical vector with components $(\frac{\partial U_p}{\partial t}, \frac{\partial \Phi_p}{\partial t})$
$\mathcal{F}$	Adimensional friction function

Appendix D
Pontryagin Stability Conditions

This appendix presents two results developed by Lev Semenovich Pontryagin (1908–1988) in [232] about the conditions under which transcendental equations have only zeros with negative real part.

D.1 Stability Theorems

Theorem D.1.1 *Let* $\Delta(\lambda) = p(\lambda, e^{\lambda})$ *where* $p(\lambda, w)$ *is a polynomial with a principal term. Suppose* $\Delta(j\delta)$, $\delta \in \mathbb{R}$ *is separated into its real and imaginary parts* $\Delta(j\delta) = L(\delta) + jM(\delta)$. *If all zeros of* $\Delta(\lambda)$ *have negative real parts, then* $L(\delta)$ *and* $M(\delta)$ *are real, simple, alternate and*

$$M'(\delta)L(\delta) - M(\delta)L'(\delta) > 0 \tag{D.1}$$

for $\delta \in \mathbb{R}$. *Conversely, all zeros of* $\Delta(\lambda)$ *will be in the left half-plane provided that either of the following conditions is satisfied:*

(i) *All zeros of* $L(\delta)$ *and* $M(\delta)$ *are real, simple and alternate and inequality* (D.1) *is satisfied for at least one* δ.
(ii) *All zeros of* $L(\delta)$ *are real, and for each zero, inequality* (D.1) *is satisfied.*
(iii) *All zeros of* $M(\delta)$ *are real, and for each zero, inequality* (D.1) *is satisfied.*

M.B. Saldivar Márquez et al., *Analysis and Control of Oilwell Drilling Vibrations*, Advances in Industrial Control, DOI 10.1007/978-3-319-15747-4

Theorem D.1.2 *Let*

$$f(\lambda, u, v) = \sum_{m=0}^{r}\sum_{n=0}^{s} \lambda^m \varphi_m^{(n)}(u, v),$$

where $\varphi_m^{(n)}(u, v)$ *is a homogeneous polynomial of degree n in u and v. Assume that* $\lambda^r \varphi_r^{(*)}(u, v)$ *is the principal term of* f *and let* $\varphi_*^{(s)}(u, v) = \sum_{n=0}^{s} \varphi_r^{(n)}(u, v)$ *and* $\varphi_*^{(s)}(\lambda) = \varphi_*^{(s)}(\cos\lambda, \sin\lambda)$.

If ε *is such that* $\varphi_*^{(s)}(\varepsilon + j\delta) \neq 0$, $\delta \in \mathbb{R}$, *then for sufficiently large integers k, the function* $N(\lambda) = f(\lambda, \cos\lambda, \sin\lambda)$ *will have exactly* $4ks + r$ *zeros in the strip* $-2k\pi + \varepsilon \leqslant Re\lambda \leqslant 2k\pi + \varepsilon$. *Consequently, the function* $N(\lambda)$ *will have only real roots if and only if, for sufficiently large integers k, it has exactly* $4ks + r$ *roots in the strip* $-2k\pi + \varepsilon \leqslant Re\lambda \leqslant 2k\pi + \varepsilon$.

References

1. Aarrestad, T. V., & Kyllingstad, A. (1988). An experimental and theoretical study of a coupling mechanism between longitudinal and torsional drillstring vibrations at the bit. *SPE Drilling Engineering*, *3*(1), 12–18.
2. Abdulgalil, F., & Siguerdidjane, H. (2004). Nonlinear control design for suppressing stick-slip in oil well drillstrings. In *Proceedings of the 5th Asian Control Conference* (pp. 1276–1281).
3. Abolinia, V. E., & Myshkis, A. D. (1960). A mixed problem for an almost linear hyperbolic system in the plane. *Matematicheskii Sbornik*, *50*(92), 423–442.
4. Adachi, J. I., Detournay, E., & Drescher, A. (1996). Determination of rock strength parameters from cutting tests. In *Proceedings 2nd North American Rock Mechanics Symposium* (pp. 1517–1523). Rotterdam: Balkema.
5. Adejo, A. O., Onumanyi, A. J., Anyanya, J. M., & Oyewobi, S. O. (2013). Oil and gas process monitoring through wireless sensor networks: A survey. *Ozean Journal of Applied Sciences*, *6*(2), 39–43.
6. Ait, Babram M., Arino, O., & Hbid, M. L. (2001). Computational scheme of a center manifold for neutral functional differential equations. *Journal of Mathematical Analysis and Applications*, *258*(2), 396–414.
7. Anabtawi, M. (2011). Practical stability of nonlinear stochastic hybrid parabolic systems of ito-type: Vector Lyapunov functions approach. *Nonlinear Analysis: Real World Applications*, *12*(1), 1386–1400.
8. Andrade, B. C. C., Fosenca, C. A. L. L., & Weber, H. I. (2013). Experimental and numerical drill string modeling friction induced stick-slip. In Z. Dimitrovová et al., *11th International Conference on Vibration Problems* (pp. 1–9). Portugal: Lisbon.
9. Angeli, D., Ferrell, J. E., & Sontag, E. D. (2004). Detection of multistability, bifurcations, and hysteresis in a large class of biological positive-feedback systems. *Proceedings of the National Academy of Sciences of the United States of America*, *101*(7), 1822–1827.
10. Antman, S. (1991). *Non-linear problems in elasticity*. Applied Mathematical Sciences (Vol. 107). Berlin: Springer.
11. Armstrong-Hélouvry, B. (1990). Stick-slip arising from Stribeck friction. In *IEEE International Conference on Robotics and Automation* (pp. 1377–1382). Cincinnati Ohio US.
12. Armstrong-Hélouvry, B. (1991). *Control of machines with friction*. Boston: Kluwer Academic Publishers.
13. Armstrong-Hélouvry, B. (1993). Stick slip and control in low speed motion. *IEEE Transactions on Automatic Control*, *38*(10), 1483–1496.
14. Armstrong-Hélouvry, B., Dupont, P., & Canudas-de-Wit, C. (1994). A survey of models, analysis tools and compensation methods for the control of machines with friction. *Automatica*, *30*(7), 1083–1138.
15. Armstrong, B., & Canudas-de-Wit, C. (1995). *Friction modeling and compensation*. Boca Raton: The control handbook, CRC Press.

M.B. Saldivar Márquez et al., *Analysis and Control of Oilwell Drilling Vibrations*,
Advances in Industrial Control, DOI 10.1007/978-3-319-15747-4

16. Åström K. J. (1998). Control of systems with friction. In *Proceedings of the Fourth International Conference on Motion and Vibration Control MOVIC '98*. Zurich, Switzerland.
17. Awrejcewicz, J., & Olejnik, P. (2005). Analysis of dynamic systems with various friction laws. *Applied Mechanics Reviews*, *58*, 389–411.
18. Azar, J. J., & Robello, S. G. (2007). *Drilling engineering*. Okhlahoma: PennWell Corporation.
19. Balanov, A. G., Janson, N. B., McClintock, P. V. E., & Wang, C. H. T. (2002). Bifurcation analysis of a neutral delay differential equation modelling the torsional motion of a driven drill-string. *Chaos, Solitons and Fractals*, *15*(2), 381–394.
20. Bailey, J. J., & Finnie, I. (1960). An analytical study of drillstring vibration. *Journal of Engineering for Industry, Transactions of the ASME*, *82*(2), 122–128.
21. Bailey, J. R., & Remmert, S. M. (2009). Managing drilling vibrations through BHAdesign optimization. In *Proceeding of the International PetroleumTechnology Conference (IPTC '09)* (pp. 921–930). Doha, Qatar.
22. Barnett, S. (1983). *Polynomials and linear control systems*. New York, USA: Marcel Dekker.
23. Barreto-Jijon, R., Canudas-de-Wit, C., Niculescu, S. I., & Dumon, J. (2010). Adaptive observer design under low data rate transmission with applications to oil well drill-string. In *Proceedings American Control Conference*. Baltimore, Maryland, USA.
24. Barton, D. A. W., Krauskopf, B., & Wilson, R. E. (2007). Nonlinear dynamics of torsional waves in a drill string model with spacial extent. *Journal of Vibration and Control*, *16*, 1049–1065.
25. Bauer, A. (1996). Utilisation of chaotic signals for radar and sonar purposes. *Norwegian Signal Processing Society NORSIG*, *96*, 33–6.
26. Bekiaris-Liberis, N., & Krstic, M. (2014). Compensation of wave actuator dynamics for nonlinear systems. *IEEE Transactions on Automatic Control*, *59*(6), 1555–1570.
27. Berger, E. J. (2002). Friction modeling for dynamic system simulation. *Applied Mechanics Reviews*, *55*(6), 535–577.
28. Bertsekas, D. P., & Rhodes, I. B. (1971). On the minmax reachability of target set and target tubes. *Automatica*, *7*(2), 233–247.
29. Bertsekas, D. P. (1972). Infinite-time reachability of state-space regions by using feedback control. *IEEE Transactions on Automatic Control*, *17*(5), 604–613.
30. Besaisow, A. A., Ng, F. W., & Close, D. A. (1990). *Application of ADAMS (Advanced Drill-string Analysis and Measurement System) and Improved Drilling Performance*. SPE 19998; SPE/IADC Drilling Conference.
31. Besselink, B., van de Wouw, N., & Nijmeijer, H. (2011). A semi-analytical of stick-slip oscillations in drilling systems. *ASME Journal of Computational and Nonlinear Dynamics*, *6*(2), 021006-1-021006-9.
32. Bhushan, B. (1996). *Tribology and mechanics of magnetic storage devices* (2nd ed.). New York, USA: Springer.
33. Bhushan, B. (2003). Adhesion and stiction: mechanisms, measurement techniques, and methods for reduction. *Journal of Vacuum Science & Technology B: Microelectronics and Nanometer Structures*, *21*(6), 2262–2296.
34. Bi, X. L., Wang, J., & Sun, S. H. (2011). Finite element analysis on axial-torsional coupled 108 vibration of drill string. In Y. Li, P. Wang, L. Ai, X. Sang, & J. Bu (Eds.) *Advanced materials research, materials processing technology. Vibration, noise analysis and control* (Vol. 291–294, pp. 1952–1956). Switzerland: Trans Tech Publications.
35. Blanchard, P., Devaney, R. L., & Hall, G. R. (2006). *Differential equations*. London: Thompson.
36. Blanchini, F. (1999). Set invariance in control–a survey. *Automatica*, *35*(11), 1747–1768.
37. Boukas, E. K., & Rodrigues, L. (2005). Inventory control of switched production systems: LMI approach. In *Analysis, control and optimization of complex dynamic systems* (pp. 25–42). USA: Springer.
38. Boussaada, I., & Niculescu, S. I. (2014). Computing the codimension of the singularity at the origin for delay systems: The missing link with Birkhoff incidence matrices. In *21st International Symposium on Mathematical Theory of Networks and Systems* (pp. 1699–1706). July 7–11. Groningen, The Netherlands.

39. Boussaada, I., Mounier, H., Niculescu, S. I., & Cela, A. (2012). Analysis of drilling vibrations: a time delay system approach. In *20th Mediterranean Conference on Control and Automation MED*. Barcelona, Spain.
40. Boussaada, I., Cela, A., Mounier, H., & Niculescu, S. I. (2013). Control of drilling vibrations: A time-delay system-based approach. In 11th Workshop on Time-Delay Systems Part of,. (2013). *IFAC Joint Conference SSSC*. France: Grenoble.
41. Boyd, S., El Ghaoui, L., Feron, E., & Balakrishnan, V. (1994). *Linear matrix inequalities in system and control theory*. Philadelphia: SIAM Pubilcations, Society for Industrial and Applied Mathematics.
42. Brayton, R. K. (1966). Bifurcation of periodic solutions in a non-linear difference-differential equation of neutral type. *Quarterly of Applied Mathematics, 24*, 215–224.
43. Brett, J. F. (1992). The genesis of torsional drillstring vibrations. *SPE Drilling Engineering, 7*(3), 168–174.
44. Brown, P., Byrne, G., & Hindmarsh, A. (1989). Vode: A variable-coefficient ode solver. *SIAM Journal on Scientific and Statistical Computing, 10*(5), 1038–1051.
45. Brunet, J., & Desplans, J. P. (1995). Exploitation du modèle de forage en torsion explicatif du stick-slip–estimation des variables de fond pour la réalisation de l'aide au foreur. Rapport ADERSA–avenant 1 à la convention 17 632.
46. Brunet, J., Desplans, J. P., Rey-Fabret I., & Mabile, I. (1996). Etat d'avancement de l'étude du phénomène de bit-bouncing. Note IFP 97–068.
47. Byrnes, C. I., Spong, M. W., & Tarn, T. J. (1984). A several complex variables approach to feedback stabilization of neutral delay-differential systems. *Mathematical Systems Theory, 17*(1), 97–133.
48. Campbell, S. A. (2009). *Calculating centre manifolds for delay differential equations using Maple. Delay Differential equations: Recent advances and new directions*. New York: Springer.
49. Canudas-de-Wit, C., Olsson, H., Åström, K. J., & Lischinsky, P. A. (1995). New model for control of systems with friction. *IEEE Transactions on Automatic Control, 40*(3), 419–425.
50. Canudas-de-Wit, C., Siciliano, B., & Bastin, G. (1996). *Theory of robot control* (1st ed.). New York: Springer.
51. Canudas-de-Wit, C., Rubio, F. R., & Corchero, M. A. (2008). D-OSKIL: A new mechanism for controlling stick-slip oscillations in oil well drillstrings. *IEEE Transactions on Control Systems Technology, 16*(6), 1177–1191.
52. Carr, J. (1981). *Applications of centre manifold theory*. Applied Mathematical Sciences (Vol. 35). New York: Springer.
53. Challamel, N. (2000). Rock destruction effect on the stability of a drilling structure. *Journal of Sound and Vibration, 233*(2), 235–254.
54. Chapman, C. D., Sánchez-Flores, J. L., de-León-Pérez, R., & Yu, H. (2012). Automated closed-loop drilling with ROP optimization algorithm significantly reduces drilling time and improves downhole tool reliability. In *Proceedings of the IADC/SPE Drilling Conference and Exhibition, SPE 151736* (pp. 1323–1329). San Diego, California, USA.
55. Chen, J., Gu, G., & Nett, C. N. (1995). A new method for computing delay margins for stability of linear delay systems. *Systems and Control Letters, 26*, 101–117.
56. Christoforou, A. P., & Yigit, A. S. (2003). Fully coupled vibrations of actively controlled drillstrings. *Journal of Sound and Vibration, 267*(5), 1029–1045.
57. Chunjie, H., & Tie, Y. (2009). The research on axial vibration of drill string with delphi. In: *International Conference on Computational Intelligence and Natural Computing* (pp. 78–481).
58. Clayer, F., Heneusse, & H., Sancho, J. (1992). Procédé de transmission acoustique de données de forage d'un puits. World Intellectual Property Organization, No. WO 92/04644.
59. Cobern, M. E., & Wassell, M. E. (2004). Drilling vibration monitoring & control system. In: *National Gas Technology Conference II*. Phoenix, Arizona, USA.
60. Cobern, M. E., Perry, C. A., Burgess, D. E., Barbely, J. R., & Wassell, M. (2007). Drilling tests of an active vibration damper. In: *SPE/IADC Drilling Conference, SPE-105400-MS*. Amsterdam, The Netherlands.

61. Committee on Advanced Drilling Technologies. (1994). *Drilling and excavation technologies for the future*. Washington, DC, USA: National Academy Press.
62. Cooke, K., & Krumme, D. (1968). Differential-difference equations and nonlinear initial boundary value problems for linear hyperbolic partial differential equations. *Journal of Mathematical Analysis and Applications*, *24*, 372–387.
63. Cooke, R. L., Nicholson, J. W., Sheppard, M. C., & Westlake, W. (1989). First real time measurements of downhole vibrations, forces, and pressures used to monitor directional drilling operations. In: *SPE/IADC 18651 Drilling Conference*.
64. Cull, S. J., & Tucker, R. W. (1999). On the modelling of Coulomb friction. *Journal of Physics A: Mathematical and General*, *32*(11), 2103–2113.
65. Dahl, P. R. (1968). A solid friction model. In Technical Report (Ed.), *TOR-0158H3107181-1*. El Segundo, CA: The Aerospace Corporation.
66. Dareing, D., Tlusty, J., & Zamudio, C. (1990). Self-excited vibrations induced by drag bits. *Journal of Energy Resources Technology, Transactions of the ASME*, *112*(1), 54–61.
67. Davis, J. E., Smyth, G. F., Bolivar, N., & Pastusek, P. E. (2012). Eliminating stick slip by management bit depth of cut and minizing variable torque in the drillstring. In: *Proceedings of the SPE/IADC Drilling Conference*, SPE 151133. San Diego, California, USA.
68. Debeljkovic, D. L., Buzurovic, I. M., Simeunovic, G. V., & Misic, M. (2012). Asymptotic practical stability of time delay systems. In: *IEEE 10th Jubilee International Symposium on Intelligent Systems and Informatics (SISY)* (pp. 379–384).
69. Deen, C. A., Wedel, R. J., Nayan A., Mathison, S. K., & Hightower, G. (2011). Application of a torsional impact hammer to improve drilling efficiency. In: *Proceedings of the SPE Annual Technical Conference and Exhibition*, SPE 147193. Denver, Colorado, USA.
70. Deily, F. H., Dareing, D. W., Paff, G. H., Ortloff, J. E., & Lynn, R. D. (1968). Downhole measurements of drillstring forces and motions. *Journal of Engineering for Industry, Transactions of the ASME, Series B*, *90*(2), 217–225.
71. Detournay, E., & Atkinson, C. (2000). Influence of pore pressure on the drilling response in low-permeability shear-dilatant rocks. *International Journal of Rock Mechanics and Mining Sciences*, *37*(7), 1091–1101.
72. Detournay, E., & Defourny, P. (1992). A phenomenological model for the drilling action of drags bits. *International Journal of Rock Mechanics, Mining Science and Geomechanical Abstracts*, *29*, 13–23.
73. Detournay, E. & Tan, C. P. (2002). Dependence of drilling specific energy on bottom-hole pressure in shales. In: *SPE/ISRM 78221, Rock Mechanics Conference*. Irving, Texas.
74. Drouin, A., Cunha, S. S., Ramos, B. A. C., & Mora-Camino, F. (2011). Differential flatness and control of nonlinear systems. In: *30th Chinese Control Conference (CCC)* (pp. 643–648).
75. Drumheller, D. (1989). Acoustical properties of drill strings. *Journal of the Acoustical Society of America*, *85*(3), 1048–1064.
76. Drumheller, D. (1992). An overview of acoustic telemetry. In: *Sandia Research Report*, Sand-92-0677c.
77. Drumheller, D., & Knudsen, S. (1995). The propagation of sound waves in drill strings. *Journal of the Acoustical Society of America*, *97*(4), 2116–2125.
78. Dubinsky, V. S. H., Henneuse, H. P., & Kirkman, M. A. (1992). *Surface monitoring of downhole vibrations: Russian, European, and American approaches European Petroleum Conference* (pp. 16–18). France: Cannes.
79. Dunayevsky, V., Abbassian, F., & Judzis, A. (1993). Dynamic stability of drillstrings under fluctuating weight on bit. *SPE Drilling and Completion*, *8*(2), 84–92.
80. Dupont, P. E. (1994). Avoiding stick-slip through PD control. *IEEE Transactions on Automatic Control*, *39*(5), 1094–1097.
81. Duran, E., David, A., Rasmus, J. C., Dorel, A., López-Flores, H. R., & Azizi, T. (2013). Utilizing wired drill pipe technology during managed pressure drilling operations to maintain direction control, constant bottom-hole pressures, and well-bore integrity in a deep, ultra-depleted reservoir. In: *SPE/IADC Drilling Conference*, SPE-163501-MS. Amsterdam, The Netherlands.

82. El-Farra, N. H., Mhaskar, P., & Christofides, P. D. (2005). Output feedback control of switched nonlinear systems using multiple Lyapunov functions. *Systems & Control Letters, 54*, 1163–1182.
83. Elsayed, M. A., & Aissi, C. (2006). Analysis of shock absorber characteristics for drillstrings. In: *ASME 8th Biennial Conference on Engineering Systems Design and Analysis: Dynamic Systems and Controls, Symposium on Design and Analysis of Advanced Structures, and Tribology* (Vol. 3, pp. 93–101), ESDA2006-95141. Torino, Italy.
84. Ersoy, A. (2003). Automatic drilling control based on minimum drilling specific energy using PDC and WC bits. *Transactions of the Institution of Mining and Metallurgy, Section A: Mining Technology, 112*, 86–96.
85. Ertas, D., Bailey, J. R., Wang, L., & Pastusek, P. E. (2013). Drillstring mechanics model for surveillance, root cause analysis, and mitigation of torsional and axial vibrations. In: *Proceedings of the SPE/IADC Drilling Conference and Exhibition* (pp. 203–216). Amsterdam, The Netherlands.
86. Fang, H., Lin, Z., & Hu, T. (2004). Analysis of linear systems in the presence of actuator saturation and L_2-disturbances. *Automatica, 40*, 1229–1238.
87. Faria, T., & Magalhães, L. T. (1995). Normal forms for retarded functional differential equations with parameters and applications to Hopf bifurcation. *Journal of Differential Equations, 122*(2), 81–200.
88. Faria, T., & Magãles, L. T. (2001). *Topics in functional differential and difference equations*. Providence: American Mathematical Society, Fields Institute Communications.
89. Finnie, I., & Bailey, J. J. (1960). An experimental study of drill-string vibration. *Journal of Engineering for Industry, Transactions of the ASME, 82*(2), 129–135.
90. Fliess, M., Lévine, J. L., Martin, P., & Rouchon, P. (1995). Flatness and defect of non-linear systems: Introductory theory and examples. *International Journal of Control, 61*(6), 1327–1361.
91. Fliess, M., Mounier, H., Rouchon, P., & Rudolph, J. (1995). Controllability and motion planning for linear delay systems with an application to a flexible rod. In: *Proceedings of the 34th Conference on Decision & Control*, TA16 10:40. New Orleans, LA.
92. Fliess, M., & Mounier, H. (1999). Tracking control and π-freeness of infinite dimensional linear system. In G. Picci & D. S. Gilliam (Eds.), *Dynamical systems, control coding, computer vision* (pp. 45–68). Boston: Birkhäuser.
93. Fliess, M., Join, C., & Sira-Ramirez, H. (2004). Robust residual generation for linear fault diagnosis: An algebraic setting with examples. *International Journal of Control, 77*, 1223–1242.
94. Fliess, M., Join, C., & Sira-Ramirez, H. (2008). Non-linear estimation is easy. *International Journal of Modelling, Identification and Control, 4*(1), 12–27.
95. Fliess, M., & Sira-Ramírez, H. (2003). An algebraic framework for linear identification. *ESAIM Control, Optimisation and Calculus of Variations, 9*, 151–168.
96. Fliess, M., & Sira-Ramírez, H. (2004). Reconstructeurs d'état. Comptes rendus de l'Académie des Sciences Paris (Ser. I, Vol. 338, pp. 91–96).
97. Fridman, E. (2001). New Lyapunov-Krasovskii functionals for stability of linear retarded and neutral type systems. *Systems & Control Letters, 43*(4), 309–319.
98. Fridman, E., Pila, A., & Shaked, U. (2003). Regional stabilization and $\mathscr{H}_\infty$ control of time-delay systems with saturating actuators. *International Journal of Robust and Nonlinear Control, 13*(9), 885–907.
99. Fridman, E. (2005). A new Lyapunov technique for robust control of systems with uncertain non-small delays. *IMA Journal of Mathematical Control and Information, 23*(2), 165–179.
100. Fridman, E., & Orlov, Y. (2009). An LMI approach to H_∞ boundary control of semilinear parabolic and hyperbolic systems. *Automatica, 45*(9), 2060–2066.
101. Fridman, E., Dambrine, M., & Yeganefar, N. (2008). Input to state stability of systems with time-delay: A matrix inequalities approach. *Automatica, 44*(9), 2364–2369.
102. Fridman, E., & Dambrine, M. (2009). Control under quantization, saturation and delay: A LMI approach. *Automatica, 45*(10), 2258–2264.

103. Fridman, E., Mondié, S., & Saldivar, M. B. (2010). Bounds on the response of a drilling pipe model. *Special issue on Time-Delay Systems, IMA Journal of Mathematical Control & Information, 27*(4), 513–526.
104. Fu, P., Chen, J., & Niculescu, S. I. (2006). Stability of linear neutral time-delay systems: Exact conditions via matrix pencil solutions. *IEEE Transactions on Automatic Control, 51*(6), 1063–1069.
105. Fu, P., Chen, J., & Niculescu, S. I. (2006). Generalized eigenvalue-based stability tests for 2-D linear systems: Necessary and sufficient conditions. *Automatica, 42*(9), 1569–1576.
106. Fubin, S., Linxiu, S., Lin, L., & Qizhi, Z. (2010). Adaptive PID control of rotary drilling system with stick slip oscillation. In: *The 2nd International Conference on Signal Processing Systems, ICSPS 2010* (pp. 289–292).
107. Gamboa-Ritto, T. (2010). *Numerical analysis of the nonlinear dynamics of a drill-string with uncertainty modeling*. PhD Thesis, Postgraduate Program in Mechanical Engineering. Brazil: Pontifícia Universidade Católica do Rio de Janeiro.
108. García-Collado, F. A., D'Andréa-Novel, B., Fliess, M., & Mounier, H. (2009). *Analyse fréquentielle des dérivateurs algébriques*. Dijon, France: XXIIe Colloque GRETSI.
109. Gensior, A., Woywode, O., Rudolph, J., & Guldner, H. (2006). On differential flatness, trajectory planning, observers, and stabilization for DC-DC converters. *IEEE Transactions on Circuits and Systems, 53*(9), 2000–2010.
110. Germay, C. (2009). *Modeling and analysis of self-excited drill bit vibrations*. PhD Dissertation. Belgium: University of Liège.
111. Germay, C., Van De Wouw, N., Nijmeijer, H., & Sepulchre, R. (2005). Nonlinear drilling dynamics analysis. *SIAM Journal on Applied Dynamical Systems, 8*(2), 527–553.
112. Ghasemloonia, A., Rideout, D. G., & Butt, S. D. (2013). Vibration analysis of a drillstring in vibration-assisted rotary drilling: Finite element modeling with analytical validation. *Journal of Energy Resources Technology, 135*(3), Article ID 032902.
113. Ghanes, M., De Leon, J., & Barbot, J. P. (2013). Observer design for nonlinear systems under unknown time-varying delays. *IEEE Transactions on Automatic Control, 58*(6), 1529–1534.
114. Ghasemloonia, A., Rideout, D. G., Butt, S. D., & Hajnayeb, A. (2014). Elastodynamic and finite element vibration analysis of a drillstring with a downhole vibration generator tool and a shock sub. *Proceedings of the Institution of Mechanical Engineers, Part C: Journal of Mechanical Engineering Science* 0954406214543491.
115. Gjerding, K., Heisig, G., Hernández, M., Macpherson, J., Reeves, M., Wolter, H., & Zaeper, R. (2007). The first offshore use of an ultra high speed drillstring telemetry network involving a full LWD logging suite and rotary steerable drilling system. In: *SPE Annual Technical Conference and Exhibition*. Anaheim, California.
116. Glover, D., & Schweppe, F. (1971). Control of linear dynamic systems with set constrained disturbances. *IEEE Transactions on Automatic Control, 16*(5), 411–423.
117. Gonzalez-Garcia, S., Polyakov, A. E., & Poznyak, A. S. (2011). Using the method of invariant ellipsoids for linear robust output stabilization of spacecraft. *Automation and Remote Control, 72*(3), 540–555.
118. Grujic, L. T. (1973). On practical stability. *International Journal of Control, 17*(4), 881–887.
119. Gu, K., Kharitonov, V. L., & Chen, J. (2003). *Stability of time-delay systems*. Boston, USA: Birkhäuser.
120. Guay, M. (2005). Real-time dynamic optimization of nonlinear systems: A flatness-based approach. In: *Proceedings of the 44th IEEE Conference on Decision and Control, and the European Control Conference*. Seville, Spain.
121. Guckenheimer, J., & Holmes, P. (1983). *Nonlinear oscillations, dynamical systems, and bifurcation of vector fields*. Applied Mathematical Sciences (Vol. 42). New York: Springer.
122. Halanay, A. (1966). *Differential equations: Stability, oscillations, time lags*. New York: Academic.
123. Hale, J. K. (1971). Functional differential equations. In: *Analytic theory of differential equations*. Lecture Notes in Mathematics (Vol. 183, pp. 9–22). Berlin: Springer.

124. Hale, J.K. (1977). Theory of functional differential equations. Applied Mathematical Sciences, Vol. 3, Springer, New York.
125. Hale, J. K., Infante, E. F., & Tsen, F. S. P. (1985). Stability in linear delay equations. *Journal of Mathematical Analysis and Applications*, *105*, 533–555.
126. Hale, J. K., & Verduyn-Lunel, S. M. (1993). *Introduction to functional differential equations*. Applied Mathematical Sciences (Vol. 99). New York: Springer.
127. Hale, J. K., & Huang, W. (1994). Period doubling in singularly perturbed delay equations. *Journal of Differential Equations*, *114*, 1–23.
128. Halsey, G. W., Kyllingstad, A., & Kylling, A. (1988). Torque feedback used to cure slip-stick motion. In: *Proceedings of the 63rd Society of Petroleum Engineers Drilling Engineering, Annual Technical Conference and Exhibition* (pp. 277–282). Houston, TX.
129. Han, J. H., Kim, Y. J., & Karkoub, M. (2013). Modeling of wave propagation in drill strings using vibration transfer matrix methods. *Journal of the Acoustical Society of America*, *134*(3), 1920–1931.
130. Hensen, R. H. A., van de Molengraft, M. J. G., & Steinbuch, M. (2003). Friction induced hunting limit cycles: A comparison between the LuGre and switch friction model. *Automatica*, *39*, 2131–2137.
131. Hernández-Suárez, R., Puebla, H., Aguilar-López, R., & Hernández-Martínez, E. (2009). An integral high-order sliding mode control approach for stick-slip suppression in oil drillstrings. *Petroleum Science and Technology*, *27*(8), 788–800.
132. Horn, R. A., & Johnson, C. A. (1991). *Topics in Matrix Analysis*. Cambridge, UK: Cambridge Univ. press.
133. Hutchinson, M., Burgess, D., Thompson, F., & Kopfstein, A. (2013). Self-adapting bottom hole assembly vibration suppression. In: *Proceedings of the SPE Annual Technical Conference and Exhibition*, SPE 66071. New Orleans, USA.
134. Iserles, A. (1993). On the generalised pantograph functional-differential-equation. *European Journal of Applied Mathematics*, *4*(1), 1–38.
135. Jain, J. R., Ledgerwood, L. W. III, Hoffmann, O. J., Schwefe, T., & Fuselier, D. M. (2011). Mitigation of torsional stick-slip vibrations in oil well drilling through PDC bit design: Putting theories to the test. In: *Proceedings of the SPE Annual Technical Conference and Exhibition (ATCE 11)* (pp. 1815–1828), SPE. Denver, Colorado, USA.
136. Jamaleddine, R., & Vinet, A. (1999). Role of gap junction resistance in rate-induced delay in conduction in a cable model of the atrioventricular node. *Journal of Biological Systems*, *7*(4), 475–490.
137. Jansen, J. D. (1993). *Nonlinear dynamics of oilwell drillstrings*. PhD Dissertation, Delft University of Technology. The Netherlands: Delft University Press.
138. Jansen, J. D., & van den Steen, L. (1995). Active damping of self-excited torsional vibrations in oil well drillstrings. *Journal of Sound and Vibration*, *179*(4), 647–668.
139. Javanmardi, K., & Gaspard, D. (1992). *Application of soft torque rotary table in mobile bay*. IADC/SPE 23913. Paper presented at the Drilling Conference held in New Orleans.
140. Jeffryes, B., Moriarty, K., & Reyes, S. (2001). Method and apparatus for enhanced acoustic mud-pulse telemetry. In: *World Intellectual Property*. Organization, No. WO 01/66912 A1.
141. Jones, E., Oliphant, T., Peterson, P. et al. (2001). SciPy: Open source scientific tools for *Python*.
142. Karkoub, M., Abdel-Magid, Y. L., & Balachandran, B. (2009). Drillstring torsional vibration suppression using GA optimized controllers. *Journal of Canadian PetroleumTechnology*, *48*(12), 32–38.
143. Karkoub, M., Zribi, M., Elchaar, L., & Lamont, L. (2010). *Robust μ-synthesis controllers for suppressing stick-slip induced vibrations in oil well drill strings*. Multibody System Dynamics (pp. 191–207). New York: Springer.
144. Kappel, F. (2006). *Linear autonomous functional differential equations*. Delay Differential Equations and Applications (Vol. 205, Part I, pp. 41–139). Netherlands: Springer.
145. Karnopp, D. (1985). Computer simulation of stick-slip friction in mechanical dynamic systems. *Journal of Dynamic Systems Measurement and Control*, *107*(1), 100–103.

146. Kazemi, R., Jafari A. A., & Mahyari, M. F. (2010). The effect of drilling mud flow on the lateral and axial vibrations of drill string. In: *ASME 2010 10th Biennial Conference on Engineering Systems Design and Analysis* (Vol. 5). Istanbul, Turkey.
147. Kermani, P., & Kleinrock, L. (1980). A tradeoff study of switching systems in computer communication networks. *IEEE Transactions on Computers, C, 29*(12), 1052–1060.
148. Khalil, H. (1992). *Nonlinear Systems*. New York: Macmillan Publishing Company.
149. Kharitonov, V. (1998). Robust stability analysis of time delay systems: A survey. In: *Fourth IFAC Conference on System Structure and Control*, Nantes, France, plenary lecture (pp. 1–12).
150. Kharitonov, V. L. (2013). *Time-delay systems: Lyapunov functionals and matrices*. Boston: Control Engineering, Birkhäuser.
151. Kharitonov, V. L., & Niculescu, S. I. (2003). On the stability of linear systems with uncertain delay. *IEEE Transactions on Automatic Control, 48*(1), 127–132.
152. Kharitonov, V., Collado, J., & Mondié, S. (2006). Exponential estimates for neutral time delay systems with multiple delays. *International Journal of Robust and Nonlinear Control, 16*(2), 71–84.
153. Kim, S., Campbell, S. A., & Liu, X. (2006). Stability of a class of linear switching systems with time delay. *IEEE Transactions on Circuits and Systems, 53*(2), 384–393.
154. Kirillov, O. N., & Pelinovsky, D. E. (2013). *Nonlinear physical systems: Spectral analysis, stability and bifurcations*. London: Wiley.
155. Knüppel, T., Woittennek, F., Boussaada, I., Mounier, H., & Niculescu, S. I. (2013). Flatness-based control for a non-linear spatially distributed model of a drilling system. In: *Low complexity controllers for time delay systems*. Advances in Delays and Dynamics. Berlin: Springer.
156. Kolar B., & Schlacher K. (2013). Flatness based control of a gantry crane. In: *9th IFAC Symposium on Nonlinear Control Systems* (pp. 487–492). Toulouse, France.
157. Kolmanovskii, V. B., & Nosov, V. R. (1986). *Stability of functional differential equations*. London: Academic Press.
158. Kolmanovskii, V. B., & Myshkis, A. (1992). *Applied theory of functional differential equations*. Mathematics and Applications (Vol. 85). Dordrecht: Kluwer Academy.
159. Kolmanovskii, V. B., Niculescu, S. I., & Gu, K. (1999). Delay effects on stability: A survey. In: *38th IEEE Conference on Decision and Control (CDC)* (pp. 1993–1998). Phoenix, AZ.
160. Kreisle, L. F., & Vance, J. M. (1970). Mathematical analysis of the effect of a shock sub on the longitudinal vibrations of an oilwell drill string. *Society of Petroleum Engineers Journal, SPE-2778-PA, 10*(4), 349–356.
161. Kreuzer, E., & Steidl, M. (2012). Controlling torsional vibrations of drill strings via decomposition of traveling waves. *Archive of Applied Mechanics, 82*, 515–531.
162. Kriesels, P. C., Keultjes, W. J. G., Dumont, P., Huneidi, I., Furat, A., Owoeye, O. O., & Hartmann, R. A. (1999). Cost savings through an integrated approach to drillstring vibration control. In: *SPE/IADC Middle East Driling Technology Conference, SPE/IADC 57555*. Abu Dhabi.
163. Krstic, M. (2009). Delay compensation for nonlinear, adaptive, and PDE systems. Systems & Control: Foundations & Applications, Birkhäuser Boston, a part of Springer Science+Business Media LLC.
164. Kurzhanski, A. B., & Veliov, V. M. (1994). *Modeling techniques for uncertain systems*. Progress in Systems and Control Theory (Vol. 18). Boston: Birkhäuser.
165. Kuznetsov, Y. (1998). *Elements of applied bifurcation theory* (2nd ed., Vol. 112). Applied Mathematics Sciences. New York: Springer.
166. Kyllingstad, A., & Haisey, G. W. (1988). A study of slip/stick motion of the bit. *SPE Drilling Engineering, 3*, 369–373.
167. Lakshmikantham, V., Leela, S., & Martynyuk, A. A. (1990). *Practical stability of nonlinear systems*. Singapore: World Scientific Publishing, Company Private Limited.
168. La Salle, J., & Lefschetz, S. (1961). *Stability by Lyapunov's direct method: With applications*. London: Academic Press Inc.
169. Lear, W., & Dareing, D. (1990). Effect of drillstring vibrations on MWD pressure pulse signals. *Journal of Energy Resources Technology, 112*(2), 84–89.

170. Ledgerwood, L. W., Jain, J. R., Hoffmann, O. J., & Spencer, R. W. (2013). Downhole measurement and monitoring lead to an enhanced understanding of drilling vibrations and polycrystalline diamond compact bit damage. *SPE Drill Completion, 28*(3), 254–262.
171. Leine, R. I., Van Campen, D. H., & De Kraker, A. (1998). Stick-slip vibrations induced by alternate friction models. *Nonlinear Dynamics, 16*, 41–54.
172. Lévine, J., Lottin, J., & Ponsart, J. C. (1996). A nonlinear approach to the control of magnetic bearings. *IEEE Transactions on Control Systems Technology, 4*(5), 524–544.
173. Li, B., & Zou, J. (2011). The design of oil drilling wireless data acquisition system. In: *International Conference on Electrical and Control Engineering* (pp. 1810–1813). Yichang, China.
174. Li, L., Zhang, Q. Z., & Rasol, N. (2011). Time-varying sliding mode adaptive control for rotary drilling system. *Journal of Computers, 6*(3), 564–570.
175. Li, Z., Gaob, H., Agarwalc, R., & Kaynakd, O. (2013). H_∞ control of switched delayed systems with average dwell time. *International Journal of Control, 86*(12), 2146–2158.
176. Lin, A. Y. M., & Silvester, J. A. (1991). Priority queueing strategies for traffic control at an ATM integrated broadband switching system. *IEEE Pacific Rim Conference on Communications, Computers and Signal Processing, 2*, 429–432.
177. Liao, C., Balachandran, B., Karkoub, M., & Abdel-Magid, Y. L. (2011). Drill-string dynamics: reduced-order models and experimental studies. Journal of Vibration and Acoustics, 133(4), 8, ID 041008.
178. Lin, Y., & Wang, Y. (1991). Stick-slip vibration of drill strings. *Transactions of ASME, 113*, 38–43.
179. Liu, D. Y., Liu, X. Z., & Zhong, S. M. (2008). Delay-dependent robust stability and control synthesis for uncertain switched neutral systems with mixed delays. *Applied Mathematics and Computation, 202*(2), 828–839.
180. Liu, X., Li, B., & Yue, Y. (2007). Transmission behavior of mud-pressure pulse along well bore. *Journal of Hydrodinamics, 19*(2), 236–240.
181. Liu, X., Vlajic, N., Long, X., Meng, G., & Balachandran, B. (2013). Nonlinear motions of a flexible rotor with a drill bit: Stick-slip and delay effects. *Nonlinear Dynamics, 72*(1–2), 61–77.
182. Liu, X., Vlajic, N., Long, X., Meng, G., & Balachandran, B. (2014). State-dependent delay influenced drill-string oscillations and stability analysis. *Journal of Vibration and Acoustics, 136*(5), 051008.
183. Liu, X., Vlajic, N., Long, X., Meng, G., & Balachandran, B. (2014). Multiple regenerative effects in cutting process and nonlinear oscillations. *International Journal of Dynamics and Control, 2*, 86–101.
184. Liu, W., & Zhou, Y. (2009). Vibration control management to secure safety and fast drilling. In: *International Conference on Management and Service Science* (pp. 1–4). Wuhan, China.
185. Loiseau, J. J. (1998). Algebraic tools for the control and stabilization of time-delay systems. In: *First IFAC Workshop on Linear Time Delay Systems* (pp. 234–249). Grenoble, France, plenary lecture.
186. Loiseau, J. J., Cardelli, M., & Dusser, X. (2001). Neutral type time-delay systems that are not formally stable are not BIBO stabilizable. *IMA Journal of Mathematical Control and Information, 19*, 217–227.
187. López, I., & Nijmeijer, H. (2009). Prediction and validation of the energy dissipation of a friction damper. *Journal of Sound and Vibration, 328*, 396–410.
188. Louembet, C., Cazaurang, F., Zolghadri, A., Charbonnel, C., & Pittet, C. (2009). Path planning for satellite slew manoeuvres: A combined flatness and collocation-based approach. *Control Theory & Applications, IET, 3*(4), 481–491.
189. Lu, H., Dumon, J., & Canudas-de-Wit, C. (2009). Experimental study of the D-OSKIL mechanism for controlling the stick-slip oscillations in a drilling laboratory testbed. In: *2009 IEEE Control Applications, (CCA) & Intelligent Control, (ISIC)* (pp. 1551–1556). Russia: St. Petersburg.

190. Macpherson, J. D., Mason, J. S., & Kingman, J. E. E. (1993). Surface measurement and analysis of drillstring vibrations while drilling. In: *SPE/IADC paper 25777*. German: Society of Petroleum Engineers.
191. McCartney, C., Allen, S., Hernández, M., MacFarlane, D., Baksh, A., & Reeves, M. E. (2009). Step-change improvements with wired-pipe telemetry. In: *SPE/IADC Drilling Conference and Exhibition, SPE-119570-MS*. Amsterdam, The Netherlands.
192. Mahyari, M. F., Behzad, M., & Rashed, G. R. (2010). Drill string instability reduction by optimum positioning of stabilizers. *Proceedings of the Institution of Mechanical Engineers, Part C: Journal of Mechanical Engineering Science*, *224*(3), 647–653.
193. Mason, J. S., & Sprawls, B. M. (1998). Addressing BHA whirl: The culprit in mobile bay. *SPE Drilling and Completion*, *13*(4), 231–236.
194. Martin, P., & Rouchon, P. (1996). Flatness and sampling control of induction motors. In: *IFAC World Congress* (pp. 389–394). San Francisco.
195. Michiels, W., Niculescu, S. I., & Moreau, L. (2004). Using delays and time-varying gains to improve the output feedback stabilizability of linear systems: A comparison. *IMA Journal of Mathematical Control and Information*, *21*(4), 393–418.
196. Michiels, W., & Niculescu, S. I. (2007). *Stability and stabilization of time-delay systems: an eigenvalue-based approach*. Advances in Design and Control. Philadelphia, PA: Society for Industrial and Applied Mathematics (SIAM).
197. Mitchell, R. F., & Allen, M. B. (1987). Case studies of BHA vibration failure. In: *SPE 16675*. Society of Petroleum Engineers.
198. Moreau, L., & Aeyels, D. (2000). Practical stability and stabilization. *IEEE Transactions on Automatic Control*, *45*(8), 1554–1558.
199. Mounier, H. (1995). Propriétés structurelles des systèmes linéaires à retard: aspects théorique et pratique. *Thèse de l'Université Paris-Sud*, (3888), 148.
200. Mounier, H. (1996). Planification de trajectoires du système de forage utilisant des équations des ondes. Note IFP 97.
201. Mounier, H. (1997). Dynamique du système de forage: description de la modélisation et de l'estimation de variables de fond. *Rapport IFP*, *43*, 953.
202. Mounier, H., & Rudolph, J. (1998). Flatness based control of nonlinear delay systems: A chemical reactor example. *International Journal of Control*, *71*(5), 871–890.
203. Mounier, H., & Rudolph, J. (2008). Flatness and quasi-static state feedback in nonlinear delay systems. *International Journal of Control*, *81*(3), 445–456.
204. Murray, R. M., Sastry, S. S., & Zexiang, L. (1994). *A Mathematical introduction to robotic manipulation*. Boca Raton, FL, USA: CRC Press Inc.
205. Murray, R. M., Rathinam, M., & Sluis, W. (1995). Differential flatness of mechanical control systems—a catalog of prototype systems. In: *ASME International Mechanical Engineering Congress and Exposition*. San Francisco, California, USA.
206. Murray, R. M. (1996). Trajectory generation for a towed cable system using differential flatness. In: *IFAC World Congress* (pp. 395–400). San Francisco.
207. Nandakumar, K., & Wiercigroch, M. (2013). Stability analysis of a state dependent delayed, coupled two DOF model of drill-string vibration. *Journal of Sound and Vibration*, *332*, 2575–2592.
208. Navarro-López, E., & Suárez, R. (2004). Modelling and analysis of stick-slip behaviour in a drillstring under dry friction. *Congreso Nacional de Control Automático* (pp. 330–335). México.
209. Navarro-López, E., & Suárez, R. (2004). Practical approach to modelling and controlling stick-slip oscillations in oilwell drillstrings. In: *Proceedings of the 2004 IEEE International Conference on Control Applications* (pp. 1454–1460).
210. Navarro-López, E., & Cortés, D. (2007). Sliding-mode control of a multi-DOF oilwell drillstring with stick-slip oscillations. In: *Proceedings of the 2007 American Control Conference* (pp. 3837–3842). New York City, USA.
211. Navarro-López, E. M., & Cortés, D. (2007). Avoiding harmful oscillations in a drillstring through dynamical analysis. *Journal of Sound and Vibrations*, *307*(1–2), 152–171.

212. Navarro-López, E. (2009). An alternative characterization of bit-sticking phenomena in a multi-degree-of-freedom controlled drillstring. *Nonlinear Analysis: Real World Applications*, *10*(5), 3162–3174.
213. Navarro-López, E., & Licéaga-Castro, E. (2009). Non-desired transitions and sliding-mode control of a multi-DOF mechanical system with stick-slip oscillations. *Chaos, Solitons and Fractals*, *41*, 2035–2044.
214. Navarro-López, E. (2010). *Bit-sticking phenomena in a multidegree-of-freedom controlled drill strings*. Technical Report : University of Manchester.
215. Nazin, A., Polyak, B., & Topunov, M. (2007). Rejection of bounded exogenous disturbances by the method of invariant ellipsoids. *Automation and Remote Control*, *68*(3), 467–486.
216. Nicaise, S., & Pignotti, C. (2008). Stabilization of the wave equations with variable coefficients and boundary conditions of memory type. *Asymptotic Analysis*, *50*(1–2), 31–67.
217. Niculescu, S. I. (2001). *Delay effects on stability, a robust control approach*. Lecture Notes in Control and Information Sciences (Vol. 269, p. XVI). London: Springer.
218. Nijmeijer, H., & van der Schaft, A. J. (1990). *Nonlinear dynamical control systems*. New York: Springer.
219. Nussbaum, R. D. (2002). *Handbook of dynamical systems* (Vol. 2, B. Fiedler (Ed.)). New York: Elsevier Science B.V.
220. Ollivier, F., & Sedoglavic, A. (2001). A generalization of flatness to nonlinear systems of partial differential equations—application to the command of a flexible rod. In: *Proceedings of the 5th IFAC Symposium "Nonlinear Control Systems"* (Vol. 1, pp. 196–200). Russia: Saint Petersburg.
221. Olsson, H., & Åström, K. J. (2001). Friction generated limit cycles. *IEEE Transactions on Control Systems Technology*, *9*(4), 629–636.
222. Olsson, H., Åström, K. J., Canudas-de-Wit, C., Gafvert, M., & Lischinsky, P. (1998). Friction models and friction compensation. *European Journal of Control*, *4*(3), 176–195.
223. Ottesen, J. T. (1997). Modelling of the baroflex-feedback mechanism with time-delay. *Journal of Mathematical Biology*, *36*, 41–63.
224. Patil, P. A., & Teodoriu, C. (2013). Model development of torsional drillstring and investigating parametrically the stick-slips influencing factors. *Journal of Energy Resources Technology*, *135*(1), 1–7.
225. Pavković, D., Deur, J., & Lisac, A. (2011). A torque estimator-based control strategy for oil-well drill-string torsional vibrations active damping including an auto-tuning algorithm. *Control Engineering Practice*, *19*(8), 836–850.
226. Pavone, D. R., & Desplans, J. P. (1994). Application of high sampling rate downhole measurements for analysis and cure of stick-slip in drilling. In: *SPE Annual Technical Conference and Exhibition*, SPE 28324 (pp. 335–345). New Orleans, L.A.
227. Pavone, D. R., & Desplans, J. P. (1996). Analyse et Modélisation du comportement dynamique d'un rig de forage. *IFP report*, 42208.
228. Pelfrene, G., Sellami, H., & Gerbaud, L. (2011). Mitigating stick slip in deep drilling based on optimization. In: *Proceedings of the SPE/IADC Drilling Conference and Exihibition*, SPE 139839. Amsterdam, The Netherlands.
229. Petit, N., Rouchon, P., Boueilh, J. M., Guérin, F., & Pinvidic, P. (2002). Control of an industrial polymerization reactor using flatness. *Journal of Process Control*, *12*, 659–665.
230. Pessier, R., & Damschen, M. (2011). Hybrid bits offer distinct advantages in selected roller-cone and PDC-bit applications. *SPE Drilling and Completion*, *26*(1), 96–103.
231. Poincaré, H. (1885). L'équilibre d'une masse fluide animée d'un mouvement de rotation. *Acta Mathematica*, t.7, 259–380.
232. Pontryagin, L. S. (1955). On the zeros of some elementary trascendental functions. *Transactions of the American Mathematical Society*, *2*, 95–110.
233. Popov, V. L. (2010). *Contact mechanics and friction. Physical principles and applications*. Heidelberg Dordrecht London New York: Springer.
234. Poznyak, A. S. (2008). *Advanced mathematical tools for automatic control engineers: Deterministic techniques* (Vol. 2). Amsterdam: Elsevier.

235. Rabia, H. (1982). Specific energy as a criterion for drilling performance prediction. *International Journal of Rock Mechanics and Mining Sciences*, *19*, 39–42.
236. Rahman, N. A., Mohaideen, A., Bakar, F. H., Tang, K. H., Maury, R., Cox, P., Le, P., Donald, H., Brahmanto, E., & Subroto, B. (2012). Solving stickslip dilemma: dynamic modeling system significantly reduces vibration, increases ROP by 54 %. In: *Proceedings of the Abu Dhabi International Petroleum Exhibition and Conference*, SPE 161155 (pp. 948–963). Abu Dhabi, UAE.
237. Rajnauth, J., & Jagai, T. (2012). Reduce torsional vibration and improve drilling operations. *International Journal of Applied Science and Technology*, *2*(7), 109–123.
238. Rasvan, V. (1975). A method for distributed parameter control systems and electrical networks analysis. *Revue Roumaine des Sciences Techniques, Serie Electrotechnique Energy*, *20*, 561–566.
239. Rasvan, V., & Niculescu, S. I. (2002). Oscillations in lossless propagation models: A Lyapunov-Krasovskii approach. *IMA Journal of Mathematical Control and Information*, *19*(1–2), 157–172.
240. Reddish, D. J., & Yasar, E. (1996). A new portable rock strength index test based on specific energy of drilling. *International Journal of Rock Mechanics Mining Sciences & Geomechanics.*, *33*(5), 543–548.
241. Rey-Fabret, I., Nauroy, J. F., Vincké, O., Peysson, Y., King, I., Chauvin, H., et al. (2004). Intelligent drilling surveillance through real time diagnosis. *Oil & Gas Science and Technology—Revue d'IFP Energies nouvelles*, *59*(4), 357–369.
242. Richard, J. P. (2003). Time-delay systems: An overview of some recent advances and open problems. *Automatica*, *39*(10), 1667–1694.
243. Richard, T., Germay, C., & Detournay, E. (2007). A simplified model to explore the root cause of stickslip vibrations in drilling systems with drag bits. *Journal of Sound and Vibration*, *305*, 432–456.
244. Rommetveit, R., Bjørkevoll, K. S., Halsey, G. W., Larsen, H. F., Merlo, A., Nossaman, L. N., Sweep, M. N., Silseth, K. M., & Ødegård, S. I. (2004). Drilltronics: An integrated system for real-time optimization of the drilling process. In: *IADC/SPE Drilling Conference*, SPE-87124-MS. Dallas, Texas, USA.
245. Rommetveit, R., Bjørkevoll, K. S., Ødegård, S. I., Herbert, M., Halsey, G. W., Kluge, R., & Korsvold, T. (2008). eDrilling used on Ekofisk for real-time drilling supervision, simulation, 3D visualization and diagnosis. In: *Intelligent Energy Conference and Exhibition*, SPE-112109-MS. Amsterdam, The Netherlands.
246. Rothfuss, R., Rudolph, J., & Zeitz, M. (1996). Flatness based control of a nonlinear chemical reactor model. *Automatica*, *32*(10), 1433–1439.
247. Rouchon, P. (1998). Flatness and stick-slip stabilization. *Technical Report*, *492*, 1–9.
248. Rudat, J., & Dashevskiy, D. (2011). Development of an innovative model-based stick/slip control system. In: *Proceeding of the SPE/IADC Drilling Conference and Exhibition* (pp. 585–596). Amsterdam, The Netherlands.
249. Runia, D. J., Dwars, S., & Stulemeijer, I. P. J. M. (2013). A brief history of the Shell "Soft Torque Rotary Syste" and some recent case studies. In: *SPE/IADC Drilling Conference*, SPE-163548-MS. Amsterdam, The Netherlands.
250. Sagert, C., Di Meglio, F., Krstic, M., & Rouchon, P. (2013). Backstepping and flatness approaches for stabilization of the stick-slip phenomenon for drilling. In: IFAC Symposium on System Structure and Control. Grenoble, France.
251. Saldivar, B., Mondié, S., & Fridman, E. (2009). Ultimate boundedness of the response of a drilling pipe model. In: *IFAC Workshop on Time Delay Systems (TDS 2009)*. Sinaia, Rumania.
252. Saldivar, B., Mondié, S., Loiseau, J.J. (2009). Reducing stick-slip oscillations in oilwell drillstrings. 6th International Conference on Electrical Engineering, Computing Science and Automatic Control (CCE), Toluca, México, 1–6.
253. Saldivar, B., Mondié, S., Loiseau, J. J., & Rasvan, V. (2011). Stick-slip oscillations in oillwell drilstrings: Distributed parameter and neutral type retarded model approaches. In: *18th IFAC World Congress* (pp. 284–289). Milano, Italy.

254. Saldivar, B., Mondie, S., Loiseau, J. J., & Rasvan, V. (2011). Exponential stability analysis of the drilling system described by a switched neutral type delay equation with nonlinear perturbations. In: *50th IEEE Conference on Decision and Control and European Control Conference (CDC-ECC)*. Orlando, FL, USA.
255. Saldivar, B., Seuret A., & Mondié, S. (2011). Exponential stabilization of a class of nonlinear neutral type time-delay systems, an oilwell drilling model example. In: *8th International Conference on Electrical Engineering, Computer Science and Automatic Control*. Mérida, Yucatan.
256. Saldivar, B. (2013). Analysis, modeling and control of an oilwell drilling system. Cotutelle Ph.D. Thesis. Centro de Investigación y de Estudios Avanzados del Instituto Politécnico Nacional, Mexico. Institut de Recherche en Communications et Cybernétique de Nantes, France.
257. Saldivar, B., & Mondié, S. (2013). Drilling vibration reduction via attractive ellipsoid method. *Journal of the Franklin Institute Elsevier*, *350*(3), 485–502.
258. Saldivar, B., Mondié, S., Loiseau, J. J., & Rasvan, V. (2013). Suppressing axial torsional coupled vibrations in oilwell drillstrings. *Journal of Control Engineering and Applied Informatics*, *15*(1), 3–10.
259. Saldivar, B., Boussaada, I., Mounier, H., Mondié, S., & Niculescu, S. I. (2014). An overview on the modeling of oilwell drilling vibrations. In: *19th World Congress of the International Federation of Automatic Control*. Cape Town, South Africa.
260. Saldivar, B., Knüppel, T., Woittennek, F., Boussaada, I., Mounier, H., & Niculescu, S. I. (2014). Flatness-based control of torsional-axial coupled drilling vibrations. In: *19th World Congress of the International Federation of Automatic Control*. Cape Town, South Africa.
261. Sananikone, P. (1993). *Method and apparatus for determining the torque applied to a drillstring at the surface*. U.S. Patent No. 5,205,163. Houston, Texas: Schlumberger Technology Co.
262. Sargin, M. (1971). *Stress-strain relationship for concrete and the analysis of structural concrete sections*. SM Study 4, Solid Mechanics Division. Canada: University of Waterloo.
263. Sastry, S. S., & Bodson, M. (1989). *Adaptive control: Stability, convergence, and robustness*. Englewood Cliffs: Prentice-Hall.
264. Schiesser, W. E. (1991). *The numerical method of lines: integration of partial differential equations*. San Diego: Academic Press.
265. Seidel, H., & Herzel, H. (1998). Bifurcations in a non-linear model of the baroreceptor-cardiac reflex. *Physica D: Nonlinear Phenomena*, *115*, 145–160.
266. Serrarens, A. F. A., van de Molengraft, M. J. G., Kok, J. J., & van den Steeen, L. (1998). H_∞ control for suppressing stick-slip in oil well drillstrings. *IEEE Control Systems*, *18*(2), 19–30.
267. Seuret, A., Dambrine, M., & Richard, J. P. (2004). Robust exponential stabilization for systems with time-varying delays. In: *5th IFAC Workshop on Time Delays Systems*.
268. Seuret, A. (2006). Commande et observation des systèmes à retard variables, théorie et applications (pp. 57–94). Ph.D. Thesis, Ecole Centrale de Lille, Univiersité des Sciences et Technologies de Lille.
269. Shewalla, M. (2007). *Evaluation of shear strength parameters of shale and siltstone using single point cutter tests*. M.Sc. Thesis. Graduate Faculty of the Louisiana State University and Agricultural and Mechanical College.
270. Shyu, R. J. (1989). *Bending vibration of rotating drill strings*. Ph.D Thesis, Massachusetts Institute of Technology.
271. Sieber, J., & Krauskopf, B. (2004). Bifurcation analysis of an inverted pendulum with delayed feedback control near a triple-zero eigenvalue singularity. *Nonlinearity*, *17*, 85–103.
272. Sinanovic, S., Johnson, D. H., Shah, V. V., & Gardner, W. R. (2004). Data communication along the drill string using acoustic waves. *IEEE International Conference on Acoustics, Speech, and Signal Processing ICASSP '04* (Vol. 4, pp. 909–912). Montreal, Quebec, Canada.
273. Skaugen, E. (1987). *The effects of quasi-random drill bit vibrations upon drillstring dynamic behavior*. SPE 16660, Society of Petroleum Engineers.

274. Smyshlyaev, A., & Krstic, M. (2009). Boundary control of an antistable wave equation with anti-damping on the uncontrolled boundary. In: *American Control Conference* (pp. 1511–1516).
275. Soendervik, K. (2013). Autonomous robotic drilling systems. In: *SPE/IADC Drilling Conference*, SPE-163466-MS. Amsterdam, The Netherlands.
276. Sowers, S. F., Dupriest, F. E., Bailey J. R., & Wang, L. (2009). Roller reamers improve drilling performance in wells limited by bit and bottomhole assembly vibrations. In: *Proceedings of the SPE/IADC Drilling Conference and Exhibition*, SPE 119375. Amsterdam, The Netherlands.
277. Stamova, I. M. (2007). Vector Lyapunov functions for practical stability of nonlinear impulsive functional differential equations. *Journal of Mathematical Analysis and Applications*, *325*(1), 612–623.
278. Stribeck, R. (1902). Die wesentlichen Eigenschaften der Gleit- und Rollenlager The key qualities of sliding and roller bearings. *Zeitschrift des Vereines Seutscher Ingenieure*, *46*(38, 39), 1342–1348, 1432–1437.
279. Stroud, D., Bird, N., Norton, P., Kennedy, M. J., & Greener, M. (2012). Roller reamer fulcrum in point the bit rotary steerable system reduces stick slip and backward whirl. In: *Proceedings of the IADC/SPE Drilling Conference and Exhibition*, SPE 151603. San Diego, California, USA.
280. Suplin, V., Fridman, E., & Shaked, U. (2006). H^∞ control of linear uncertain time-delay systems—projection approach. *IEEE Transactions on Automatic Control*, *51*(4), 680–685.
281. Tariku, F. (1992). *Simulation of dynamic mechanical systems with stick-slip friction*. M.Sc Thesis, University of New Brunswick, Canada.
282. Teale, R. (1965). The concept of specific energy in rock drilling. *International Journal of Rock Mechanics & Mining Sciences*, *2*, 57–73.
283. Tubei, P., Bergeron, C., & Bell, S. (1992). Mud-pulser telemetry system for down hole Measurement-While-Drilling. In: *IEEE Proceedings of the 9th Instrumentation and Measurement Technology Conference* (pp. 219–223).
284. Tucker, R. W., & Wang, C. (1999). On the effective control of torsional vibrations in drilling systems. *Journal of Sound and Vibration*, *224*(1), 101–122.
285. Tucker, R. W., & Wang, C. (1999). An integrated model for drill-string dynamics. *Journal of Sound and Vibration*, *224*(1), 123–165.
286. Tucker, R. W., & Wang, C. (2000). *The excitation and control of torsional slip-stick in the presence of axial vibrations*. Citeseer Computer and Information Science Publications collection. Id: 41923656.
287. Tucker, R. W., & Wang, C. (2003). Torsional vibration control and Cosserat dynamics of a drill-rig assembly. *Mecanica*, *38*(1), 145–161.
288. VanAntwerp, J. G., & Braatz, R. D. (2000). A tutorial on linear and bilinear matrix inequalities. *Journal of Process Control*, *10*(1), 363–385.
289. Van der Heijden, G. H. M. (1994). *Nonlinear drillstring dynamics*. Ph.D Thesis. The Netherlands: University of Utrecht.
290. Van Geffen, V. (2009). *A study of friction models and friction compensation*. Traineeship report DCT 2009.118. Technische Universiteit Eindhoven, Department Mechanical Engineering, Dynamics and Control Technology Group.
291. Vandiver, J. K., Nicholson, J. W., & Shyu, R. J. (1990). *Case studies of the bending vibration and whirling motion of drill collars*. SPEDE, SPE 18652 (pp. 282–290).
292. Veeningen, D. (2011). Novel high speed telemetry system with measurements along the string mitigate drilling risk and improve drilling efficiency. In: *Proceedings of the Brazil Offshore Conference and Exhibition* (pp. 533–544). Macaé, Brazil.
293. Verriest, E. I., & Niculescu, S. I. (1998). Delay-independent stability of linear neutral systems: A Riccati equation approach. *Lecture Notes in Control and Information Sciences, Stability and Control of Time-delay Systems*, *228*, 92–100.
294. Veslin, E. Y., Slama, J., Dutra, M. S., & Lengerke, O. (2011). Motion planning on mobile robots using differential flatness. *IEEE Latin America Transactions*, *9*(7), 1006–1011.

295. Vidyasagar, M. (1993). *Nonlinear Systems Analysis* (2nd ed.). Englewood Cliffs: Prentice-Hall.
296. Villafuerte, R., Mondié, S., & Poznyak, P. (2008). Practical stability of neutral type time delay systems: LMIs approach. In: *IEEE International Conference on Electrical Engineering, Computing Science and Automatic Control (CCE)* (pp. 75–79).
297. Villafuerte, R., Mondié, S., & Poznyak, P. (2011). Practical stability of time-delay systems: An LMIs ap-proach. *European Journal of Control*, *17*(2), 127–138.
298. Villafuerte, R., Saldivar, B., & Mondié, S. (2013). Practical stability and stabilization of a class of nonlinear neutral type time delay systems with multiple delays: BMI's approaches. *International Journal of Control, Automation and Systems*, *11*(5), 859–867.
299. Vlajic, N., Liu, X., Karki, H., & Balachandran, B. (2014). Torsional oscillations of a rotor with continuous stator contact. *International Journal of Mechanical Sciences*, *83*, 65–75.
300. Vyhlídal, T., & Zítek, P. (2003). Quasipolynomial mapping based rootfinder for analysis of time delay systems. In: *4th IFAC workshop on Time Delay Systems, Rocquencourt* (pp. 227–232). France.
301. Wang, T. (1994). Stability in abstract functional-differential equations. II. Applications. *Journal of Mathematical Analysis and Applications*, *186*, 835–861.
302. Wang, C. H., Zhang, L. X., Gao, H. J., & Wu, L. G. (2005). Delay-dependent stability and stabilization of a class of linear switched time-varying delay systems. In: *Proceeding of the Fourth ICMLC* (pp. 917–922). Guangzhou.
303. Wedel, R., Mathison, S., & Hightower, G. (2011). Mitigating bit related stick slip with a torsional impact hammer. In: *Proceedings of the AADE National Technical Conference and Exhibition*. Houston, Tex, USA.
304. Weedermann, M. (2001). *Normal forms for neutral functional differential equations*. Topics in functional differential and difference equations (Vol. 29) of Fields Inst. Commun. (pp. 361–368). Amer. Math. Soc., Providence, RI.
305. Weedermann, M. (2006). Hopf Bifurcation calculations for scalar delay differential equations. *Nonlinearity*, *19*, 2091–2102.
306. Wolter, H., Gjerding, K., Reeves, M. E., Hernández, M., Duncan-Macpherson, J., Heisig, G., & Zaeper, R. (2007). The first offshore use of an ultra high speed drillstring telemetry network involving a full LWD logging suite and rotary steerable drilling system. In: *SPE Annual Technical Conference and Exhibition*, SPE-110939-MS. Anaheim, California, USA.
307. Wu, X., Karuppiah, V., Nagaraj, M., Partin, U. T., Machado, M., Franco, M., & Duvvuru, H. K. (2012). Identifying the root cause of drilling vibration and stick slip enables fit for purpose solutions. In: *Proceedings of the IADC/SPE Drilling Conference and Exhibition*, SPE no. 151347. San Diego, California, USA.
308. Wu, S. X., Paez, L. C., Partin, U. T., & Agnihotri, M. (2010). Decoupling stick-slip and whirl to achieve breakthrough in drilling performance. In: *Proceedings of the IADC/SPE Drilling Conference and Exhibition* (pp. 966–978). New Orleans, La, USA.
309. Xia, Y., & Jia, Y. (2003). Robust control of state delayed systems with polytopic type uncertainties via parameter-dependent Lyapunov functionals. *Systems & Control Letters*, *50*(3), 183–193.
310. Yu, L. (2004). Comments and improvement on "Robust control of state delayed systems with polytopic type uncertainties via parameter-dependent Lyapunov functionals". *Systems & Control Letters*, *53*(3–4), 321–323.
311. Xiong, L., Zhong, S., & Mao, Y. (2009). New stability analysis for switched neutral systems. *Second International Symposium on Computational Intelligence and Design*, *1*, 180–183.
312. Zakuan, A., Junaida, A., Subroto, B., Hermawan, H., Fatakh, A., & Halim, A. (2011). Stick slip mitigation plan to improve drilling. In: *Proceedings of the SPE Asia Pacific Oil and Gas Conference and Exhibition*, SPE 141988 (pp. 65–69). Jakarta, Indonesia.
313. Zamanian, M., Khadem, S. E., & Ghazavi, M. R. (2007). Stick-slip oscillations of drag bits by considering damping of drilling mud and active damping system. *Journal of Petroleum Science and Engineering*, *59*, 289–299.

314. Zhu, X., Tang, L., & Yang, Q. (2014). A literature review of approaches for stick-slip vibration suppression in oilwell drillstring. *Advances in Mechanical Engineering*. Hindawi Publishing Corporation, Article ID 967952.

315. Zribi, M., Karkoub, M., & Huang, C. C. (2011). Control of stick-slip oscillations in oil well drill strings using the back-stepping technique. *International Journal of Acoustics and Vibrations*, *16*(3), 134–143.

Index

M.B. Saldivar Márquez et al., *Analysis and Control of Oilwell Drilling Vibrations*, Advances in Industrial Control, DOI 10.1007/978-3-319-15747-4

MIX
Papier aus verantwortungsvollen Quellen
Paper from responsible sources
FSC® C105338

If you have any concerns about our products,
you can contact us on
ProductSafety@springernature.com

In case Publisher is established outside the EU,
the EU authorized representative is:
Springer Nature Customer Service Center GmbH
Europaplatz 3, 69115 Heidelberg, Germany

Printed by Libri Plureos GmbH
in Hamburg, Germany